T0299120

The fungal colony

Fungi are amongst the simplest of eukaryotes. Their study has provided useful paradigms for processes that are fundamental to the way in which higher cells grow, divide, establish form and shape, and communicate with one another. The majority of work has been carried out on the budding yeast *Saccharomyces cerevisiae*, but in nature unicellular fungi are greatly outnumbered by filamentous forms, for which our knowledge is much less well developed. This volume focuses on the analysis of the filamentous life style, particularly on the hyphae which constitute the fungal mycelial colony. It provides the most recent insights into the molecular genetics and physiological mechanisms underlying the elaboration of the branching mycelium and the interactions between individual fungal mycelia. As such it offers much to interest mycologists and, equally, those working in the fields of cell biology, developmental biology, physiology and biochemistry.

The fungal colony

SYMPOSIUM OF
THE BRITISH MYCOLOGICAL SOCIETY
HELD AT THE SCIENTIFIC SOCIETIES' LECTURE
THEATRE, LONDON
SEPTEMBER 1997

EDITED BY

N. A. R. GOW, G. D. ROBSON
AND G. M. GADD

Published for the British Mycological Society

CAMBRIDGE
UNIVERSITY PRESS

University Printing House, Cambridge CB2 8BS, United Kingdom

Cambridge University Press is part of the University of Cambridge.

It furthers the University's mission by disseminating knowledge in the pursuit of education, learning and research at the highest international levels of excellence.

www.cambridge.org
Information on this title: www.cambridge.org/9780521621175

First published 1999

A catalogue record for this publication is available from the British Library

ISBN 978-0-521-62117-5 Hardback

Contents

Contributors

J. R. Beeching
School of Biology and Biochemistry, University of Bath, Claverton Down, Bath BA2 7AY, UK

S. Bocking
School of Biological Sciences, 1.800 Stopford Building, University of Manchester, Manchester M13 9PT, UK

J. W. Crawford
Soil Plant Dynamics Unit, Cellular and Environmental Physiology Department, Scottish Crop Research Institute, Invergowrie, Dundee DD2 5DA, UK

H. J. P. Dalstra
Department of Plant Biology, Groningen Biotechnology Institute, University of Groningen, Kerklaan 30, 9751 NN, Haren, The Netherlands

G. M. Gadd
Department of Biological Sciences, University of Dundee, Dundee DD1 4HN, UK

G. W. Gooday
Department of Molecular and Cell Biology, University of Aberdeen, Institute of Medical Sciences, Foresterhill, Aberdeen AB25 2ZD, UK

N. A. R. Gow
Department of Molecular and Cell Biology, Institute of Medical Sciences, Foresterhill, University of Aberdeen, Aberdeen AB25 2ZD, UK

J. E. Hamer
Department of Biological Sciences, Purdue University, West Lafayette, Indiana 47907-1968, USA

L. Hamer
Department of Biological Sciences, Purdue University, West Lafayette, Indiana 47907-1968, USA

S. D. Harris
Department of Microbiology, The University of Connecticut Health Center, Farmington CT 06030-3205, USA

M. Momany
Department of Botany, University of Georgia, Athens, GA 30602, USA

J. A. Morrell
Department of Cell Biology, Vanderbilt University, Nashville, Tennessee, USA

S. Olsson
The Royal Veterinary and Agricultural University, Department of Ecology and Molecular Biology (Microbiology), Rolighedsvej 21, DK-1958 Frederiksberg C, Copenhagen, Denmark

L. M. Ramsay
Department of Biological Sciences, University of Dundee, Dundee DD1 4HN, UK

M. Ramsdale
Department of Zoology, University of Cambridge, Cambridge CB2 3EJ, UK

A. D. M. Rayner
School of Biology and Biochemistry, University of Bath, Claverton Down, Bath BA2 7AY, UK

K. Ritz
Soil Plant Dynamics Unit, Cellular and Environmental Physiology Department, Scottish Crop Research Institute, Invergowrie, Dundee DD2 5DA, UK

G. D. Robson
School of Biological Sciences, 1.800 Stopford Building, University of Manchester, Manchester M13 9PT, UK

J. A. Sayer
Department of Biological Sciences, University of Dundee, Dundee DD1 4HN, UK

J. M. J. Scheer
Department of Plant Biology, Groningen Biotechnology Institute, University of Groningen, Kerklaan 30, 9751 NN, Haren, The Netherlands

T. A. Schuurs
Department of Plant Biology, Groningen Biotechnology Institute, University of Groningen, Kerklaan 30, 9751 NN, Haren, The Netherlands

M. L. Smith
Biology Department, Carleton University, Ottawa, K1S 5B6 Canada

R. J. Swift
School of Biological Sciences, 1.800 Stopford Building, University of Manchester, Manchester M13 9PT, UK

A. P. J. Trinci
School of Biological Sciences, 1.800 Stopford Building, University of Manchester, Manchester M13 9PT, UK

G. Turner
Department of Molecular Biology and Biotechnology, The University of Sheffield, Firth Court, Western Bank, Sheffield S10 2TN, UK

Z. R. Watkins
School of Biology and Biochemistry, University of Bath, Claverton Down, Bath BA2 7AY, UK

S. Watkinson
Department of Plant Sciences, University of Oxford, South Parks Road, Oxford OX1 3RB, UK

J. G. H. Wessels
Department of Plant Biology, Groningen Biotechnology Institute, University of Groningen, Kerklaan 30, 9751 NN, Haren, The Netherlands

M. G. Wiebe
School of Biological Sciences, 1.800 Stopford Building, University of Manchester, Manchester M13 9PT, UK

J. M. Withers
School of Biological Sciences, 1.800 Stopford Building, University of Manchester, Manchester M13 9PT, UK

T. Wolkow
Department of Biological Sciences, Purdue University, West Lafayette, Indiana 47907-1968, USA

Preface

Fungi are amongst the simplest of eukaryotes and have become useful paradigms for processes that are fundamental to the way in which higher cells grow, divide, establish form and shape and communicate with one another. Leading the way has been the budding yeast *Saccharomyces cerevisiae* whose ease of manipulation and accessible systems for sexual and molecular genetics have spearheaded basic investigations into fundamental processes as diverse as the analysis of the cell cycle, to investigations of longevity. Although unicellular fungi are greatly outnumbered in nature by the moulds our knowledge of them is much less developed than in this single yeast species. The true hallmark of the filamentous fungus is the hypha – tubular tip-growing cells that are the constituent components of the fungal mycelial colony. This work is dedicated to the analysis of the filamentous life style of fungi, the elaboration of the branching mycelium and the interactions between fungal mycelia. Mycelial fungi also offer major and exciting opportunities for cell and developmental biologists, physiologists, biochemists and developmental biologists. For example, the fungal hypha and the branching mycelium is an excellent system in which to explore the regulation of polarized cell growth, intracellular transport and signalling, how nuclei interact within a common cytoplasm, how genetically similar and dissimilar species interact and recognize one another and how growth responses can be coordinated as an organism explores and infiltrates a heterogeneous environment.

These themes form the rationale for this work. The chapters, all written by leading authorities in their fields, represent the frontiers in research into the molecular and cell biology of moulds, their physiology and the relationship between their filamentous growth habit and their ecology. The first four chapters deal with the properties of intact mycelia, their response to their environment and the signalling systems that articulate environmental sensing. Chapter 4 to 8 deal with metabolism, biochemistry, enzyme secretion, and responses to toxic metal stress in fungi that are important to humans in the manufacture of single cell protein or as wood-rotting fungi or potential remediators of environmental pollutants. Three chapters deal with recent developments in the molecular analysis of tip growth, branching and associated cell cycle regulation. The final three contributions then consider the ways in which genetic information is exchanged during mating, stabilized within large ceonocytic compartments of long-lived individual mycelia and expressed in dikaryotic or monokaryotic hyphae. Together these chapters illustrate how the fungal colony represents a highly adaptive and successful growth form in natural environments and a structure that offers many insights into how unicellular life ramifies into multicellularity so that the properties of the whole mycelium exceeds the sum of its hyphal parts.

The ecology and exploitability of fungi give many of these fundamental studies an important applied dimension. Curious, then, to consider that the vast majority of fungal species remain undescribed, uncharacterized, and that support for mycology is under increasing threat.

Neil A.R. Gow
Geoff D. Robson
Geoffrey M. Gadd

1

Self-integration – an emerging concept from the fungal mycelium

A. D. M. RAYNER, Z. R. WATKINS
AND J. R. BEECHING

Introduction

For so long neglected in the development and promulgation of evolutionary theory, there are increasing signs that mycelial fungi can bring new insights into the origins of phenotypic diversity and change. They challenge some of our most fundamental assumptions about natural selection and its significance relative to other processes in determining the direction of evolutionary pathways. This is because of the way mycelia are physically organized as versatile systems of interconnected tubes that can span heterogeneous environments in which energy is often in very variable supply (Rayner 1994; Rayner, Griffith & Ainsworth, 1995*a*).

Current models of evolutionary change effectively treat the boundaries of living systems and their components as fixed (that is, determinate). Consequently, the dynamic processes underlying change are assumed to be driven by purely external forces acting on discrete objects – genes and individuals (see Dawkins, 1995). However, such discretist models of evolutionary and ecological dynamics are potentially very misleading because all known life forms, from single cells to communities, are dynamic systems which assimilate supplies of free energy from their surroundings and distribute this energy into growth, development, reproduction and movement. They achieve this by possessing boundaries through which they regulate energy exchange with their surroundings and other life forms (Rayner, 1997*a*). For life forms to thrive and survive as energy supplies wax and wane, these boundaries have to be capable of enhancing gains through the proliferation of assimilative free surface in energy-rich environments whilst minimizing losses by various means of containment in inhospitable environments. This requires the configuration and properties of boundaries to change according to circumstances.

Living system boundaries cannot therefore be absolutely fixed any more than they can be absolutely sealed: rather they are dynamic, reactive interfaces, forever in some degree of flux. Their properties both define and are defined by the properties of the interactive arenas – the 'dynamic contexts' – incorporated by living systems. (Please note that in this sense 'context' does not equate with external environment as such, but is rather the 'domain', 'territory' or 'field' occupied by and including a life form as it develops through space and time.) Many of these properties derive from materials or energy sources – for example, water, air, minerals and light – that are not encoded in DNA, and are overlooked in much evolutionary theory, but are nevertheless salient in moulding the dynamic interplay between genetic information and environment into diverse phenotypic forms and behaviours. The dynamic boundaries of living systems therefore define both the sites and mode of action of natural selection as an interactive, channelling process rather than a mechanical sifting of particle-like units. They enable living systems to respond to the environmental heterogeneity that these systems both interact with and help to generate.

This chapter aims to show how the interconnectedness and versatility of mycelial organization uphold a 'systemic' evolutionary approach which explains phenotypic diversity in terms of how the properties of dynamic contextual boundaries regulate processes of input, throughput and output of energy. Special emphasis will be given to the way mycelia epitomize the integrational processes of boundary-sealing, boundary-redistribution and boundary-fusion. These much neglected processes counteract the tendency for living systems to subdivide into discrete, competitive, energy-dissipating units.

Order, organization and chaos – the mycelial example

Recent decades have witnessed significant developments in the way that pattern-generating processes in dynamic physical systems can be understood and modelled mathematically. These developments are encompassed within an array of interrelated concepts, variously described as non-linearity, chaos, complexity, fractal geometry and self-organization. Using fungal mycelia for cross-reference, we will try here to clarify the biological relevance of these developments. We hope to show how the concepts extend beyond rather than negate discretist paradigms, and so open up new prospects for future understanding.

'Self-differentiation' – the route to 'incoherence'

Many of the developments just referred to arise from consideration of the effects of two kinds of feedback and their counteraction. Positive feedback – autocatalysis – arises from the ability of a system to amplify itself using energy input from its local environment. This ability generates an expansive drive which, if unconstrained, causes the system to increase exponentially. Negative feedback damps down expansive drive by directly or indirectly increasing resistance or dissipation as input increases.

The counteraction between positive and negative feedback causes systems to be non-linear (non-additive) and to become unstable if the rate of input exceeds a critical threshold or 'throughput capacity'. Below this threshold, the counteraction causes a smooth build-up to a dynamic equilibrium at which there is no net increase in the system's expansion: the system then remains, in effect, self-contained. Above this threshold, the system becomes 'forced' and hence prone to subdivide, by means of a series of bifurcations, into increasing numbers of subdomains or states. These subdomains may be manifest as increasingly complex, but nonetheless recurrent and predictable, oscillations, countercurrents or branches. Above a yet higher threshold, the subdivisions cease to occur recurrently. Instead, the system traverses what approaches an infinite variety of states in a manner which is apparently erratic and extremely sensitive to initial conditions, and therefore unpredictable in the long term. This is deterministic chaos.

An implicit feature of physical systems that exhibit non-linear dynamics is the presence of one or more dynamic boundaries. The very term 'feedback' implies a reactive interface that mediates this influence. Without a boundary, whether of attraction or constraint, that allows assimilation but prevents instantaneous dispersion, there can be no autocatalysis and no containment. The fact that the importance of dynamic boundaries is often overlooked, has led consciously or unconsciously to discretist interpretations of non-linear systems. These interpretations arise because attention is focused on the behaviour of individual components of the systems rather than the boundaries which shape and are shaped by these behaviours. An example occurs in what has been termed 'self-organization theory' and its attendant metaphor of 'order out of chaos' (Prigogine & Stengers, 1984). The most generally accepted idea of self-organization is that it involves the production of potentially complex patterns or structures through the operation of simple calculational procedures (algorithms) in a many-bodied system. Since the algorithms do not themselves directly encode the patterns or structures, generation of the latter is described as an 'emergent property' of the system (see Bonabeau *et al.*, 1997).

For self-organization to occur, it has been considered necessary for the systems to be thermodynamically open and far from equilibrium, so that they can be sustained by high rates of input and dissipation of energy. Consequently, the emergent structures or patterns they produce are described as 'dissipative', maximizing the conversion of free energy input to entropy (Prigogine & Stengers, 1984). Since emergence of dissipative structures occurs in what appears to be a previously patternless or structureless domain, it is assumed to originate from chaos or even randomness. Examples commonly used to illustrate this idea include 'random' mixtures of autocatalytic ('activator') and constraining ('inhibitor') chemicals, and 'random' arrays of social organisms (for example, slime mould amoebae, ants). These systems generate annular and spiral patterns if suitably prompted by local perturbations (e.g. Goodwin, 1994).

The assumption of a chaotic or random origin for self-organizing patterns may, however, be inappropriate. In fact, it is thought more apt to describe chaos as an extreme form of order which emerges as a consequence of high rates of input of free energy into an initially coherent (self-contained) system (Rayner, 1997*b*). Here, it is important to understand what is implied systemically by coherence, randomness, homogeneity and heterogeneity, and how these terms relate to concepts of order, organization, chaos and entropy. The systemic application of all these terms and concepts depends on the way that systems both define and are defined by their dynamic boundaries, and so differs in some important respects from conventional analytical usage. To begin with, it is vital to realize that randomness is the converse of homogeneity. This fact is often overlooked because for purposes of calculation, random assemblies are assumed analytically to be sets of independent (discrete) data points whose density can be treated on average as homogeneous – the same in different samples – provided that sufficient numbers are accounted for. When sample sizes are small, however, random distributions exhibit extreme heterogeneity. Furthermore, although the distribution of data between set intervals does not imply that these data are interdependent and so non-random from an analytical perspective, in which boundaries are absolute, the same cannot be said from a systemic viewpoint. What randomness, that is, total 'incoherence', implies systemically is the lack of a containing boundary, so the components of a system can be anywhere, anytime and incapable of concerted action. Such absolute disorder is equatable with entropy. By contrast, homogeneity implies being the same everywhere, at all scales, that is, absolute order. Whilst a fully random system is incapable of concerted action, a fully coherent system is incapable of change. Dynamic systems therefore operate between these extremes – that is, with increasing degrees of freedom from relative coherence to relative incoherence. The boundaries of these systems represent sites of relative order which, when in disordered surroundings, tend to lose coherence.

These considerations focus attention, at any particular scale of reference, on the boundary of a system as the expression of its relative order and dynamic state. The effect of introducing free energy into a dynamically bounded system is, directly or indirectly, to cause an expansion of the system's boundary. If the rate of input to the system is below the 'throughput capacity' defined by the resistances imposed by the system's boundary (see above), the boundary expands smoothly, retaining its symmetry and minimizing its surface area and consequent dissipation to its surroundings. A germinating fungal spore exhibiting initial spherical growth exemplifies this. However, if the rate of input exceeds the throughput capacity, the system begins to lose coherence by 'breaking its symmetry' and generating emergent structure. It first polarizes and then subdivides to produce more and more dissipative (and assimilative) free surface – as epitomized by the emergence and subsequent branching of a germ tube (see Fig. 1.1). All this emergent, increasingly complex structure, the most extreme form of which is chaotically distributed, represents proliferated boundary – and hence as presently defined, increased order. However, the origin of this order is not disorder, but a highly integrated, coherent initial state. We view this initial state as more highly organized.

Like the packaging that is used to enclose all kinds of commodities, the order invested in boundaries is energetically costly, for two reasons. Firstly, a high rate of energy input is required to cause systems to become unstable and break symmetry. Secondly, proliferating boundaries present an increased dissipative free surface which renders the system more susceptible to random environmental influences and counteracts the input of free energy, so that more erratic but less labile structures emerge.

Given that boundary-proliferation can only be sustained by continuing energy input, an important question is what happens to dissipative struc-

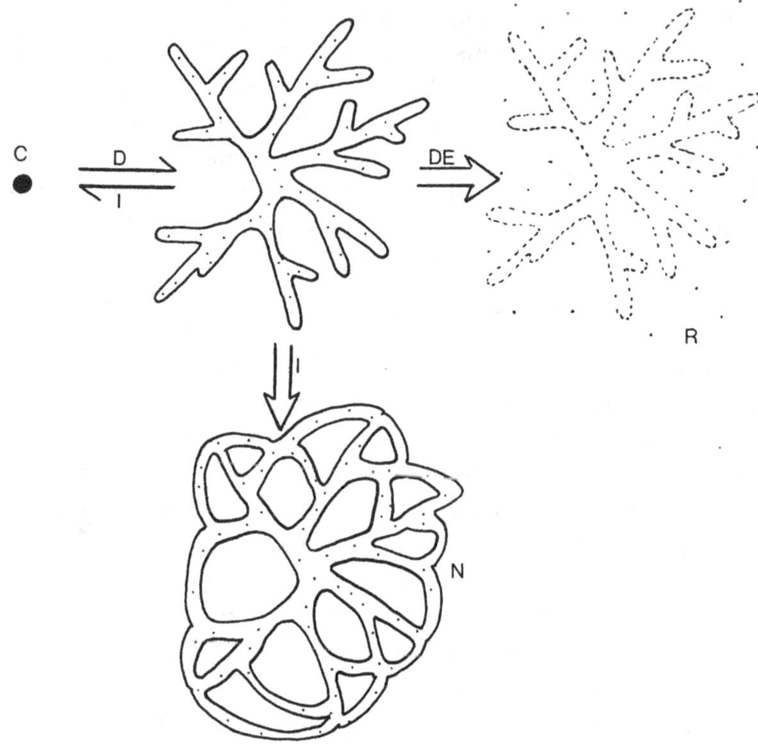

Fig. 1.1. The role of dynamic boundaries in the production of dissipative structure ('order') and coherent 'organization'. Assimilation of free energy into a coherent initial state (C) results in the proliferation and subdivision of boundary (dissipative structure) by 'self-differentiation' (D). Irreversible decay or degeneration (DE) of this structure in the absence of energy replenishment leads to random disorder (R). 'Self-integration' of this structure by boundary-fusion, boundary-sealing and boundary-redistribution minimizes its dissipative free surface, enabling it to reconfigure into coherent initial states or persistent networks (N). (From Rayner, 1997*b*.)

tures when external supplies of free energy are restricted? A related, fundamentally important, question is what is the origin of the initial coherent state from which dissipative structures emerge in energetically unrestricted environments? Essentially, if external energy supplies are withdrawn from a dissipatively structured system, the long-term survival of the system (or part of it) rests on a stark alternative (Fig. 1.1). The system may continue to dissipate, or it undergoes processes that minimize exposure of free surface. The first option leads to dissolution, an irreversible decay into an entropic state. The second option results in a reduction of order and an increase in organization. The sustainability and persistence of life forms in energetically variable environments depends on this second option, which involves three dissipation-minimizing processes: boundary-fusion, boundary-sealing and boundary-redistribution. We term this second option 'self-integration', as distinct from the emergence of dissipative structures, which we term 'self-differentiation'.

'Self-integration' – retaining and regaining coherence

Since dissipative free surface is energetically costly, any process that minimizes this surface is energy-saving and even energy-yielding. The three self-integrational processes depicted in Figs. 1.1 and 1.2 all have these effects. Operating separately or in concert, these processes enable living systems to conserve, explore for and recycle resources by means of fundamentally simple adjustments in their boundary properties that accord with local circumstances. They are well illustrated by fungal mycelia.

Boundary-fusion both lessens the amount of dissipative surface and releases energy that was previously contained in this surface through its dissolution. It is most obvious amongst mycelial fungi in the process of anastomosis. Anastomosis can occur both between individual hyphae, and between hyphal aggregates such as mycelial cords (Thompson & Rayner, 1983; Dowson, Rayner & Boddy, 1988*a*). It converts a dendritic branching system with resistances in series, to a more coherent network with resistances at least partially in parallel. It thereby makes the system more retentive and less prone to proliferate branches. At the same time it enables the system to amplify its organizational scale, through enhanced delivery to sites of emergence of distributive or reproductive structures, for example rhizomorphs and fruit bodies, on its boundary.

Boundary-sealing involves various ways of reducing permeability and hence increasing the 'insulation' of a system. Sealing a fixed boundary results in the production of survival structures, as in various kinds of constitutively dormant spores, sclerotia and pseudosclerotia by fungal mycelia. Sealing a deformable boundary results in the emergence of distributive structures that serve reproductive or explorative/migratory functions. Since the sites of input to these structures are distal to their sites of proliferation, their branching pattern will be distributary- or fountain-like, contrasting with the tributary-like branching pattern of

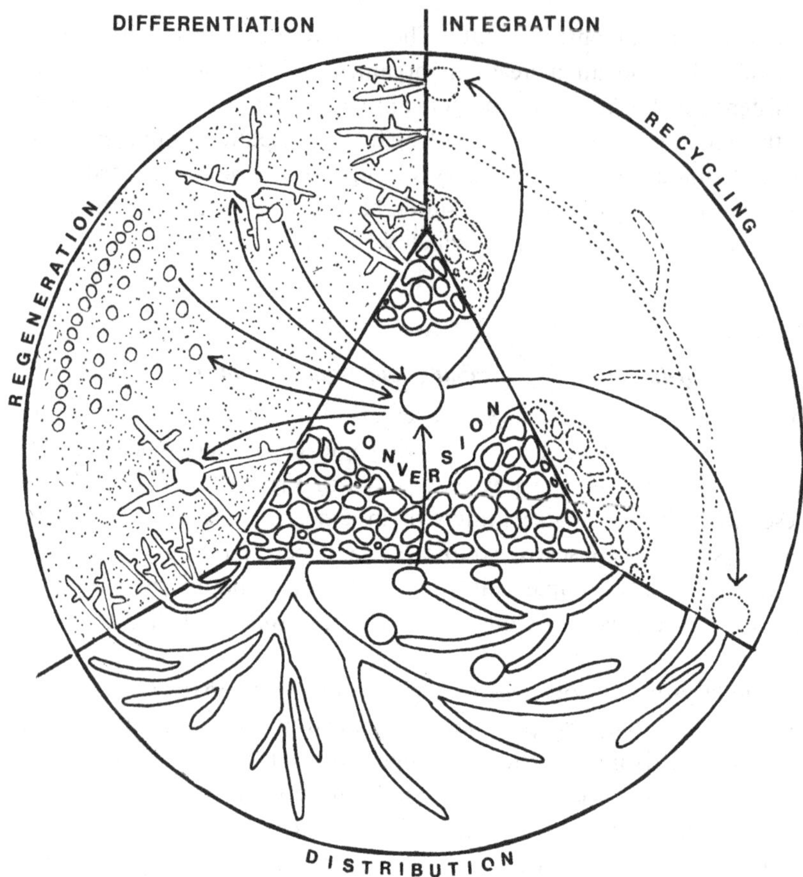

Fig. 1.2. The interplay between self-integration and self-differentiation to produce distinctive organizational states in resource-rich (stippled) and resource-restricted environments. The interplay enables energy to be assimilated (allowing regeneration of boundaries), conserved (by conversion of boundaries into impermeable form), explored for (through internal distribution of energy sources) or recycled (via redistribution of boundaries) according to circumstances. Fine lines indicate permeable contextual boundaries, bold lines impermeable boundaries and dotted lines degenerating boundaries. (From Rayner, 1997*a*.)

assimilative structures. Moreover, sealing the lateral boundary of a tube or channel whilst maintaining its apical boundary in a deformable, responsive state enables a much more focused response to a directional stimulus – much like blinkering a horse! This correlates with the observation that migratory structures in fungal mycelia are more prone to exhibit tropic responses than assimilative structures (Rayner & Boddy, 1988).

Such responses greatly enhance the energetic efficiency of any distributive system.

Boundary-redistribution involves the transfer of resources from degenerative to generative sites. In mycelia, it is evident in various examples of 'autolysis' and 'self-parasitism' (e.g. Rayner, 1977*a*; Rayner & Boddy, 1988).

Boundary properties and life history patterns

From a discretist standpoint, life cycles are commonly thought of as reproductive cycles – the means by which genes make more of themselves. The ability of a particular genotype, given a particular set of niche parameters, to make use of these cycles in delivering as many as possible of its own offspring into succeeding generations defines its adaptive fitness.

Even from this standpoint, however, it has long been appreciated that reproduction is subject to biotic or abiotic environmental constraints. Recognition of the effects of these constraints on population dynamics has given rise to classical theories of *r*- and K-selection, based on the reproductive rate (r) and equilibrium or carrying capacity ($K = 1 - 1/r$) terms of the non-linear logistic equation (see, for example, Andrews, 1992). Correspondingly, high rates of reproduction are associated with *r*-selection in unrestrictive environments, whilst lower rates of reproduction occur as a result of K-selection in restrictive environments.

Whilst *r*–K-selection theories explain why, in an adaptational sense, reproductive rates in unrestrictive and restrictive environments differ, how, in an organizational sense, life forms and life cycle stages are attuned physically to their surroundings has attracted less attention. Consequently, important insights into the origins and versatility of phenotypic form in response to inconstant environmental circumstances may have been missed.

From the systemic perspective illustrated in Figs. 1.1 and 1.2, far from defining the beginnings and endings of discontinuous generations of discrete individuals within finite niches, life cycles represent a means of generating and maintaining a continuous dynamic context in changeable surroundings. This is the answer to the riddle of the chicken and the egg (or the spore and the sporophore): neither came first – rather they represent distinctive boundary configurations of the same dynamic system!

The continuity of context that is ensured by life cycles is an expression of the fundamental indeterminacy of living systems, that is, their capacity for ongoing production and reconfiguration of boundaries, and obscures the discretist distinction between growth and reproduction. Examples of this indeterminacy can be found from molecular to social scales of biological organization, with boundaries being defined anywhere from intracellular to extra-organism locations (Rayner, 1997*a*). Correspondingly, the dynamic contexts of motile organisms – such as many animals – are not defined by where the body boundaries of these organisms are at a particular instant. Rather, they are defined topographically by the trajectories that these organisms map out as they use their powers of locomotion to follow and create paths of least resistance, and regionally by the territories within which the trajectories are confined. By contrast, the topographical and regional contexts of organisms, such as many plants and mycelial fungi, that grow rather than move bodily from place to place, coincide directly with the proliferation and overall extent of their body boundaries.

The concept that life forms inhabit and generate indeterminate contexts introduces the need to develop a more dynamic view of niches not as fixed but as fluid and variably interconnected space–time–energy domains. This in turn has important implications for the way *r*–K-selection theory can be used to understand the relationship between life form and life cycle in the generation of exploitative and/or persistent organizations.

When supplies of readily accessible resources are temporarily plentiful – that is, under *r*-selective conditions – following destructive or enrichment disturbance of natural habitats, the self-differentiation or regenerative processes depicted in Figs. 1.1 and 1.2 are promoted. These processes result in rapid proliferation, associated with high metabolic rates, but produce highly dissipative structures that are only sustainable as long as there is continual enrichment. In the absence of replenishment, conditions in any habitat are prone to become more restrictive – that is, to change from *r*- to *K*-selective – due to increasing competition or abiotic stress (including resource depletion). This necessitates self-integration into a more coherent organization if total dissipation is to be avoided.

Systems in which boundary-redistribution into relatively discrete dispersal and survival units predominates are characteristically strongly exploitative and somatically non-persistent – for example, many mitosporic (asexual) fungi. By contrast, boundary-sealing, accompanied by fusion and redistribution, produces a more retentive, coherent organization that allows resources to be conserved or distributed within a protective or explorative context, for example within sclerotia or rhizomorphs.

These processes therefore enable either 'stress management' through the production of survival and emigratory structures, or the development of overtly territorial, invasive and resistive organizations.

These propositions are supported by the general (but not invariable) differences evident between life styles typical amongst basidiomycetes and those of other fungi. Basidiomycetes are generally regarded as having more K-selected properties than other fungi in that they become dominant at late stages of fungal succession, associated with greater powers of persistence, invasiveness and reproduction via meiotic pathways (Frankland, 1992). Persistence is due to the production of resilient, often sclerotized and anastomosed boundaries (for example, pseudosclerotial crusts) which sequester rather than release hydrophobic metabolites. Invasiveness is associated with the formation of cable-like mycelial cords and rhizomorphs, and meiospores are produced in macroscopic basidiomes.

Dynamic networks – varying scale and pattern

The way that self-differentiation and self-integration processes are coordinated in a system capable of changing its scale and mode of operation according to circumstances can be demonstrated by growing mycelia, especially those of basidiomycetes, in heterogeneous culture systems. For example, experiments in which certain wood-decay fungi are grown between colonized inocula and uncolonized 'baits' in trays of soil have revealed a variety of long-range and short-range 'foraging strategies' that produce patterns extraordinarily similar to, for example, the raid swarms of army ants and the roots and stoloniferous systems of plants (e.g. Boddy, 1993; Dowson, Rayner & Boddy, 1986, 1988*b*).

Especially revealing are patterns produced by mycelia grown in matrix systems of the kind illustrated in Fig. 1.3. Here it is possible to see how, purely by responding to local circumstances and without any central administration, a mycelium can generate a persistent network which is reinforced along avenues of successful exploration and capable of producing coherent, scaled-up outgrowths with reproductive and migratory functions.

These patterns can be explained by the organization of mycelia as nonlinear hydrodynamic systems (Rayner, Ramsdale & Watkins, 1995*b*; Rayner, 1994, 1996*a*, *b*, *c*). The uptake of water and nutrients generates an expansive drive that results in the hydraulic displacement of deformable components of hyphal boundaries. Whenever its 'through-

Fig. 1.3. Two examples of development of mycelium of the basidiomycete, *Coprinus radians*, when grown through a matrix of 25×4 cm^2 chambers. The chambers alternately contain 2% (w/v) malt agar and distilled water agar and are interconnected by narrow channels cut in the plastic partitions just above the level of the medium. Notice the diffuse proliferation in the high nutrient chambers and production of fruit bodies and fruit body initials in the low nutrient chambers. (Photograph by Timothy Jones.)

put capacity', due to the resistance to displacement to existing sites of boundary deformation, becomes exceeded by the rate of uptake, the non-linear system becomes prone to branch. In purely assimilative hyphae, generated by self-differentiation, the branches will form in a tributary-like pattern – due to the fact that the sites of uptake and proliferation coincide. In hyphae where uptake is distal to the site of proliferation, the branches form in a distributary-like pattern.

The feasibility of the theory that the versatility of mycelial systems can be explained in terms of varied hydraulic resistances to uptake, throughput and discharge of resources, can be tested using non-linear mathematical models. A simple reaction-diffusion model of this kind has been developed by Davidson *et al.* (1996, 1997). This model has four fundamental components: (1) a diffusible substrate, which is the energy source of the system; (2) replenishment of this substrate at a constant specific rate; (3) an autocatalytic activator, which facilitates conversion of the substrate into energy, drives the proliferation of biomass and decays at a constant specific rate; (4) diffusion of the activator at a rate which is inversely related to the system's resistance to throughput.

As shown in Fig. 1.4, the model used by Davidson *et al.* (1996,1997) was capable of reproducing many of the macroscopic patterns of biomass distribution actually observed in fungal mycelia. These included the production of smooth, annular and irregular biomass density profiles depending on the resistance to throughput. Smooth profiles were produced by low resistance (that is, highly networked) systems that ceased to expand radially unless replenishment was prevented – simulating the effect of internal degeneration – whence a 'fairy-ring'-like travelling wave front was propagated.

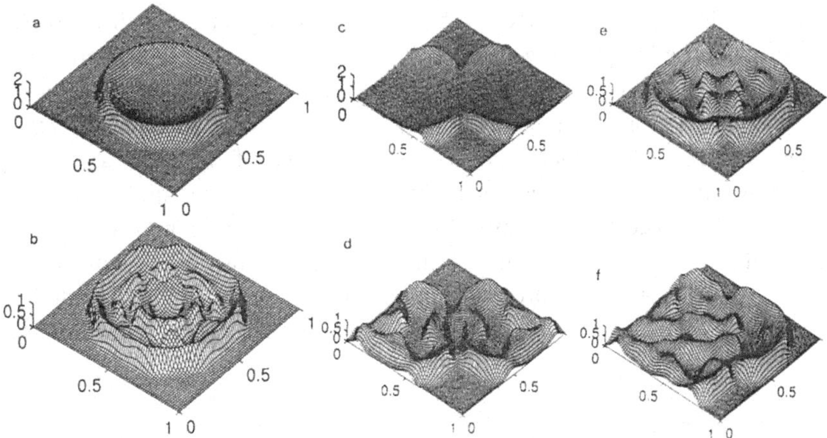

Fig. 1.4. Use of a reaction-diffusion model to predict energy assimilation patterns in growing and interacting mycelial networks with varied resistance to throughput and replenishment of substrate. Numbers on the vertical and horizontal axes respectively represent units of activator concentration (and hence, biomass-generating capacity) and spatial intervals. (a) Fairy-ring-like travelling-wave solution produced when replenishment is low or absent. (b) Irregularly lobed pattern produced in a relatively high resistance system when replenishment is sufficient to offset decay. In low resistance systems, the heterogeneity is reduced and expansion ceases (but can be resumed as a travelling wave if replenishment is prevented). (c) Mutual extinction of the interface between colliding travelling waves, as is observed in natural fairy rings. (d) Formation of a demarcation zone between established, replenished systems – as commonly exhibited by self-inhibiting cultures. (e) Coalescence of immature, replenished systems – as commonly seen in self-pairings between mycelia. (f) Formation of ridges protruding from an established into an immature replenished system – simulating the commonly observed penetration of mycelial cords from one colony into another (from Davidson *et al.*, 1996).

It should be noted, however, that the model could not sustain either an evenly growing margin without a decline in the capacity for substrate utilization in the interior, or the extension and expansion of activator peaks. The latter would correspond with the emergence of distributive structures such as mycelial cords and fruit bodies, due to enhanced delivery to local sites on the boundary of an integrated system. These inabilities may be related to the fact that as a first step, the model used was unable to seal its boundaries and/or alter its resistance as a consequence of throughput and so was incapable of effective conservation or distribution.

A more recent model based on circuitry has, however, succeeded in generating emergent structures (D. Marles, S.C. Harris, D. Williams & A.D.M. Rayner, unpublished). This model consists of a set of distributive channels emerging from a resource base and travelling along a gradient towards a resource source. The channels were allowed to branch and to anastomose at random. Their extension rate was dependent on their conductivity, which increased as a consequence of throughput. Under appropriate circumstances, anastomosis and consequent increased conductivity enabled the rapid proliferation of fan-like branching systems in a manner strongly reminiscent of the slow/dense – fast/effuse transitions in fungal mycelia that have been implicated in switches from predominantly assimilative to explorative growth patterns (e.g. Rayner & Coates, 1987).

Feedback mechanisms – the primary role of oxidative stress?

Having proposed roles for self-differentiation and self-integration in the dynamics of mycelia, the next step is to identify the specific feedback mechanisms responsible for bringing these processes into operation. Given that their physical properties of hyphal boundaries depend on their chemistry, the key questions are how does this chemistry affect, and how is it influenced by the external and internal environments of hyphae?

Since the energy which drives proliferation derives most fundamentally from external sources of reducing and oxidizing power, a primary consideration is how the availability of these sources affects boundary chemistry. Here, the ability of oxygen to serve constructive and destructive as well as energy-releasing roles may be important. These contrasting roles arise from the affinity of oxygen for electrons, which it accepts one at a time in the course of its reduction to water or other compounds. Such reduction releases chemical energy but also generates reactive oxygen species (ROS) capable of bringing about oxidative cross-linking as well as destroying the chemical order of living protoplasm through the generation of free radicals (species with one or more unpaired electrons; Halliwell & Gutteridge, 1989).

Here, an important possibility has been suggested by Hansberg & Aguirre (1990), who found that the development of a 'hyperoxidant state' is a necessary prelude to aerial mycelium and spore formation in the ascomycete, *Neurospora crassa*. This is a state in which the capacity of protoplasm to neutralize ROS is exceeded. Unless mitigated in some way, it can lead to protoplasmic degeneration. It is promoted by any factors that diminish availability of reducing power, enhance exposure to oxygen (especially in the gaseous phase, as in terrestrial habitats) or impede oxidative phosphorylation. On the other hand it is attenuated by any mechanisms that maximize resource uptake whilst minimizing intracellular oxygen concentrations.

Mycelia have four main ways of responding to this threat and promise of oxygen, each with contrasting effects on boundary chemistry. Firstly, they can assimilate nutrients in solution from plentiful external supplies through hydrophilic, permeable boundaries: in so doing they acquire respirable substrate and the ability to proliferate as dissipative systems – but only as long as nutrients are replenished. Secondly, they can neutralize intracellular ROS and molecular oxygen by means of antioxidant enzymes, pathways and metabolites, many of which are currently classified under the general heading of 'secondary metabolism'. Thirdly, they can produce a relatively oxygen- (and thereby also solute- and water-) impermeable boundary. This can be achieved by anastomosis (so restricting proliferation of branches), aggregation and the generation and oxidative cross-linking of hydrophobic phenolic, proteinaceous and lipid compounds in the presence of phenol oxidase and peroxidase enzymes. Fourthly, they can actively enable or passively allow a hyperoxidant state to arise and lead to protoplasmic disorder and cell degeneration (Rayner, 1996*b*, 1997*a*).

Onset of the hyperoxidant state, due to an inability to reduce intracellular oxygen fully to water via the respiratory chain, may therefore be an important, and possibly the most fundamental cue for self-integration. Moreover, degenerative processes would be initiated above a high oxidative stress threshold, whereas protective mechanisms would come into play above a lower threshold (Rayner, 1996*b*). A recent study, which indicated that initiation of sclerotia in the Basidiomycete, *Sclerotium rolfsii*, is characterized by a high degree of peroxidation in its total lipids, accords with this interpretation. Lipid peroxidation is characteristic of oxidative stress, and factors that in the past have been shown to inhibit or promote sclerotium initiation respectively act as free radical scavengers or pro-oxidants (Georgiou, 1997).

A general hypothesis that arises from these considerations is that when the ratio of external to internal supplies of resources providing reducing power exceeds a threshold, mycelia differentiate as assimilative, dissipative structures. However, when the ratio falls below this threshold, self-integration is induced – accompanied to varying degrees by the production, sequestration and release of extracellular compounds. Where these compounds are sequestered, they reduce the permeability of hyphal boundaries. This hypothesis accords both with the changes in mycelial organization associated with resource enrichment or attenuation (for example, in matrix plates), and with the onset of 'secondary metabolism' (e.g. Bushell, 1989*a*,*b*). Moreover, it suggests why organization and metabolite production are interrelated. These propositions are also consistent with data obtained from three phenotypically distinctive strains of the basidiomycete, *Hypholoma fasciculare* (Crowe, 1997). One of these strains was a typical dikaryon, which produced silky, cord-forming mycelium with distributary-like branching. Several major hydrophobic components were present in extracts from the mycelium as opposed to the growth medium of this dikaryon. These hydrophobic components were also consistently produced by other dikaryotic genotypes. By contrast, monokaryons of *H. fasciculare*, which have predominantly tributary-like branching, as well as 'flat' dikaryons, which did not produce mycelial cords and aerial mycelium, lacked these components. The importance of hydrophobic compounds in the formation of emergent mycelium has been demonstrated more explicitly for an increasingly wide range of fungi that produce the cysteine-rich proteins known as 'hydrophobins'. These proteins coat the walls of emergent hyphae, but are released into the medium from submerged hyphae (Wessels, 1994).

An important adjunct of the general hypothesis is that the nature of boundary-sealing compounds may be expected to vary with habitat – notably with respect to the degree of aeration and availability of nitrogen sources. Correspondingly, whereas relatively aquatic systems often produce abundant mucilage at their external boundaries, terrestrial systems undergo sclerotization. Sclerotization characteristically involves the oxidative cross-linking of reducing compounds, notably phenolics, lipids and cysteine-rich proteins. It not only makes outer boundaries impermeable to water, but also consumes and reduces permeability to oxygen.

That basidiomycetes generally display the most overtly terrestrial life styles amongst fungi is consistent with this proposition. Moreover, a striking example of the transition from mucilaginous to sclerotized boundaries within a single invasive structure is found in the rhizomorphs of honey fungus (*Armillaria* spp.), which both create and grow down oxygen gradients. The apex of a growing rhizomorph is coated by a thick layer of mucilage, which both lubricates the structure and makes it deformable (Rayner *et al.*, 1985). Behind the apex, the outermost layer progressively darkens and rigidifies due to the oxidative cross-linking of phenolic compounds in the presence of the extracellular phenoloxidase, laccase (Worrall, Chet & Hüttermann, 1986; Rehmann & Thurston, 1992). Within the core of the structure is a medulla, which probably acts as an air channel (Griffin, 1972). Surrounding the medulla is a cortex, the innermost layer of which, forming a dome over the medulla, contains hyphae rich in mitochondria and bundles of microfilaments (Rayner *et al.*, 1985). The rhizomorph is remarkable for its ability to invade anoxic environments and to penetrate the intact bark and vascular cambium of woody roots, with increased aeration in its wake. If it emerges into air, it ceases extension and its tip sclerotizes. The mitochondria-rich hyphae presumably serve both to generate the energy necessary to drive extension and to use up the oxygen supplied through the medulla that would otherwise sclerotize the tip. The same basic principle of progressive sclerotization is demonstrated by many other examples of the 'hardening off' of maturing boundaries, ranging from insect cuticles to tree bark, and also links in with the general principle, described earlier, of lateral boundary-sealing in distributive structures.

Interactive boundaries – mycelia in conflict and partnership

Given their proposed significance and sensitivity, boundary properties will both influence and be influenced by the outcome of encounters between fungi and between fungi and other organisms. Here we will signpost the ways we think self-integration theory may help to clarify how these influences are brought to bear.

Competition

Competition occurs when the boundaries of systems with similar resource requirements occupy the same contextual domain but fail to integrate: the outcome may then be mutual dissolution, mutual restriction or the overtaking of one system by another (Rayner, 1996*c*, 1997*a*). An indication of how these distinctive outcomes are embedded in the physical organization of mycelia as dynamic, assimilative systems is provided by observations of collisions between the mathematical model systems of Davidson *et al.* (1996, 1997; Fig. 1.4). Collisions between travelling waves resulted in mutual extinction of their interactive interface, as with natural fairy rings (Dowson, Rayner & Boddy, 1989). Collisions between irregular systems resulted in coalescence, formation of demarcation zones and reflective waves or incursion from one system into the other, depending on the distance apart and relative timing. All these patterns occur in real mycelial interactions.

In understanding the mechanisms and consequences of competition, it is important to appreciate the distinction between 'primary resource capture', primacy in locating and assimilating suitable external supplies of free energy, and combat, the gaining or denial of access to resources already incorporated by a system (Cooke & Rayner, 1984). Whereas primary resource capture is maximized through assimilative free surface, combative prowess is enhanced by increased coherence and a boundary that is both resilient to and capable of serving as a secretory surface for allelopathic (damaging/inhibitory) substances.

Numerous observations point to the importance of combat in fungal successions (Frankland, 1992). Here, relatively more *r*-selected and/or aquatic forms with permeable boundaries might be expected to lack combative properties altogether, or release diffusible inhibitors/antibiotics. By contrast, forms with oxidatively polymerized boundaries would be more capable of producing resistive or invasive structures. These structures would not only be relatively immune to, but could benefit territorially from, the secretion of reactive molecules, such as hydrogen peroxide (which in the presence of peroxidase could also provide a mechanism for eliminating excess oxygen and boundary-sealing).

The dominance of basidiomycetes late in succession, and the characteristic way in which they produce emergent mycelial phases and induce cell death following contact with one another and other organisms accords with these expectations. So too does evidence that the release of hydrophobic metabolites by these fungi is suppressed when they interact with one another or metabolic inhibitors such as 2,4-dichlorophenol, associated with enhanced phenol oxidase and peroxidase activity (Griffith, Rayner & Wildman, 1994*a*,*b*,*c*). The production of hydrogen peroxide, accompanied by the formation of invasive mycelial cords and lysis of opposing mycelium, has been demonstrated in *Phanerochaete velutina* (Z.R. Watkins & A.D.M. Rayner, unpublished). Also, of the three phenotypes of *H. fasciculare* described earlier, the typical dikaryon was more invasive than the monokaryon and flat dikaryon when paired with another fungus (Crowe, 1997).

Mating

In those fungi where plasmogamy involves integration of systems that are genetically non-identical (i.e. 'non-self'), a context is thereby produced for the encounter of disparate nuclear and mitochondrial genomic organelles. The boundaries and activities of these organelles may be 'compatible' – capable of integration, or 'incompatible' – incapable of integration.

Since any mechanism that compromises a cell's ability to maintain reactive oxygen species below a critical level will induce oxidative stress, incompatibility and consequent interference between gene products that directly or indirectly influence electron flows could be critical to survival. For example, disparity between nuclei with regard to the specification of key components of mitochondria could induce dysfunction of respiratory pathways and protoplasmic degeneration (e.g. Rayner & Ross, 1991). This could explain the *dynamic origin* of post-fusion somatic incompatibility reactions amongst higher fungi (Rayner, 1991, 1996*c*) and the general difference between ascomycetes and basidiomycetes with respect to their ability to form stable heterokaryons in strongly aerated conditions (Rayner, 1996*c*). Ascomycetes may generally lack the latter ability because they do not, except in their fruit bodies, produce distributive hyphae with strongly insulated boundaries. The relatively terrestrial, independently growing heterokaryotic states of many Basidiomycetes, do, however, form such boundaries.

In Basidiomycetes, the smoothness of the transition between homokaryotic and heterokaryotic stages may depend on the rapidity with which an insulated boundary is formed by heterokaryotic states resulting from hyphal fusion, *before* cellular degeneration. The sensitivity of this transition is evident in matings of *Stereum* from different geographical regions (that is, 'allopatric' matings), where takeover, degeneracy and subdivision into conflicting local domains have all been observed (Ainsworth & Rayner, 1989; Ainsworth *et al.*, 1990, 1992) accompanied by the release of secondary metabolites. The latter include the sesquiterpene, (+)-torreyol, and can be suppressed by the antioxidant, N-acetyl-L-cysteine (Z.R. Watkins & A.D.M. Rayner, unpublished).

Parasitic symbiosis

Parasitism occurs when the boundary of a host entity provides a context for, but does not coalesce with, the boundary of an invasive entity, so establishing an interface across which the invader can unilaterally assimilate resources. The nature and sustainability of the parasitism then depends on the integrity and responsiveness of host and parasite system boundaries. Necrotrophic parasitism requires the destruction of host boundaries before they can be sealed, so that their contents are released and made available to the parasite. Biotrophic parasitism depends on the host boundary remaining intact, but permeable, so that the parasite can assimilate resources from living host cells or tissues. Any mechanism that directly or indirectly disrupts the boundary of the parasite or seals the boundary of the host obviates the parasitism. This fact probably underlies the varied mechanisms and processes that underlie resistance and susceptibility to infection, including the production of reactive oxygen species during hypersensitive, cell death, responses (Baker & Orlandi, 1995).

Mutualistic symbiosis

Mutualistic symbiosis depends on the boundaries of disparate entities associating in such ways as to produce an interactive interface for reciprocal exchange of resources. In mycorrhiza-forming fungi, the provision of a linking mycelium between an array of such interfaces can serve to interconnect separate host plants in a manner analogous to computer networks (Read, 1997; Simard *et al.*, 1997). However, degenerative incompatible responses can be induced and need to be suppressed by specific host genes if successful integration is to occur (Gianinazzi-Pearson *et al.*, 1995).

Conclusions – the wider context

We have sought to show how the behaviour of life forms in general and fungi in particular as dynamic systems within a changeable context affects the way they proliferate, evolve and interact. From this systemic perspective, the way that energy flows are maintained and channelled by the differentiation and integration of dynamic boundaries is key to a complex interplay between genetic information and physical context. This interplay both generates and sustains phenotypic diversity by enabling organisms to distribute their activities – in effect as controlled water flows – through highly variable niches.

By opening their boundaries to enable assimilation, fungi enhance their ability to proliferate and compete in primary resource capture (Cooke & Rayner, 1984), but render themselves susceptible to the vicissitudes of external biotic and abiotic influences. By sealing and anastomosing their boundaries, they produce resilient, obstructive organizations that maintain their physiological integrity. When appropriate, these organizations are capable of delivering considerable power to local sites of emergence of explorative and invasive structures, that, aided by production of extracellular enzymes and reactive molecules, create and follow paths of least resistance. By undergoing local degeneration, fungi can limit incursion of pathogens, competitors and incompatible genetic information, and overcome the retentiveness of excessively interconnected and over-centralized organizations.

The resultant organizational patterns reflect themes evident in all kinds of indeterminate systems. The dynamic origins of these patterns beg a rich variety of questions to fuel future research. Ultimately, these questions focus on dynamic boundaries that can neither be absolutely *in*side nor *out*side, but instead represent the reactive interface for feedback between the two.

Acknowledgements

We thank GlaxoWellcome and the Biotechnology and Biological Sciences Research Council for financial support of some of the research described in this chapter, and Jackie Constable for technical assistance.

References

Ainsworth, A. M., Beeching, J. R., Broxholme, S. J., Hunt, B. A., Rayner, A. D. M. & Scard, P. T. (1992). Complex outcome of reciprocal exchange of nuclear DNA between two members of the basidiomycete genus *Stereum*. *Journal of General Microbiology*, **138**, 1147–57.

Ainsworth, A. M. & Rayner, A. D. M. (1989). Hyphal and mycelial responses associated with genetic exchange within and between species of the basidiomycete genus *Stereum*. *Journal of General Microbiology*, **135**, 1643–59.

Ainsworth, A. M., Rayner, A. D. M., Broxholme, S. J., Beeching, J. R., Pryke, J. A., Scard, P. T., Berriman, J., Powell, K. A., Floyd, A. J. & Branch, S. K. (1990). Production and properties of the sesquiterpene, (+)-torreyol, in degenerative mycelial interactions between strains of *Stereum*. *Mycological Research*, **94**, 799–809.

Andrews, J. H. (1992). Fungal life-history strategies. In *The Fungal Community. Its Organization and Role in the Ecosystem*, ed. G. C. Carroll & D. T. Wicklow, pp. 119–45. New York: Marcel Dekker.

Baker, C. J. & Orlandi, E. W. (1995). Active oxygen in plant pathogenesis. *Annual Review of Phytopathology*, **33**, 299–321.

Boddy, L. (1993). Saprotrophic cord-forming fungi: warfare strategies and other ecological aspects. *Mycological Research*, **94**, 641–55.

Bonabeau, E., Theralauz, G., Deneuborg, J-L., Aron, S. & Camazine, S. (1997). Self-organization in social insects. *Trends in Ecology and Evolution*, **12**, 183–93.

Bushell, M. E. (1989*a*). Biowars in the bioreactor. *New Scientist*, **124**, 42–5.

Bushell, M. E. (1989*b*). The process physiology of secondary metabolite production. *Symposia of the Society for General Microbiology*, **44**, 95–120.

Cooke, R. C. & Rayner, A. D. M. (1984). *Ecology of Saprotrophic Fungi*. London and New York: Longman

Crowe, J. D. (1997).Origins of heterogeneity in *Hypholoma fasciculare*. PhD Thesis, University of Bath

Davidson, F. A., Sleeman, B. D., Rayner, A. D. M., Crawford, J. W. & Ritz, K. (1996).Context-dependent macroscopic patterns in growing and interacting fungal networks. *Proceedings of the Royal Society of London, Series B*, **263**, 873–80.

Davidson, F. A. Sleeman, B. D. Rayner, A. D. M., Crawford, J. W. & Ritz, K. (1997). Travelling waves and pattern formation in a model for fungal development. *Journal of Mathematical Biology*, **35**, 589–608.

Dawkins, R. (1995). *River Out of Eden*. London: Weidenfeld & Nicolson

Dowson, C. G., Rayner, A. D. M. & Boddy, L. (1986). Outgrowth patterns of mycelial cord-forming basidiomycetes from and between woody resource units in soil. *Journal of General Microbiology*, **121**, 203–11.

Dowson, C. G., Rayner, A. D. M. & Boddy, L. (1988*a*). The form and outcome of mycelial interactions involving cord-forming decomposer basidiomycetes in homogeneous and heterogeneous environments. *New Phytologist*, **109**, 423–32.

Dowson, C. G., Rayner, A. D. M. & Boddy, L. (1988*b*). Foraging patterns of *Phallus impudicus*, *Phanerochaete laevis* and *Steccherinum fimbriatum* between discontinuous resource units in soil. *FEMS Microbiology Ecology*, **53**, 291–8.

Dowson, C. G., Rayner, A. D. M. & Boddy, L. (1989). Spatial dynamics and interactions of the woodland fairy ring fungus, *Clitocybe nebularis*. *New Phytologist*, **111**, 501–9.

Frankland, J. C. (1992). Mechanisms in fungal succession. In *The Fungal Community. Its Organization and Role in the Ecosystem*, ed. G. C. Carroll and D. T. Wicklow, pp. 383–401. New York: Marcel Dekker.

Georgiou, C. D. (1997). Lipid peroxidation in *Sclerotium rolfsii*: a new look into the mechanism of sclerotial biogenesis in fungi. *Mycological Research*, **101**, 460–4.

Gianinazzi-Pearson, V., Gollotte, A., Lherminier, J., Tisserant, B., Franken, P., Dumas-Gaudot, E., Lemoine, M-C., van Tuinen, D. & Gianinazzi, S. (1995). Cellular and molecular approaches in the characterization of symbiotic events in functional arbuscular mycorrhizal associations. *Canadian Journal of Botany*, **73**, S526–32.

Goodwin, B. (1994). *How the Leopard Changed its Spots*. London: Weidenfeld & Nicolson.

Griffin, D. M. (1972). *Ecology of Soil Fungi*. London: Chapman & Hall.

Griffith, G. S., Rayner, A. D. M. & Wildman, H. G. (1994*a*). Interspecific interactions and mycelial morphogenesis of *Hypholoma fasciculare* (Agaricaceae). *Nova Hedwigia*, **59**, 47–75.

Griffith, G. S., Rayner, A. D. M. & Wildman, H. G. (1994*b*). Interspecific interactions, mycelial morphogenesis and extracellular metabolite production in *Phlebia radiata* (Aphyllophorales). *Nova Hedwigia*, **59**, 331–44.

Griffith, G. S., Rayner, A. D. M. & Wildman, H. G. (1994*c*). Extracellular metabolites and mycelial morphogenesis of *Hypholoma fasciculare* and *Phlebia radiata* (Hymenomycetes). *Nova Hedwigia*, **59**, 311–29.

Halliwell, B. & Gutteridge, J. M. C. (1989). *Free Radicals in Biology and Medicine*, 2nd edn. Oxford: Clarendon Press.

Hansberg, W. & Aguirre, J. (1990). Hyperoxidant states cause microbial cell differentiation by cell isolation from dioxygen. *Journal of Theoretical Biology*, **142**, 201–21.

Prigogine, I. & Stengers, I. (1984). *Order out of Chaos*. London: Heinemann.

Rayner, A. D. M. (1991). The challenge of the individualistic mycelium. *Mycologia*, **83**, 48–71.

Rayner, A. D. M. (1994). Pattern-generating processes in fungal communities. In *Beyond the Biomass*, ed. K. Ritz, J. Dighton & K. E. Giller, pp. 247–58. Chichester: Wiley.

Rayner, A. D. M. (1996*a*). Has chaos theory a place in environmental mycology? In *Fungi and Environmental Change*, ed. J. C. Frankland, N. Magan & G. M. Gadd, pp. 317–41. Cambridge: Cambridge University Press.

Rayner, A. D. M. (1996*b*). Antagonism and synergism in the plant surface colonization strategies of fungi. In *Microbiology of Aerial Plant Surfaces*, ed. C. E. Morris, P. Nicot & C. Nguyen-The, pp. 139–54. New York and London: Plenum Press.

Rayner, A. D. M. (1996*c*). Interconnectedness and individualism in fungal mycelia. In *A Century of Mycology*, ed. B. C. Sutton, pp. 193–232. Cambridge: Cambridge University Press.

Rayner, A. D. M. (1997*a*). *Degrees of Freedom – Living in Dynamic Boundaries*. London: Imperial College Press.

Rayner, A. D. M. (1997*b*). Evolving boundaries – the systemic origin of phenotypic diversity. *Journal of Transfigural Mathematics*, **3** (2), 13–22.

Rayner, A. D. M. & Boddy, L. (1988) *Fungal Decomposition of Wood*. Chichester: John Wiley.

Rayner, A. D. M. & Coates, D. (1987). Regulation of mycelial organization and responses. In *Evolutionary Biology of the Fungi*, ed. A. D. M. Rayner,

C. M. Brasier & D. Moore, pp.115–35. Cambridge: Cambridge University Press.

Rayner, A. D. M., Griffith, G. S. & Ainsworth, A. M. (1995*a*). Mycelial interconnectedness. In *The Growing Fungus*, ed. N. A. R. Gow & G. M. Gadd, pp. 21–40. London: Chapman & Hall.

Rayner, A. D. M., Powell, K. A., Thompson, W. & Jennings, D. H. (1985). Morphogenesis of vegetative organs. In *Developmental Biology of Higher Fungi*, ed. D. Moore, L. A. Casselton, D. A. Wood & J. C. Frankland, pp. 249–79. Cambridge: Cambridge University Press.

Rayner, A. D. M., Ramsdale, M. & Watkins, Z. R. (1995*b*). Origins and significance of genetic and epigenetic instability in mycelial systems. *Canadian Journal of Botany*, **73**, S1241–58.

Rayner, A. D. M. & Ross, I. K. (1991). Sexual politics in the cell. *New Scientist*, **129**, 30–33.

Read, D. (1997). The ties that bind. *Nature*, **388**, 517–18.

Rehmann, A. U. & Thurston, C. F. (1992). Purification of laccase I from *Armillaria mellea*. *Journal of General Microbiology*, **138**, 1251–7.

Simard, S. W., Perry, D. A., Jones, M. D., Myrold, D. D., Durall, D. M. & Molina, R. (1997). Net transfer of carbon between ectomycorrhizal tree species in the field. *Nature*, **388**, 579–82.

Thompson, W. & Rayner, A. D. M. (1983). Extent, development and functioning of mycelial cord systems in soil. *Transactions of the British Mycological Society*, **81**, 333–45.

Wessels, J. G. H. (1994). Developmental regulation of fungal cell wall formation. *Annual Review of Phytopathology*, **32**, 413–37.

Worrall, J. J., Chet, I. & Hüttermann, A. (1986). Association of rhizomorph formation with laccase activity in *Armillaria* spp. *Journal of General Microbiology*, **132**, 2527–33.

2

Nutrient translocation and electrical signalling in mycelia

S. OLSSON

Nutrient translocation in fungi has recently been covered in several reviews (Cairney, 1992; Boddy, 1993; Jennings, 1994; Cairney & Burke, 1996). This chapter adds some new results and discusses nutrient translocation and electric signalling as parts of the integration of activities within the fungal mycelium. Some parts of the chapter are rather speculative but this is meant both to provoke the reader to come up with better ideas and to encourage more people to start research into the subjects of this chapter.

Discussion of definitions: colony and mycelium. A suggestion for using the terms Functional Mycelium Unit and Genetic Mycelium Unit

What is a fungal colony and what is a fungal mycelium? According to *Ainsworth & Bisby's Dictionary of Fungi* (Hawksworth *et al.*, 1995) a fungal mycelium is 'a mass of hyphae' and a fungal colony is 'a group of hyphae which, if from one spore, may be one individual'. The definitions given by Jennings & Lysek (1996) are 'Mycelium = network of hyphae' and 'Colony = the coherent mycelium of one origin'. However, Rayner argues that the mycelium is a functional unit, an individual (Rayner, 1991). It seems we ought to see the mycelium and the fungal colony as they have been discussed historically, as a mixture of two distinctly different concepts.

1. Genetically identical hyphae occupying a continuous space, the Genetic Mycelium Unit (GMU).
2. The network of functionally integrated hyphae that forms an individualistic organism, the Functional Mycelium Unit (FMU).

GMU and FMU seem similar to, but is not the same as, genet and ramet (Brasier & Rayner, 1987). GMU and FMU describe the functional organization of fungal mycelia and genet and ramet are terms describing the genetic relationships between isolates of a fungal species. As genet and ramet have been used in the literature on clonal plants a genet can be a collection of isolates in test tubes with the same genetic constitution and a ramet can be physically linked to and exchange nutrients with other ramets (Abrahamson, Anderson & McCrea, 1991; Caraco & Kelly, 1991; Kelly, 1995; Cain, Dudle & Evans, 1996). A GMU can consist of many FMUs (Fig 2.1) and there might be a dynamic relationship between the GMU and the FMU in that the FMUs might join each other through anastomoses (Rayner, 1991) and form a big FMU the same size as the GMU. The maximum size of the FMU is thus the whole GMU. It has been shown that large FMUs have a better capacity to spread in an environment and to compete with other fungal species (Dowson, Rayner & Boddy, 1988; Bolton, Morris & Boddy, 1991; Holmer & Stenlid, 1993). This behaviour resembles a feudal state where warlords, tied to each other by intermarriage and each with their own stronghold, join forces into one organized army to attack and conquer other strongholds (and thus resources) held by other states.

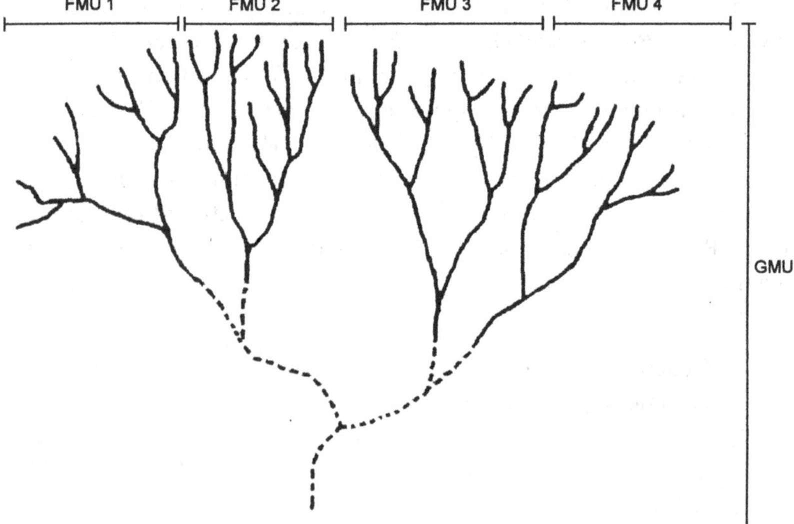

Fig. 2.1. A GMU might contain several FMUs. Drawing adapted from Carlile (1995).

The actor in this drama is the FMU and the study of the FMU should thus be set in context when trying to understand the behaviour of fungi in nature. Consequently the behaviour of FMUs of different species might vary considerably since this behaviour is likely to be a result of the interactions between the fungus and the environment. It is also the FMU that reproduces and forms spore-bearing structures. Seen from a bacterial point of view, fungi are all alike since they have comparatively similar cell physiology and thus ecologically significant species differences are more likely to be found at the mycelium organizational and behavioural level.

Studies of the Functional Mycelium Unit (FMU)

A FMU has one essential feature, which is communication between the different parts of the FMU such that the FMU exhibits some kind of integrated behaviour (Chapter 6, S. Watkinson). Communication can in principle occur in two ways, by sending goods or by sending messages. The sending of goods is normally termed nutrient translocation, organelle movement, protoplasmic streaming, etc. and the sending of messages, signalling.

Nutrient translocation

Four different mechanisms for nutrient translocation can be identified.

1. Passive

Nutrients are taken up by the hyphae according to the needs of the hyphae in the region of uptake. Translocation inside the hyphae occurs by simple diffusion as in any water phase. In a water-saturated growth system there would be no or little difference between diffusion inside the hyphae or outside the hyphae. The difference lies in that diffusion inside the hyphae is mainly one-dimensional, which compared to three-dimensional diffusion gives higher fluxes (Olsson & Jennings, 1991). In a non-saturated soil filled with air gaps the hyphae forms bridges so that nutrients can spread by symplastic or apoplastic diffusion. This type of translocation would not require any extra energy.

2. Passive + active uptake

This mechanism is similar to the passive but nutrients are taken up far in excess of the local needs (Olsson & Jennings, 1991), creating a steep nutrient gradient inside the hyphae. This results in a higher nutrient flux in the diffusion process and thus in a faster delivery of nutrients to remote parts of the mycelium. This type of translocation requires extra energy for the excess uptake but not for the actual translocation process, which is simple diffusion.

3. Active, cytoplasmic

Nutrient translocation has often been associated with active movements in the cytoplasm, either through movements of organelles (Schütte, 1956; Cooper & Tinker, 1981) or through peristaltic vacuole systems (Shepherd, Orlovich & Ashford, 1993*a*,*b*). This general mechanism involves a compartmentalization of the nutrients followed either by a translocation of the compartment itself or by a translocation of nutrients driven by peristaltic movements in a tubular compartment (Shepherd *et al.*, 1993*a*,*b*). In this type of translocation energy is required both for the compartmentalisation and for the actual translocation process.

4. Active, pressure-driven bulk flow

If the active uptake mechanism outlined above in (2) results in a high osmotic pressure inside the hyphae, water will flow in through the semi-permeable cell membrane. This will create a higher turgor inside the hyphae and water and nutrients will flow inside the mycelium in the direction of least resistance. Resistance comes from either the cell wall resistance to deformation or from resistance against water flow within the hyphae (friction). This type of translocation requires energy for both the active uptake and to overcome the resistance to water flow within the hyphae. Seen from this perspective it is no wonder that special organs, cords and rhizomorphs, have developed in fungi forming large mycelia (see Watkinson, Chapter 6). These organs often contain large thick-walled vessel hyphae free from cytoplasm, which offer low resistance to water flow and little wall deformation (Cairney, 1992).

It is important to note that all these mechanisms of nutrient translocation can be in operation simultaneously in a mycelium. Mechanisms (1) to (3) can be bi-directional in the same hyphae whereas it is difficult to envisage this occurring in mechanism (4) (Cairney, 1992). Only in mechanism (1) is extra energy not required for the net translocation of nutrients between a source and a sink. In mechanisms (3) and (4) the translocation process itself requires energy.

Examples of nutrient translocation

Plate cultures with diffusion barriers

Most studies of nutrient translocation in agar systems have been carried out in plates and in mycelia growing from a source of nutrients over a diffusion barrier onto a non-nutrient area (e.g. Girvin & Thain, 1987 and Thain & Girvin, 1987). The measured translocation has in most cases been in the direction of growth although in some early experiments translocation in the reverse direction was noted (Lyon & Lucas, 1969). More recently, in a series of papers on nutrient translocation in *Morchella esculenta* (Amir *et al.*, 1992, 1993, 1995*a*,*b*) demonstrated that nutrients could be translocated towards developing sclerotia in the older mycelium. They also showed by direct and indirect turgor potential measurements that translocation to the sclerotia takes place via the mechanism of turgor-driven mass flow (Amir *et al.*, 1992, 1995*b*). Interestingly, they also showed by thc use of inhibitors that translocation was affected by activities in both the source and the sink (Amir *et al.*, 1995*a*).

Reallocation patterns and movements in one-dimensional growth systems

Growth systems can be made to be in principle one-dimensional as in the race-tubes of Ryan familiar to most mycologists (Fig. 4.2) (Ryan, Beadle & Tatum, 1943). Studying nutrient reallocation in one-dimensional growth systems is advantageous in that the interpretation of the reallocation pattern and the analysis can be made simpler. Another advantage is that it is easy to create reproducible nutrient gradients. We have employed such a system in the study of nutrient reallocation in three moulds (Olsson & Jennings, 1991). The system consisted of glass fibre strips filled with nutrient solutions and it was possible to create gradients between a glucose side and a mineral nutrient side (Fig. 2.2). Thus the fungus growing out from the centre experienced a carbon/energy surplus at one end of the mycelium and a mineral surplus at the other end. It was then investigated how added label ([^{14}C]glucose or [^{32}P]orthophosphate) in different parts of the mycelium became reallocated. The label was only reallocated from its addition point if it was added at the end of the mycelium which had a surplus of the same nutrient. The reallocated nutrients became distributed as in a diffusion but all label was located inside the mycelium, thus the translocation observed was most likely to have occurred through diffusion inside the hyphae, aided by an uptake in excess of the local need in the hyphae. We also employed a similar agar medium based system for

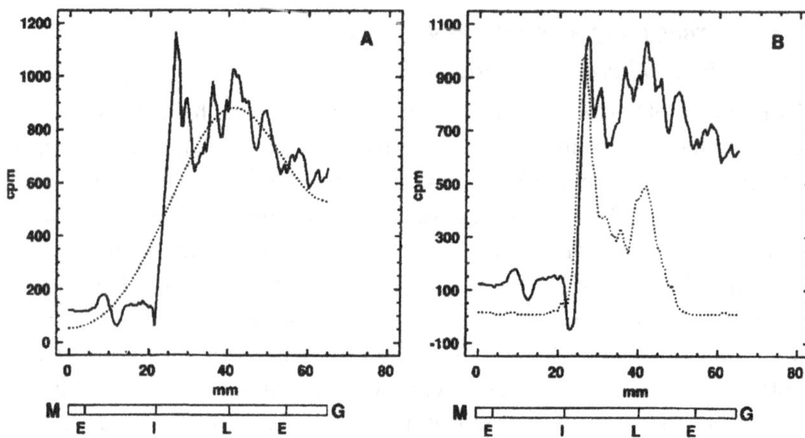

Fig. 2.2. (A) Total [^{14}C] label added as glucose (solid line) in the glass fibre strip cultures of *Rhizopus nigricans* and theoretical distribution (dotted line) if the label had been moved by diffusion. (B) Distribution of label in the medium (solid line) and in the mycelium (dotted line). Olsson & Jennings, 1991, reproduced with permission.

studying the biomass density patterns that would result if nutrients are passively or actively reallocated in fungi (Olsson, 1995). There should be marked differences in biomass density profiles if fungi are able to translocate the main nutrients needed in the mycelium by one of the active processes (Mechanism 3 or 4, above) or if they have to rely solely on diffusion. The postulated profiles can be seen in Fig. 2.3. In Fig. 2.4 a fungus with active translocation is compared with one showing a profile that is predicted for passive diffusion. In the study, 62 fungi were tested and large differences in biomass profiles were observed (Olsson, 1995). Roughly a third of the fungi showed biomass profiles that would be typical for passive translocation and another third of the fungi showed biomass profiles that indicated active translocation of all the main nutrients. The remaining fungi translocated either carbon and the most important mineral(s) passively or carbon passively and the most important mineral(s) actively.

Reallocation patterns and movements in two-dimensional growth systems

Reallocation of label in two-dimensional systems can be followed by addition of label followed by autoradiography at the end of the experiment (Finlay & Read, 1986*b*) or by repeated autoradiography (Gray *et* *al.*, 1995). A method that better captures the dynamics and the pattern of reallocation is to use a gel scanner. This has been done both for intact mycorrhizal systems (Timonen *et al.*, 1996) and for agar-grown fungal cultures (Olsson & Gray, 1998). This latter study showed that phosphorus and a non-metabolizable [^{14}C]labelled amino acid analogue ([^{14}C]2 aminoisobutyric acid, [^{14}C]AIB) suitable for tracing the amino acid pools in fungi (Lilly, Higgins & Wallweber, 1990) moved actively and showed special patterns of movements in the mycelium (Fig. 2.5). There was a quick radial movement to and from the centre of the colony along specific pathways in the mycelium. These paths were macroscopically invisible and seem to represent a functional specialization of the mycelium not followed by readily recognizable morphological changes. There was also a tangential movement of label along the hyphal front indicating that there must be a great deal of integration of the mycelium through anastomoses relatively close to the hyphal front. Label also moved with the growing hyphal front. What was mostly surprising in this study was that the same label was seen moving in opposite directions in many parts of the mycelium. This could indicate that nutrients are 'circulated' in the FMU in a similar manner as blood circulation with a sink region in the FMU removing nutrients from the stream passing by. Another important observation in the study was that label added was immobilized at the point of addition to the mycelium (Fig. 2.6a). When subtracting an earlier label distribution from a later one, it was observed that the amount of immobilized label decreases while the label at the growing edge of the mycelium increases (Fig. 2.6b). The immobilized label is thus not permanently immobilized since it obviously partakes in the reallocation of label in the mycelium.

Reallocation of nutrients in mycelial cord systems

Many studies of nutrient translocation have been centred on fungal mycelia that form specialized linear organs, cords and rhizomorphs. These linear organs have mostly been recognized as stuctures aiding nutrient reallocation within large mycelia. Several studies have employed isolated cords for the study of translocation (Brownlee & Jennings, 1982; Granlund, Jennings & Thompson, 1985). Studies of intact mycelium are more revealing in terms of the integrated behaviour of the FMU. The majority of these studies point towards a preferential translocation to physiological sinks for particular compounds. In the cord systems these sinks seem to be at sites with undifferentiated hyphal growth at the advancing mycelial front (Finlay & Read, 1986*a*,*b*; Wells,

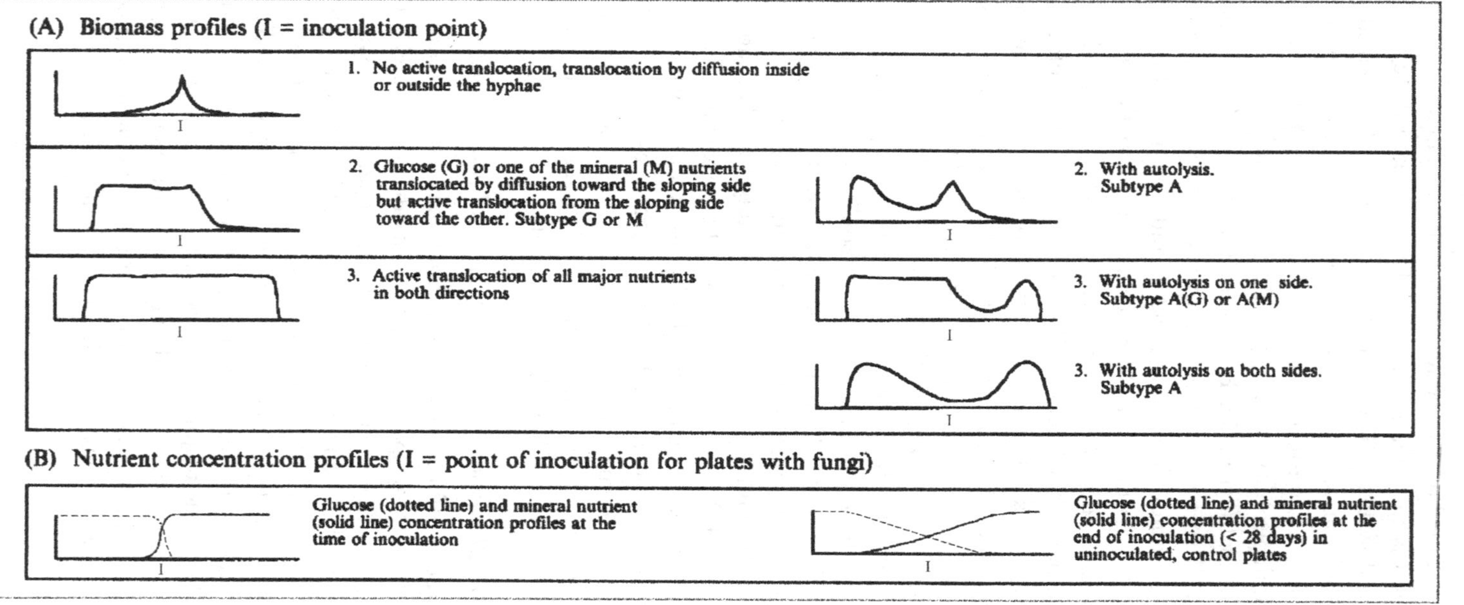

Fig. 2.3. Postulated biomass density profile types due to difference in ability to translocate nutrients when fungi grow on 20 × 2 cm agar plates with half of the plate filled with glucose medium (to the left) and the other half of the plate filled with mineral nutrient medium (to the right). Olsson, 1995, reproduced with permission.

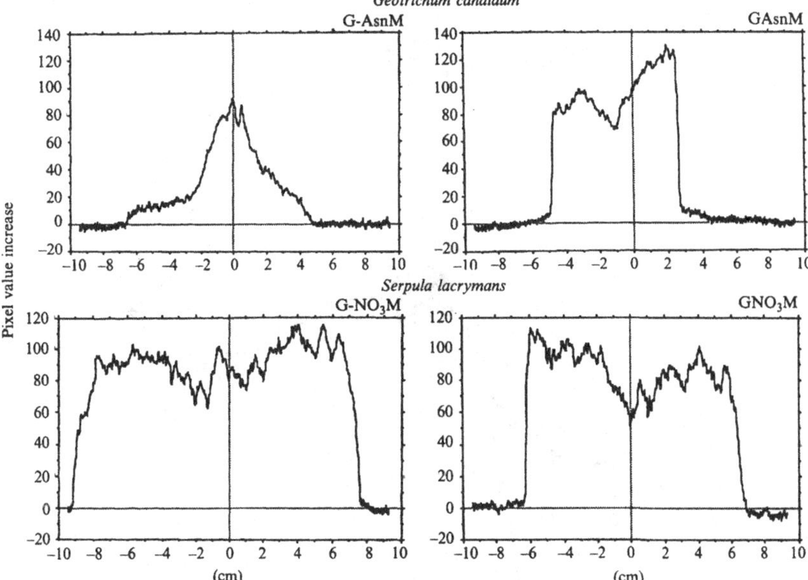

Fig. 2.4. *Geotrichum candidum* biomass density profile on normal medium (GAsnM, top right) and on gradient medium with carbon to the left and minerals to the right (G-AsnM, top left). *Serpula lacrymans* on normal medium (GNO_3M, bottom right) and on gradient medium (G-NO_3M, bottom left). Olsson, 1995, reproduced with permission.

Hughes & Boddy, 1990; Wells, Boddy & Evans, 1995) or to recently colonized pieces of wood resources (Boddy, 1993; Wells *et al.*, 1995). An especially interesting paper is one by Hughes & Boddy (1994). In this paper they studied nutrient translocation between recently colonized wood resources instead of from old resource to a new resource in the direction of mycelium growth (Wells & Boddy, 1990; Wells *et al.*, 1990). The results showed that phosphorus taken up somewhere in the FMU was distributed around the FMU and collected at sites where the demand was greatest (Hughes & Boddy, 1994). They also pointed out that since cord systems (FMUs) can extend for several metres on the forest floor they might be responsible for considerable spatial relocation of nutrients. A good example of these systems is shown in Fig. 2.7 where the distribution of labelled phosphorus added at the inoculum is shown 2 weeks after the addition (Boddy, 1993). Figure 2.7 also illustrates a possible way to map the FMU, since a practical definition of the mycelium unit is that it shares added nutrients. Phosphorus

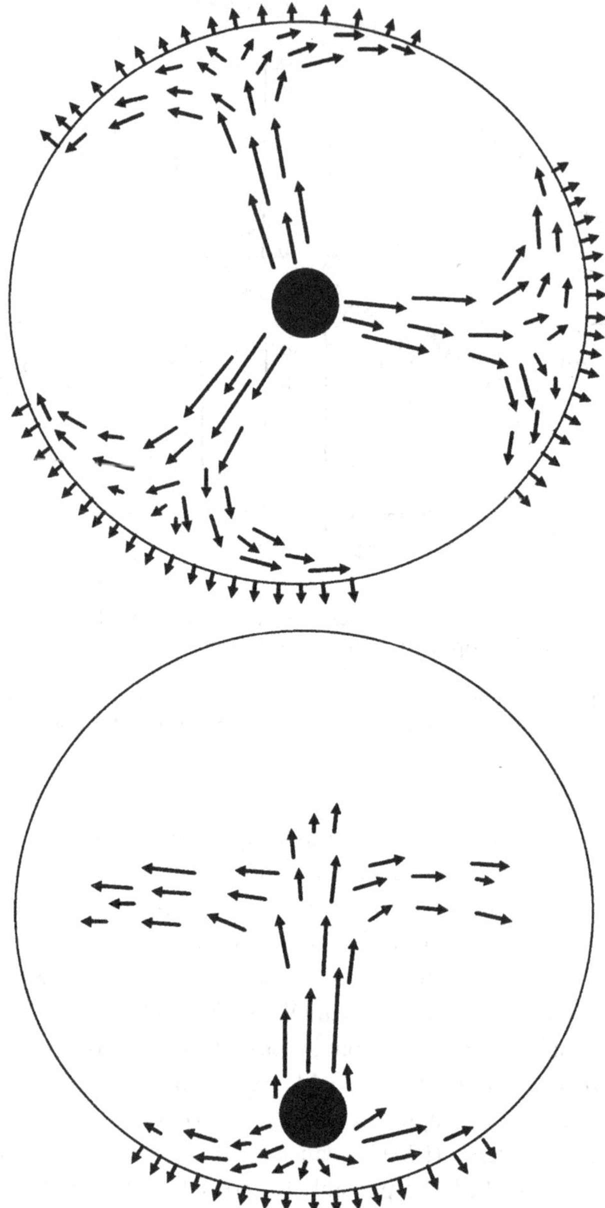

Fig. 2.5. Schematic drawing of the reallocation pattern of labelled phosphorus when added at the centre of the colony of *Pleurotus ostreatus* (above) and when added at the edge of the colony (below) of the same fungus.

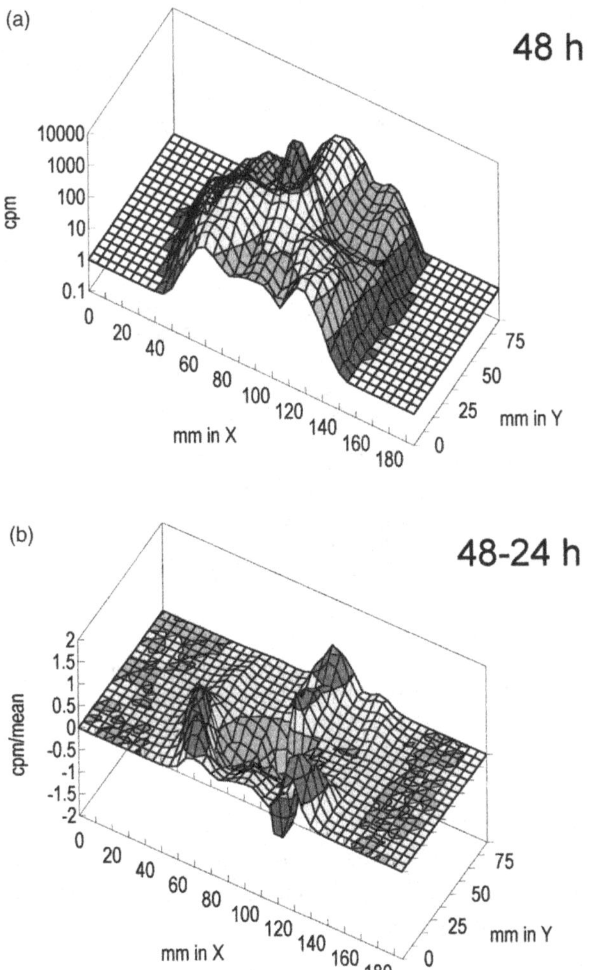

Fig. 2.6. Distribution of labelled phosphorus 48 h after addition of label at the centre of the colony (a) and difference plot showing the difference between the distribution at 48 h and 24 h (b). Olsson & Gray, 1998, reproduced with permission.

might also be the ideal tracer to outline FMUs since it does not become bound up in cell walls and seems to preferentially accumulate in growing portions of the FMU, in hyphal margins or in newly acquired resources. There are also two radioactive isotopes of phosphorus available, [^{32}P] and [^{33}P], which can readily be separated in analysis and make double labelling experiments possible.

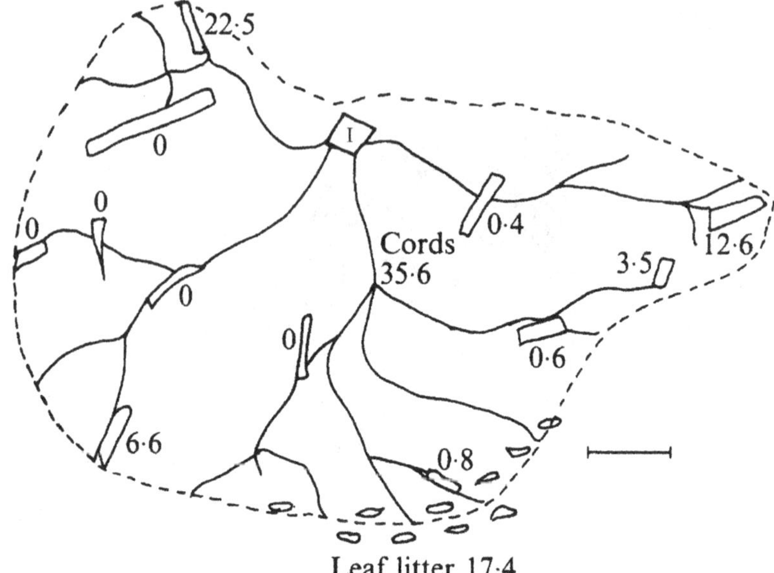

Fig. 2.7. Translocation of [^{32}P]orthophosphate, 2 weeks after addition to 'central' inoculum I, in a system of *Phanerochaete velutina*. Scale bar represents 20 cm. Figures are % translocate in different components of the system. From Boddy (1993), with permission.

Storage of nutrients and reallocation

To make a reallocation mechanism for nutrients effective there is a need for a way to store local excesses of the main nutrients, in particular for C, N and P, since none of these elements can exist in a low molecular weight and soluble form in the cytoplasm without negative consequences.

Storage of carbon

Carbon is most conveniently stored as lipids. This type of storage is common in fungi (Rosén *et al.*, 1997). Recently it was also shown that lipid bodies in *Rhizopus arrhizus* were translocated along distinct pathways that probably consisted of microtubules (Olsson, unpublished results). The translocation was bi-directional in the same hyphae and even in the same path. Thus lipid bodies were translocated in all directions in the FMU and might be a part of a mechanism for redistributing carbon/energy in the FMU and thus a part of a carbon/energy homeostasis mechanism for the whole FMU (Olsson, unpublished results).

Storage of nitrogen

Nitrogen can in principle be stored as a storage protein. In a paper by Rosén *et al.,* (1997), a lectin from the nematode-trapping fungus *Arthrobotrys oligospora* was shown to serve this purpose. The concentration of the lectin increased when the fungus was grown in nitrogen-rich media but was not formed when grown in nitrogen-poor media. After transfer from nitrogen-rich media to nitrogen-free media the fungus continued to grow while the internal pool of the lectin decreased. It took several successive transfers to nitrogen-free media before the growth rate slowed down considerably and the internal pool of lectin was fully utilized (Fig. 2.8). Immunological localization showed that the lectin was not bound to any organelles but was found free in the cytoplasm.

Immunofluorescence microscopy of the different stages during nematode capture showed that thc lectin was not formed before or during capture. It was formed in the trophic hyphae during digestion of the nematode and was also found later in the mycelium close to the nematode trap. The protein is similar in primary sequence and in binding properties to a previously described lectin in *Agaricus bisporus* (Rosén *et al.*, 1996*a,b*; Crenshaw *et al.*, 1995) and the antibodies for the *A. oligospora* lectin also recognized the *A. bisporus* lectin with high specificity (Rosén *et al.,* 1997). It might thus be that these lectins are highly conserved general nitrogen storage proteins in many fungi.

Storage of phosphorus

Previously, phosphorus was thought to be stored and translocated as insoluble polyphosphate granules (Cox *et al.,* 1980). Recently, however, Orlovich & Ashford (1993) have presented convincing evidence that these polyphosphate granules may have been artifacts of vacuoles loaded with a potassium polyphosphate solution. Thus, at present, the form in which phosphorus is stored and translocated in fungi remains to be determined.

Cell death and nutrient reallocation

There are many observations of hyphal degeneration in relation to mycelium growth. An example of this is when cords of wood-decomposing fungi grow out in all directions over a low nutrient area from a previously colonized food source. When some of the hyphae reach a new food source the cords growing in other directions quickly degenerate (Dowson *et al.*, 1989; Rayner, 1991). This degeneration seems to be con-

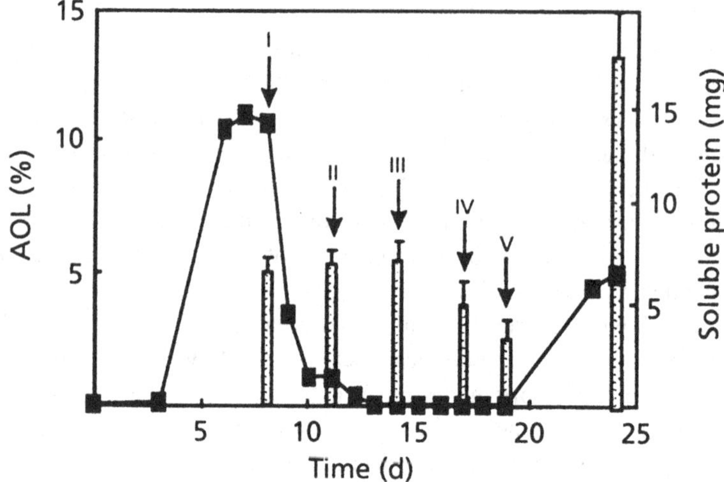

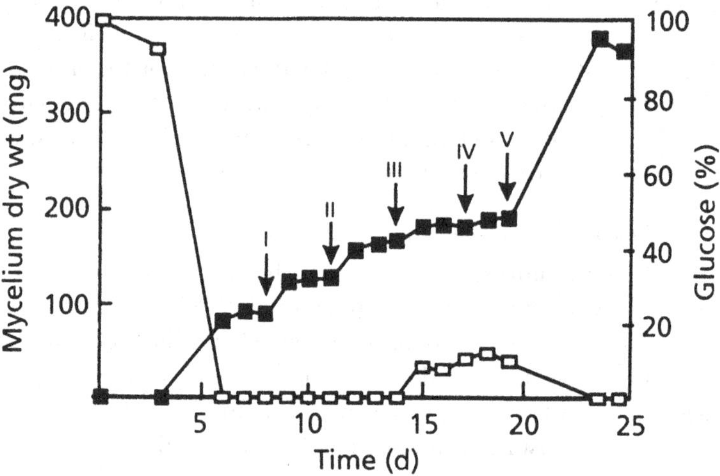

Fig. 2.8. Degradation of *Arthrobotrys oligospora* lectin (AOL) during starvation. Mycelia of *A. oligospora* grown for 8 days in a medium with C/N ratio of 6 were transferred four times (indicated by numbers I–IV) to new vessels containing medium with C/N ratio of 1000 (volume 150 ml) and thereafter to medium with C/N ratio of 6 (transfer V; volume 300 ml). (Top) AOL content, expressed as percentage of the total amount of saline-soluble proteins in the mycelium. Columns, total amount (mean) of saline-soluble proteins present in the mycelium (bars show SD with $n = 4$). (Bottom) Dry weight (filled symbols) of the transferred mycelia and the level of glucose (open symbols) in the culture filtrate (expressed as the percentage of the total level of sugars present in the fresh media). From Rosén *et al.* (1997), reproduced with permission.

trolled by the FMU in order to reallocate resources into more actively growing parts of the FMU. Cell degeneration controlled by the organism's own DNA is in other systems called apoptosis. Most eukaryotes seem to employ the same mechanism for apoptosis, including filamentous fungi (Marek *et al.*, 1997; Umar & Van Griensven, 1997). For this reason, any degeneration of hyphae that appears to be controlled by the FMU will be called apoptosis, although the strict definition of apoptosis requires experimental evidence for the appearance of several key signs. One of the best such indicators of apoptosis is that the DNA fragments into lengths which are multiples of 200 bp (due to the production of endonucleases that cleave the DNA between the regularly spaced histones) and that these changes occur without increased permeability of the cell membrane (Darzynkiewicz, Li & Gong, 1994).

Nutrient reallocation – concluding remarks

From what has been presented above it would appear that nutrients are reallocated in varying degrees in different fungi. Some fungi, notably the cord-forming basidiomycetes, appear to be equipped with the most advanced system for integrating resource use within the whole FMU and capable of doing this in large FMUs. The degree of integration of resource use in the FMU and the capacity for uptake and storage is likely to be different in different species and a reflection of the ecological niche of the particular species. Another important consideration is that nutrient translocation and net nutrient translocation need to be distinguished. Just because there was a net nutrient translocation recorded between A and B in a mycelium it does not follow that the same nutrients could not have been translocated from B to A at the same time.

The possibility for electric signalling within an FMU

Certain reactions in FMUs occur quickly and would require a fast signalling system (Rayner, Griffith & Wildman, 1994). Turgor changes have been suggested as a mechanism for fast signalling (Rayner *et al.*, 1994) although electric signalling might also be possible. The reason why electric signalling might work is that due to septal pores and anastomoses the FMU can be seen as a cytoplasmic continuum enveloped in a common membrane. In both plants and animals fast signalling is often carried by travelling depolarizations along cell membranes. Thus it should be possible for fungi to use this means of communication. In a study by Olsson

& Hansson (1995) we demonstrated high action potential-like activity in *Pleurotus ostreatus* and *Armillaria bulbosa*. This activity was also sensitive to external stimuli at centimetre distances from the intracellular insertion point of the microelectrode (Fig. 2.9) (Olsson & Hansson, 1995). Previously action potential-like activities have been measured in another filamentous fungus, *Neurospora crassa* (Slayman *et al.*, 1976) but the activity level observed was very low and was not sensitive to stimulation. We chose fungi with large long-lived mycelia that might have more complex integration of activity within their FMUs and thus a potentially greater need for a fast communication system. We inserted the microelectrodes in hyphal aggregations like cords, which stopped hyphae from sliding off the electrode tip when the electrode was

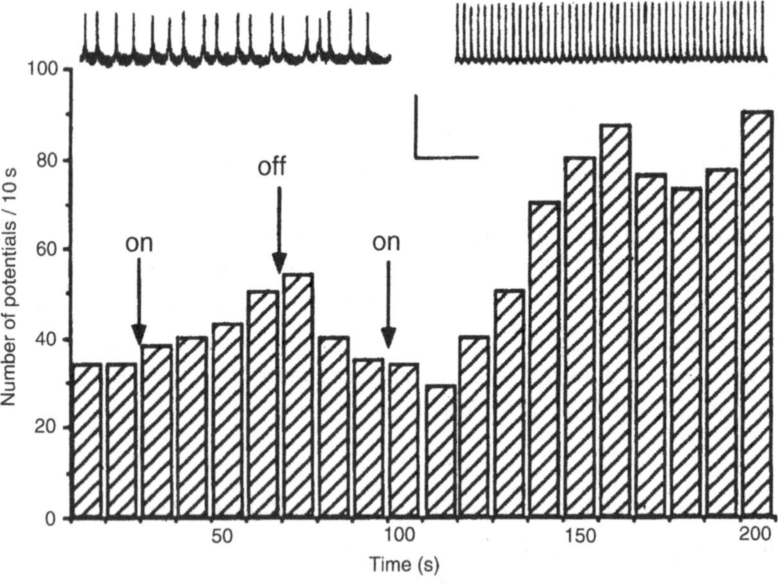

Fig. 2.9. Histogram displaying the effect of applying a piece of freshly cut beech wood to the surface of the mycelium of *Armillaria bulbosa*, 2 cm from the insertion point of the microelectrode. The wood was applied, removed, and applied again to demonstrate the reversibility of the response. Note the increase in the frequency of potentials when wood is added, and the decrease when it is removed. Above the histogram the raw data before stimulation and after is displayed. Horizontal scale bar 1 s, vertical scale bar 20 mV. From Olsson & Hansson (1995) reproduced with permission.

pressed through the cell wall. Thus, we do not know if action potentials occur in diffuse, non-differentiated mycelia. Since this study, action potential activity has been found in several other cord-forming fungi (Table 2.1), including mycorrhizal fungi in intact plant–fungal associations (Olsson & Hansson, unpublished observations).

Hypothesis for how fungi use action potential signalling

Three main hypotheses for the use of action potential signalling within a FMU can be identified.

1. The action potential may be irrelevant to the overall growth and organization of the mycelium.
2. Sensing the environment: signalling from hyphae experiencing changing external conditions to the rest of the mycelium about the external conditions.
3. Organizational: a signal used for integrating and controlling differentiation within the FMU.

Hypothesis 1 is not exciting and is very difficult to test. Hypothesis 2 requires some kind of sensory system and some way to process the information (a brain) and would also be difficult to test. Hypothesis 3 is perhaps the easiest to test and there are observations, and properties of action potential propagation, that make it the best candidate for future work. One unusual observation was that the frequency of action potential firing could be increased artificially by a slight *hyperpolarization* of the cell membrane. It has previously been shown that the fungal membrane is hyperpolarized a short distance behind an actively growing tip while depolarized at the tip creating electric currents (Gow, 1984). The hyperpolarization behind the tip also disappeared when the tip stopped extending (Gow, 1984). This might mean that the hyperpolarized portion of the membrane behind the tip triggers action potentials travelling from the tip region back into the FMU. The frequency of these potentials would increase with increasing growth rate of the tip. Another feature of action potentials is that if they meet in the same membrane they would inhibit each other since the membrane cannot depolarize a short time after a previous depolarization (Fig. 2.10). However, if two action potentials travel back from the branches into a common trunk, the frequency in the trunk would be an addition of the two branch frequencies. Thus in a branching hyphal system the hyphae or cords that connect actively growing hyphal tips would receive higher action potential signalling activity

Table 2.1. *Cord-forming basidiomycetes tested for action-potential-like activities (Olsson & Hansson, unpublished results)*

	AP present
Armillaria bulbosa	Yes
Coprimus micaceus	Yes
Hericium coralloides	Yes
Hypholoma fasciculare	Yes
Paxillus involutus	
(intact mycorrhizal association)	
cords	Yes
root tips	Yes
Phallus impudicus	Yes
Pleurotus ostreatus	Yes
Serpula lacrymans	Yes
Trechyspora vaga	No

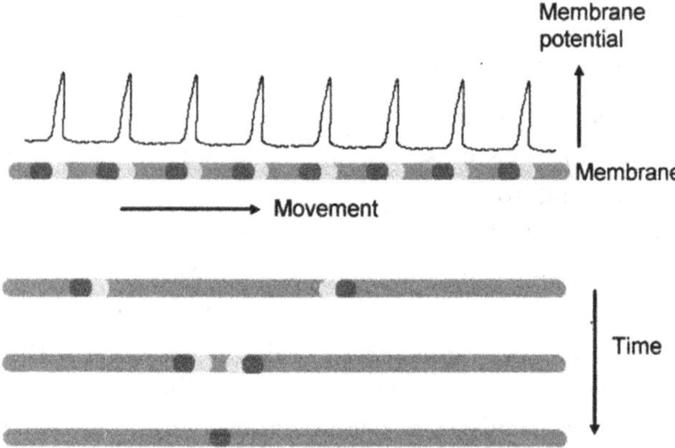

Fig. 2.10. Immediately after an action potential (depolarization, light grey) the membrane cannot trigger a new action potential (dark). Therefore the potentials will travel in one direction in the membrane. If two action potentials travel in opposite direction in a membrane and meet they will cancel each other.

than hyphae that do not connect to actively growing hyphae. In nerve and muscle cells it has been shown that an action potential input is needed from a connecting input neuron, otherwise the structures degenerate through apoptosis or apoptosis-like mechanisms (transneuronal degeneration). The early steps of the degeneration can also be reversed if the connection is restored (Rubel *et al.* 1991). Thus action potential activity in the hyphae may control growth and differentiation within large FMUs by preventing hyphae from undergoing apoptosis and instead inducing and strengthening cord formation (Fig. 2.11). Similar mechanisms could thus be controlling the relatively quick degeneration (apoptosis) of cords not connecting to new resources when some of the cords in the FMU have reached a new resource, as mentioned above.

A great advantage with the organizational hypothesis is that it should be relatively easy to test since some very clear predictions from it can be made:

1. The frequency of action potentials in a branching hypha or cord system should be lower the further out into the branches one measures.
2. Anything that increases or decreases hyphal tip growth or causes a hyperpolarization of the cell membrane should also increase or decrease the frequency of action potentials in the hyphae or cords.
3. Artificially induced action potentials should be able to inhibit or slow down degeneration in areas that normally would die.

Electric signalling within a FMU – concluding remarks

If the organizational hypothesis proves to be true then specific points in the FMU would exist where high frequency action potentials travelling in opposite direction meet. These would be perfect places for the induction of morphological structures such as fruit bodies, which are large sinks for nutrients. It would also mean that electric signals trigger gene expression (apoptosis, differentiation), which could potentially be exploited in biotechnological applications by artificially inducing signals within a FMU and so regulating gene expression.

Conclusion

Built into the organization of the animal cell is every possibility from being a unicellular organism to a human, from cells giving a solo performance to cells with defined roles in a large orchestra.

What of fungi? Why do we often seem to believe that all fungal mycelia have to be organized similarly? It should be possible to identify a large variety of radically different mycelial life-forms, if we just try.

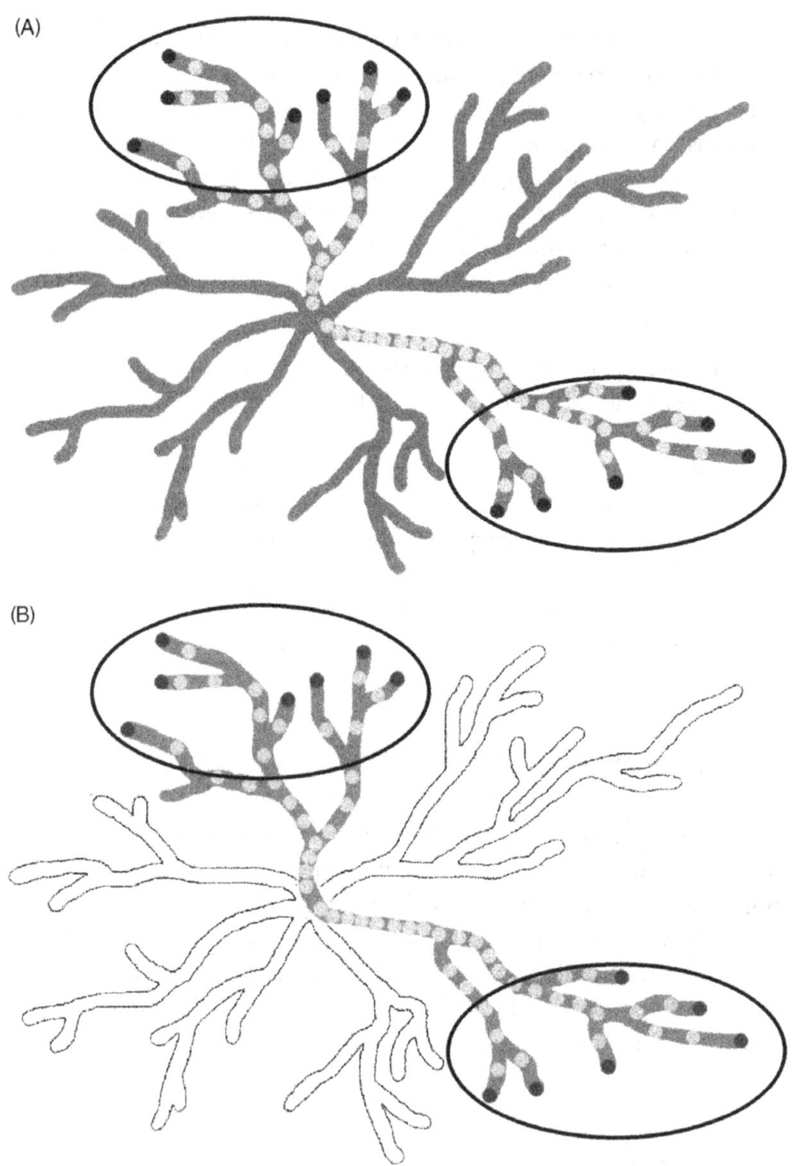

Fig. 2.11. Hypothesis for role of action potentials in the organization of mycelia. The hyperpolarization of the membranes behind actively growing hyphal tips (A) in areas with high nutrient availability (ovals) might generate action potentials travelling back into the older parts of the mycelium. The frequency of action potentials could be very high in mycelia connecting many growing hyphal tips due to additive effects. If high action potential activity stops the hyphae from degenerating (apoptosis) the hyphae not connecting with nutrients will eventually die (B).

References

Abrahamson, W. G., Anderson, S. S. & McCrea, K. D. (1991). Clonal integration nutrient sharing between sister ramets of *Solidago altissima* Compositae. *American Journal of Botany*, **78**, 1508–14.

Amir, R., Levanon, D., Hadar, Y. & Chet, I. (1992). Formation of sclerotia by *Morchella esculenta*: relationship between media composition and turgor potential in the mycelium. *Mycological Research*, **96**, 943–8.

Amir, R., Levanon, D., Hadar, Y. & Chet, I. (1993). Morphology and physiology of *Morchella esculenta* during sclerotial formation. *Mycological Research*, **97**, 683–9.

Amir, R., Levanon, D., Hadar, Y. & Chet, I. (1995*a*). Factors affecting translocation and sclerotial formation in *Morchella esculenta*. *Experimental Mycology*, **19**, 61–70.

Amir, R., Steudle, E., Levanon, D., Hadar, Y. & Chet, I. (1995*b*). Turgor changes in *Morchella esculenta* during translocation and sclerotial formation. *Experimental Mycology*, **19**, 129–36.

Boddy, L. (1993). Saprotrophic cord-forming fungi: warfare strategies and other ecological aspects. *Mycological Research*, **97**, 641–55.

Bolton, R. G. Morris, C. W. & Boddy, L. (1991). Non-destructive quantification of growth and regression of mycelial cords using image analysis. *Binary*, **3**, 127–32.

Brasier, C. M. & Rayner, A. D. M. (1987). Wither terminology below the species level in fungi? In *Evolutionary Biology of the Fungi*, ed. A. D. M. Rayner, C. M. Brasier & D. Moore, pp. 379–88. Cambridge: Cambridge University Press.

Brownlee, C. & Jennings, D. H. (1982). Long distance translocation in *Serpula lacrimans*. Velocity estimates and the continuous monitoring of induced perturbations. *Transactions of the British Mycological Society*, **79**, 143–8.

Cain, M. L., Dudle, D. A. & Evans, J. P. (1996). Spatial models of foraging in clonal plant species. *American Journal of Botany*, **83**, 76–85.

Caraco, T. & Kelly, C. K. (1991). On the adaptive value of physiological integration in clonal plants. *Ecology*, **72**, 81–93.

Cairney, J. W. G. (1992). Translocation of solutes in ectomycorrhizal and saprotrophic rhizomorphs. *Mycological Research*, **96**, 135–41.

Cairney, J. W. G. & Burke, R. M. (1996). Physiological heterogeneity within fungal mycelia: an important concept for a functional understanding of the ectomycorrhizal symbiosis. *New Phytologist*, **134**, 685–95.

Carlile, M. J. (1995). The success of the hypha and mycelium. In *The Growing Fungus*, ed. N. A. R. Gow & G. M. Gadd, pp. 3–19. London: Chapman & Hall.

Cooper, K. M. & Tinker, P. B. (1981). Translocation and transfer of nutrients in vesicular-arbuscular mycorrhizas. IV. Effect of environmental variables on movement of phosphorus. *New Phytologist*, **88**, 327–39.

Cox, G. C., Moran, K. J., Sanders, F., Nockolds, C. & Tinker, P. B. (1980). Translocation and transfer of nutrients in vesicular-arbuscular mycorrhizas. III. Polyphosphate granules and phosphorus translocation. *New Phytologist*, **84**, 649–59.

Crenshaw, R. W., Harper, S. N., Moyer, M. & Privalle, L. S. (1995). Isolation and characterization of a cDNA clone encoding a lectin gene from *Agaricus bisporus*. *Plant Physiology*, **107**, 1465–6.

Darzynkiewicz, Z., Li, X. & Gong, J. (1994). Assays of cell viability: discrimination of cells dying by apoptosis. In *Methods in Cell Biology*, Vol. 41, 2nd edn, Part A, ed. Z. Darzynkiewicz, J. P. Robinson & H. A. Crissman, pp. 16–39. San Diego: Academic Press.

Dowson, C. G., Rayner, A. D. M. & Boddy, L. (1988). The form and outcome of mycelial interactions involving cord-forming decomposer basidiomycetes in homogeneous and heterogeneous environments. *New Phytologist*, **109**, 423–32.

Dowson, C. G., Springham, P., Rayner, A. D. M. & Boddy, L. (1989). Resource relationships of foraging mycelial systems of *Phanaurochaete velutina* and *Hypholoma fasciculare* in soil. *New Phytologist*, **111**, 501–9.

Finlay, R. D. & Read, D. J. (1986*a*). The structure and function of vegetative mycelium of ectomycorrhizal plants. I. Translocation of ^{14}C-labelled carbon between plants interconnected by a common mycelium. *New Phytologist*, **103**, 143–56.

Finlay, R. D. & Read, D. J. (1986*b*). The structure and function of the vegetative mycelium of ectomycorrhizal plants. II. The uptake and distribution of phosphorus by mycelial strands interconnecting host plants. *New Phytologist*, **103**, 157–65.

Girvin, D. & Thain, J. F. (1987). Growth of and translocation in mycelium of *Neurospora crassa* on a nutrient deficient medium. *Transactions of the British Mycological Society*, **88**, 237–46.

Gow, N. A. R. (1984). Transhyphal electric currents in fungi. *Journal of General Microbiology*, **130**, 3313–18.

Granlund, H. I., Jennings, D. H. & Thompson, W. (1985). Translocation of solutes along rhizomorphs of *Armillaria mellea. Transactions of the British Mycological Society*, **84**, 111–19.

Gray, S. N., Dighton, J., Olsson, S. & Jennings, D. H. (1995). Real-time measurement of uptake and translocation of ^{137}Cs within mycelium of *Schizophyllum commune* Fr. by autoradiography followed by quantitative image analysis. *New Phytologist*, **129**, 449–65.

Hawksworth, D. L., Kirk, P. M., Sutton, B. C. & Pegler, D. N. (1995). *Ainsworth & Bisby's Dictionary of the Fungi – 8th Edition.* CAB International.

Holmer, L. & Stenlid, J. (1993). The importance of inoculum size for the competitive ability of wood decomposing fungi. *FEMS Microbiology Ecology*, **12**, 169–76.

Hughes, C. L. & Boddy, L. (1994). Translocation of ^{32}P between wood resources recently colonised by mycelial cord systems of *Phanerochaete velutina. FEMS Microbiology Ecology*, **14**, 201–12.

Jennings, D. H. (1994). Translocation in mycelia. In *The Mycota I: Growth, Differentiation and Sexuality*, ed. J. G. H. Wessels & H. Meinhardt, pp. 163–73. Berlin: Springer-Verlag.

Jennings, D. H. & Lysek, G. (1996). *Fungal Biology: Understanding the Fungal Lifestyle.* Oxford: Bios Scientific Publishers.

Kelly, C. K. (1995). Thoughts on clonal integration: Facing the evolutionary context. *Evolutionary Ecology*, **9**, 575–85.

Lilly, W. W., Higgins, S. M. & Wallweber, G. J. (1990). Uptake and translocation of 2-aminoisobutyric acid by *Schizophyllum commune. Experimental Mycology*, **14**, 169–77.

Lyon, A. J. E. & Lucas, R. L. (1969). The effect of temperature on the translocation of phosphorus by *Rhizopus stolonifera. New Phytologist*, **68**, 963–9.

Marek, S. M., Jacobson, D. J., Gilchrist, D. G. & Bostock, R. M. (1997). Investigations into apoptosis in fungi. Annual Meeting of the American Phytopathological Society, Rochester, New York, USA, August 9–13, 1997. *Phytopathology*, **87**, 63.

Olsson, S. (1995). Mycelial density profiles of fungi on heterogeneous media and their interpretation in terms of nutrient reallocation patterns. *Mycological Research*, **99**, 143–53.

Olsson, S. & Gray, S. N. (1998). Patterns and dynamics of ^{32}P-phosphate and labelled 2-aminoisobutyric acid (^{14}C-AIB) translocation in intact basidiomycete mycelia. *FEMS Microbiology Ecology*, **26**, 109–120.

Olsson, S. & Hansson, B. S. (1995). Action potential-like activity found in fungal mycelia is sensitive to stimulation. *Naturwissenschaften*, **82**, 30–1.

Olsson, S. & Jennings, D. H. (1991). Evidence for diffusion being the mechanism of translocation in the hyphae of three molds. *Experimental Mycology*, **15**, 302–9.

Orlovich, D. & Ashford, A. E. (1993). Polyphosphate granules are an artefact of specimen preparation in the ectomycorrhizal fungus *Pisolitus tinctorius*. *Protoplasma*, **173**, 91–105.

Rayner, A. D. M. (1991). The challenge of the individualistic mycelium. *Mycologia*, **83**, 48–71.

Rayner, A. D. M., Griffith, G. S. & Wildman, H. G. (1994). Differential insulation and generation of mycelial patterns. In *Linnean Society Symposium*, No 16, *Shape and Form in Plants and Fungi*, ed. D. S. Ingram & A. Hudson, pp. 291–310. London: Academic Press.

Rosén, S., Bergström, J., Karlsson, K.-A. & Tunlid, A. (1996a). A multispecific saline soluble lectin from the parasitic fungus *Arthrobotrys oligospora*. Similar binding specificities as a lectin from the mushroom *Agaricus bisporus*. *European Journal of Biochemistry*, **238**, 830–7.

Rosén, S., Kata, M., Persson, Y., Lipniunas, P. H., Wikström, M., van den Hondel, C. A. M. J. J., van den Brink, J. M., Rask, L., Hedén, L.-O. & Tunlid, A. (1996*b*). Molecular characterization of a saline soluble lectin from a parasitic fungus. Extensive sequence similarities between fungal lectins. *European Journal of Biochemistry*, **238**, 822–9.

Rosén, S., Sjollema, K., Veenhuis, M. & Tunlid, A. (1997). A cytoplasmic lectin produced by the fungus *Arthrobotrys oligospora* functions as a storage protein during saprophytic and parasitic growth. *Microbiology*, **143**, 2593–60.

Rubel, E. W., Falk, P. M., Canaday, K. S. & Steward, D. (1991). A cellular mechanism underlaying activity-dependent transneuronal degeneration: rapid but reversible destruction of neuronal ribosomes. *Brain Dysfunction*, **4**, 55–74.

Ryan, F. J., Beadle, G. W. & Tatum, E. L. (1943). The tube method of measuring the growth rate of Neurospora. *American Journal of Botany*, **30**, 789–99.

Schütte, K. H. (1956). Translocation in the fungi. *New Phytologist*, **55**, 164–82.

Shepherd, V. A., Orlovich, D. A. & Ashford, A. E. (1993*a*). A dynamic continuum of pleiomorphic tubules and vacuoles in growing hyphae of a fungus. *Journal of Cell Science*, **104**, 495–507.

Shepherd, V. A., Orlovich, D. A. & Ashford, A. E. (1993*b*). Cell-to-cell transport via motile tubules in growing hyphae of a fungus. *Journal of Cell Science*, **105**, 1173–8.

Slayman, C. L., Long, W. S. & Gradmann, D. (1976). Action potentials in *Neurospora crassa* a mycelial fungus. *Biochimica et Biophysica Acta*, **426**, 732–44.

Thain, J. F. & Girvin, D. (1987). Translocation through established mycelium of *Neurospora crassa* on a nutrient-free substrate. *Transactions of the British Mycological Society*, **89**, 45–9.

Timonen, S., Finlay, R. D., Olsson, S. and Söderström, B. (1996). Dynamics of phosphorus translocation in intact ectomycorrhizal systems: non-destructive monitoring using a β-scanner. *FEMS Microbiology Ecology*, **19**, 171–80.

Umar, M. H. & Van Griensven, L. J. L. D. (1997). Morphogenetic cell death in developing primordia of *Agaricus bisporus*. *Mycologia*, **89**, 274–7.

Wells, J. M. & Boddy, L. (1990). Wood decay and phosphorus and fungal biomass allocation in mycelial cord systems. *New Phytologist*, **116**, 285–96.

Wells, J. M., Boddy, L. & Evans, R. (1995). Carbon translocation in mycelial cord systems of *Phanerochaete velutina* (DC.: Pers.) Parmasto. *New Phytologist*, **129**, 467–76.

Wells, J. M., Hughes, C. & Boddy, L. (1990). The fate of soil-derived phosphorus in mycelial cord systems of *Phanerochaete velutina* and *Phallus impudicus*. *New Phytologist*, **114**, 595–606.

3

Colony development in nutritionally heterogeneous environments

K. RITZ AND J. W. CRAWFORD

Introduction

The majority of natural environments which fungi inhabit are heterogeneous in both space and time with respect to many factors, and this is particularly the case in the soil habitat. Purely structural non-uniformity is exemplified by the complex spatial architecture of soil, where a myriad of connecting and blind pore networks and tortuous surfaces prevail across a wide range of scales. Temperature and moisture profiles can vary markedly both seasonally and over short timescales (Marshall & Holmes, 1988), for example on a day with patchy cloud, temperatures near the surface can be highly dynamic. Biological heterogeneity is also the norm, with considerable spatio-temporal variation in microbial and faunal community structures being a characteristic of most soils. Nutrient resources are generally distributed patchily, and also may show seasonal variation, since they are often linked to plant growth cycles.

The degree of heterogeneity in these parameters is intimately linked to the scale under consideration. In spatial terms, all environments are heterogeneous at certain scales: at the atomic level, even a solution of salt is effectively non-uniform, and at the greatest of scales, it is apparent that the large-scale structure of the known universe is heterogeneous (Saunders *et al.*, 1991). In mycological terms, the spatial scale of reference is not as readily definable as it might at first seem, since mycelia are composed of operational units represented by hyphae of the order of a few microns in diameter, and thalli that may extend typically several centimetres, and occasionally several kilometres (Smith, Bruhn & Anderson, 1992). Whether heterogeneity is present as a gradient, i.e. continuously changing with respect to space, or more discrete and patchy, is also partly scale-dependent, and to an extent depends on the nature of the component under consideration. For example, from a fungal perspective, fragments of straw in an arable field, faecal deposits in a grassland, or twigs in a forest will be patchily distributed, but soluble components emanating from such nutrient depots will tend to form gradients.

The consequence of environmental heterogeneity for organisms is essentially an uncertainty in the nature of their habitat and its attendant niches: food resources must be acquired to permit growth, survival and reproduction, and physical, chemical and biological constraints dealt with when they are encountered. Organisms have developed different mechanisms for dealing with these factors, which vary according to whether they are motile or sessile. The growth form of eucarpic fungi in some ways bridges these categories, in that the mycelium is a static structure capable of longevity, but extending hyphae also provide a mechanism that effects movement of the organism into new territory. The space-filling, indeterminate nature of mycelia, coupled with the ability to translocate resources along hyphae to different zones of the colony, makes the fungal mycelium an extremely effective growth form, especially for soil-inhabiting organisms (Rayner, 1991).

Studies on fungal biology occasionally acknowledge the existence of environmental heterogeneity, but very few explicitly accommodate it. There has long been a preoccupation, as in most areas of microbiology, with the Petri dish, Erlenmeyer flask and chemostat, and consequently most of our knowledge of fungal physiology is based on fungi grown, as Rayner (Chapter 1, this volume) has colloquially put it, in 'sensory deprivation chambers'. Whilst such studies have an obviously important role in fundamental mycology, and have direct relevance to, for example, the industrial and biotechnological growth of fungi, it is becoming increasingly apparent that the nature and extent of environmental heterogeneity can play an important governing role in fungal growth and development (Olsson, 1995; Cairney & Burke, 1996; Ritz, Millar & Crawford, 1996).

This chapter reviews some of the recent experimental approaches and mathematical modelling work which have been carried out on the interactions between saprotrophic filamentous fungi and the spatial distribution of their nutrient resources, and the potential consequences of such interactions. The effects of 'biological' heterogeneity, essentially the spatial interactions between fungi and other organisms, have been reviewed elsewhere (e.g. Christensen, 1989; Rayner, 1991), whilst the effects of non-uniformity in environmental topography on colony development has barely been studied (but see Read *et al.*, 1997).

Experimental studies

Much description of fungal growth in heterogeneous environments has been based on observations of fungi growing in their natural habitats (e.g. Thomson, 1984; Rayner & Boddy, 1988; Boddy, 1993). Whilst such work provides a useful basis for the development of a 'natural history' and synthesis of concepts, a more mechanistic understanding can only be achieved through experimental control of nutritional heterogeneity, and ultimately the quantification of fungal growth and activity in such systems. This is in practice quite difficult to achieve since there is a paucity of statistical or mathematical frameworks for rigorously describing, analysing or interpreting spatial heterogeneity.

Much of the early experimental work involving the growth of fungi on contrasting media was carried out to investigate translocation, for example using radiolabelled forms of carbon, phosphorus or potassium (for a review, see Jennings, 1987). These studies generally did not involve measurement of fungal growth or biomass, rather the quantity of radiolabel present in different mycelial zones arising from local application of the isotopes to other mycelial zones. Suprisingly, the potential of substrate heterogeneity as a determinant in establishing translocation patterns was rarely considered. A common thesis was that for fungi to grow on very low nutrient domains, such as water agar or Petri dish surfaces, translocation of resources from nutrient-rich domains to permit such growth was a prerequisite (e.g. Howard, 1978). However, it is becoming increasingly apparent that many fungi are excellent scavengers of assimilable substrate and mycelia can develop under conditions of extremely low 'available' nutrients (e.g. Tribe & Mabadeje, 1972; Parkinson, Wainwright & Killham, 1989; Wainwright, 1993).

In experimental terms, most of the work done to date has tended to utilize either continuous gradients, or discrete or 'patchy' resources.

Resource gradients

Nutrient gradients in agar systems can be difficult to control precisely. One way of achieving such gradients is to pour an agar slope of one nutrient composition, and when set to place the culture vessel horizontally and overlay the slope with an agar of contrasting composition, the so-called Szybalski wedge plate (Bryson & Szybalski, 1952). The resultant gradients can change with respect to compounds that diffuse relatively slowly in the gel medium being used, but can be unpredictable when a mycelium is growing over them. Few studies using this technique have considered the development of individual mycelia over such gradients. Instead, they have been used as a means of screening populations of microbes for responses to environmental stress gradients, such as tolerance to antibiotic, osmotic, salt, nutrient or pH stresses imposed by the media (e.g. Wimpenny & Waters, 1984, 1987).

Olsson & Jennings (1991*a*) established contrasting gradients of glucose/glutamic acid and phosphorus along strips of glass fibre filter paper by simply loading such strips with nutrient solutions and allowing them to connect at their ends for a period of time. The resultant gradients were measured by using radioactive forms of C and P, and scanning the strips with a detector. Theoretical predictions of the gradients based on diffusion were found to be reliable using a one-dimensional model. The quantity of these elements in mycelia growing across such gradients was measured by washing unabsorbed phases out of the filters prior to scanning. However, fungal growth could not be quantified because the mycelia were generally invisible whilst growing through the glass fibre matrix. Some qualitative observations were possible, especially where colouration, associated with conidiation in some species, occurred. Vital staining with fluorescein diacetate and chromophoric staining with Cotton Blue were reported as feasible but no data were presented. This technique was used to establish that diffusion, as opposed to active translocation, may be the mechanism for translocation of nutrients in *Rhizopus nigricans*, *Trichoderma viride* and a *Stemphylium* sp. (Olsson & Jennings, 1991*b*).

Olsson (1995) subsequently extended this concept by abutting thin agar gels of contrasting nutrient status in place of the glass fibre filters used previously. Glucose concentration profiles (see Fig. 2.3) were measured by the innovative application of a chemical assay coupled to image analysis (Olsson, 1994), and were in good agreement with theoretical profiles calculated using a one-dimensional diffusion model. By utilizing fungal species with non-pigmented mycelia, it was possible to quantify colony growth in such media by measuring density profiles using a video capture system coupled to an image analyser. Growth of mycelia inoculated into the central portion of contrasting gradients of glucose versus mineral nutrients generally followed one of three characteristic profiles, with or without autolysis of the older mycelium (see Chapter 2, Fig. 2.3).

High-resolution imaging of hyphal distribution in mycelia is problematic because of the large difference in spatial scale between the hypha and the mycelium. Thus only a small portion of the mycelium can normally be visualized at one time, even using a dissecting microscope with low-power objectives. Transect photomicrography followed by tessellation of the resultant images is difficult and time-consuming because registration of overlapping images is rarely straightforward, although recent advances in computerized image processing make this procedure considerably easier. We have devised a method that enables resolution of all individual hyphae within large mycelia in a single image (Crawford, Ritz & Young, 1993; Ritz *et al.*, 1996). The procedure is based on the growth of mycelia on a transparent cellulose membrane, whereby the mycelium itself is placed in a photographic enlarger, and using point-source illumination the colony projected onto high-contrast film. The resultant internegative is then re-enlarged onto high-contrast paper, and a large-scale mycelial map (LSMM) can be produced. Using this approach mycelia up to an area as large as the enlarger can accommodate (typically 5 × 7 cm) can be mapped. The technique works best where mycelia are sufficiently sparse and in order that individual hyphae can be resolved, typically where nutrient concentrations are relatively low. This technique has been used to image colonies of *T. viride* growing in the presence of a nutrient gradient (Crawford *et al.*, 1993), and showed distinct preferential growth toward the nutrient resource during the early stages of development (Fig. 3.1). It was not possible in this study to determine precisely the nutrient profile in this experimental system, and the potential to combine Olsson's gradient-generating approach (Chapter 2, pp. 29–36) with the LSMM technique offers a potentially powerful combination. Such images allow the precise determination of how mycelial mass is distributed in space, and the accurate calculation of spatial measures such as the fractal dimension (see below).

Discrete resources

Experimental systems to control the distribution of spatially discrete resources which are not prone to rapid diffusion, such as wood, leaves and litter, are more straightforward to establish than those involving readily diffusible substrates. Indeed, a considerable body of knowledge about foraging strategies in wood-decomposing fungi has been assembled through the utilization of microcosms containing discrete

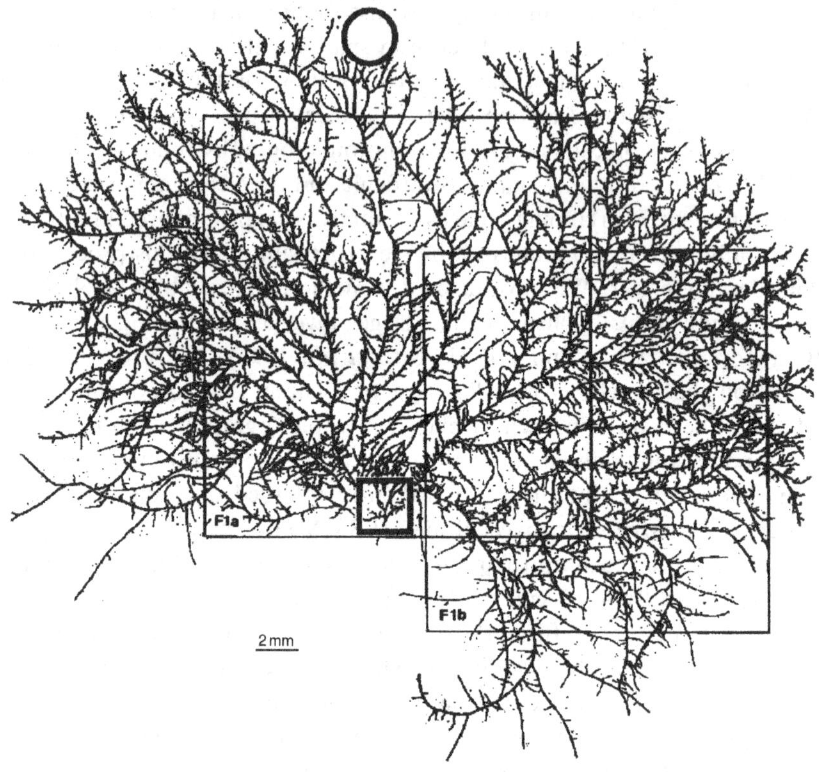

Fig. 3.1. Large-scale mycelial map of colony of *T. viride* growing from inoculation site (square) toward nutrient bait (circle). Fractal dimension of region F1a $= 2.01 \pm 0.0049$ and of region F1b $= 1.97 \pm 0.0034$. Reprinted from Crawford *et al.* (1993) with kind permission from Elsevier Science – NL, Sara Burgerharstraat 25, 1055 KV Amsterdam, The Netherlands.

placement of wood and litter bait in spatial patterns. The simplest form of such experiments is the placement of a single piece of bait in a nutritionally impoverished matrix, such as non-sterile woodland soil, distant from an inoculation point (e.g. Dowson, Rayner & Boddy, 1986), and the design can be extended to incorporate more complex spatial arrangements (e.g. Dowson, Rayner & Boddy, 1988; Dowson *et al.*, 1989). Foraging strategies expressed by different species can be classified as close-range, short-range and long-range, and are each suited to the exploitation of different types of spatial distribution in resource units (Rayner, 1994 and Chapter 1). Close-range foraging is

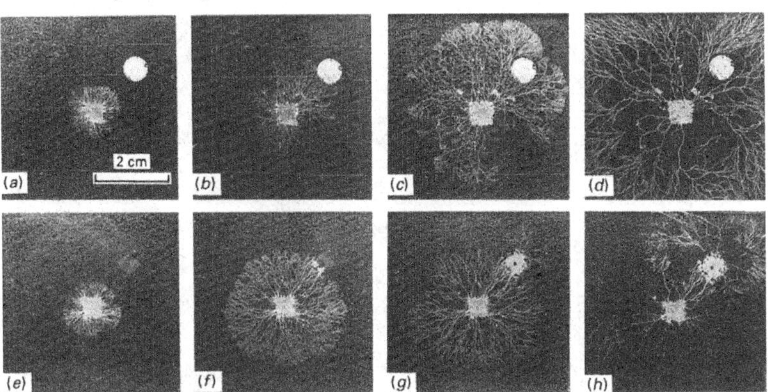

Fig. 3.2. Foraging patterns of *Hypholoma fasciculare* in resource capture experiments after 11, 25, 44 and 71 days. (a)–(d) Control dish with plastic bottle lid as bait; (e)–(h) with 2 × 2 cm beech wood block as bait. Reprinted from Dowson *et al.* (1989) with permission from Cambridge University Press.

apposite where resource units are small and spatially frequent, such as leaf litter, and is exemplified by fairy-ring formation, for example by *Clitocybe nebularis* (Dowson *et al.*, 1989). Here, short mycelial cords precede a dense, assimilative mycelial front, with a subsequent degenerative trailing edge. Short-range foraging, such as is expressed by *Hypholoma fasciculare* (Fig. 3.2) and *Steccherinum fimbriatum* (Dowson *et al.*, 1986, 1988, 1989), involves the outgrowth of dense mycelia from resource bases, which is curtailed following contact with new, previously uncolonized, resources. Degeneration of the older mycelium then occurs as the new material is colonized. This strategy is effective where resources are relatively small and frequent, but spatially separated. Long-range foraging is appropriate where resources are large and spatially distant, and is found in species such as *Phanerochaete velutina* (Fig. 3.3) (Dowson *et al.*, 1989) and *Armillaria* spp. (Rayner, 1994). Here, outgrowth of distinctly sparse, explorative, mycelial cords or rhizomorphs occurs followed by localized proliferation of assimilative hyphae when a resource is encountered. The explorative mycelium shows some degree of persistence, a feature which is pronounced in species such as *Phallus impudicus* (Dowson *et al.*, 1988), and has the consequence of allowing some ability to forage in time as well as space, since resources which may arrive in the vicinity of such cords will rapidly become colonized (Rayner, 1994). Translocation

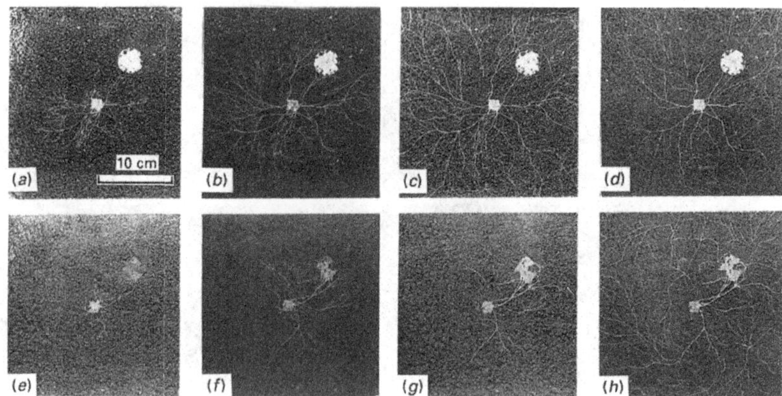

Fig. 3.3. Foraging patterns of *Phanerochaete velutina* in resource capture experiments after 15, 25, 44 and 71 days. (a)–(d) Control dish with plastic bottle lid as bait; (e)–(h) with 2 × 2 cm beech wood block as bait. Reprinted from Dowson *et al.* (1989) with permission from Cambridge University Press.

of resources to such localized woody resources has also been measured showing that larger, recently colonized units of wood act as greater translocatory sinks than smaller or more decayed ones (Wells & Boddy, 1990; Wells, Hughes & Boddy, 1990; Wells, Boddy & Evans, 1995).

Ritz (1995) devised a system whereby spatial patterns of soluble nutrient resources could be prescribed using tessellations of agar tiles of differing composition. Here, diffusion of nutrients into adjacent domains was precluded by an air gap between the tiles, which offered little barrier to the growth of hyphae. A tessellation of 3 × 3 tiles was used to establish a variety of spatial patterns of high- and low-nutrient status domains (Fig. 3.4a). The central tile in the array was inoculated with a single colony of a number of filamentous fungi, and colony development mapped by regular observations (Fig. 3.4b). Such mapping had to be qualitative, but nonetheless demonstrated a number of characteristic responses by different species. In general, hyphal density reflected the underlying nutrient status of the tile, being high and low in nutrient-rich and nutrient-poor domains respectively, although there was evidence for growth in low-nutrient tiles being greater when high-nutrient tiles were present in the tessellation. Different colony extension rates were observed for *T. viride*, being greater on low nutrient tiles than on high domains, which is evidence for a foraging strategy commensurate with

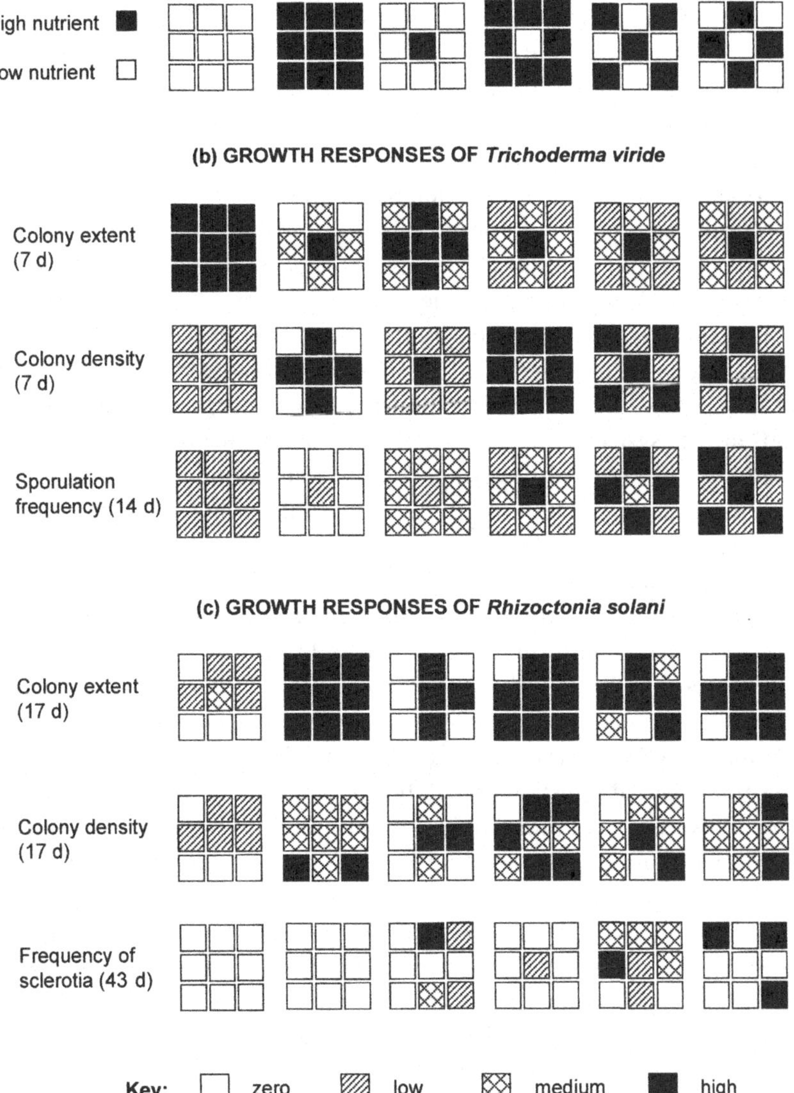

Fig. 3.4. Tessellated agar tile system for studying colony responses to spatial heterogeneity in nutrient distribution: spatial designs and some growth responses by three fungal species. Adapted from Ritz (1995).

the exploration/exploitation strategy alluded to above. Patterns of formation of reproductive structures were strongly influenced by the underlying heterogeneity in the tessellation. Such structures tended to be formed only in low-nutrient tiles by *T. viride* and *Rhizoctonia solani*, and only high-nutrient tiles by *Alternaria alternata*. Indeed, sclerotia were only produced by *R. solani* in tessellations where there was some heterogeneity, being absent in the nutritionally uniform arrangements. Furthermore, sclerotial production was strongly asymmetric in nutritionally symmetric, but heterogeneous, tessellations. A variation on this approach has been applied by Rayner *et al.* (Chapter 1), where agar domains are kept separate in the square chambers of Repli-plates, and connection between them achieved by boring small holes in the plastic walls. Again, growth patterns of cord-forming basidiomycetes in particular are shown to be strongly influenced by the heterogeneity of nutrients in such system, and are clearly further modulated by the constriction the hole imposes on the mycelia: lines of 'reinforcement' in cord formation are apparent directly along the axis joining such holes. Also, there is a general response of characteristically different mycelial forms on high- versus low-nutrient domains.

The tesselated agar tile concept was extended by Toyota, Ritz & Young (1996) with the substitution of agar tiles by soil aggregates. Here, the primary aim of the study was to examine the impact of the prevailing biotic environment on the ability of *Fusarium oxysporum* f. sp. *raphani* to colonize soil aggregates, but the effect of nutrient status on such colonizing ability was also demonstrated. It was shown that there was a relationship between the quantity and quality of nutrient resource available to the fungus and its ability to bridge air gaps and colonize adjacent non-sterile aggregates.

A more detailed visualization of colony responses to nutritional heterogeneity was made possible by a combination of LSMM with discretized gel arrays, as applied by Ritz *et al.* (1996). Here, high- and low-nutrient gels are kept apart by a perforated plastic surface overlain with a cellulose membrane (Fig. 3.5a). Whilst there is some potential in this system for diffusion between domains, this was not considered to be significant, as suggested by the highly confined growth responses of *T. viride* in these systems (Fig. 3.5b). Again, the general trend of mycelial density mirrored the underlying nutrient status of the substratum. This response was less pronounced for *R. solani*, where the overall nutrient status of the system was very low, i.e. 'high'-nutrient domains are 10%

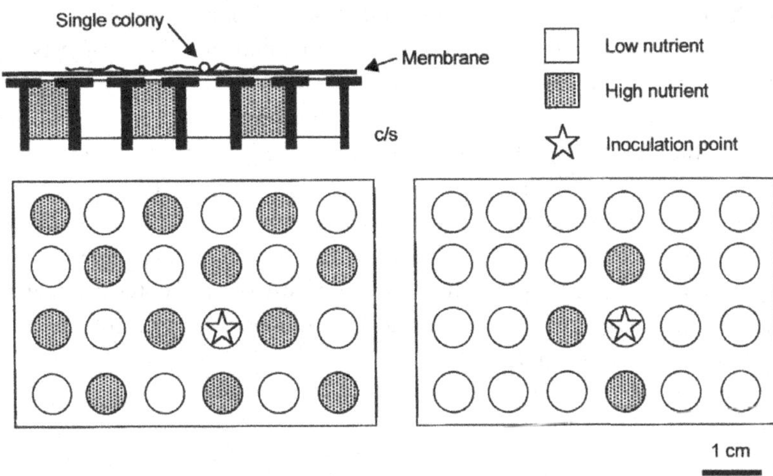

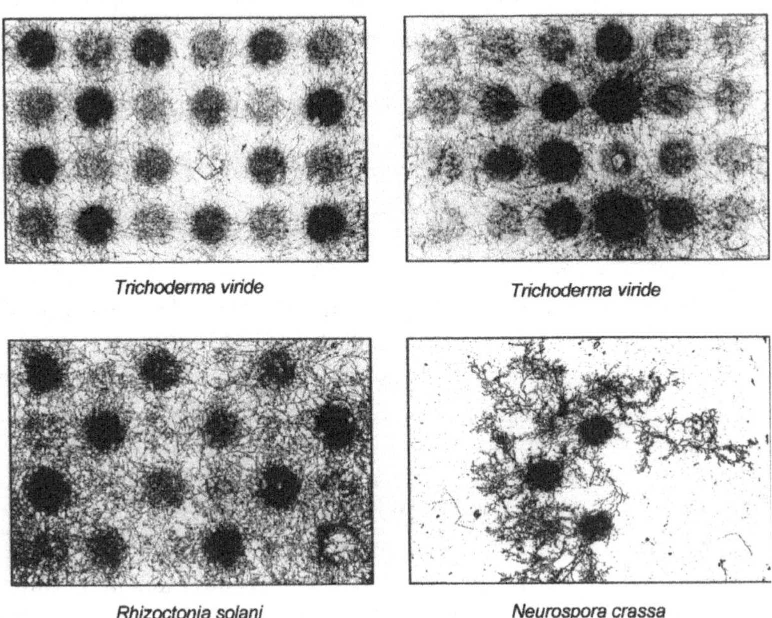

Fig. 3.5. Tessellated agar spot system for producing large-scale mycelial maps of fungi growing in nutritionally heterogeneous microcosms. Adapted from Ritz *et al.* (1996).

tap-water agar (Ritz *et al.*, 1996). Mycelial maps produced by this method for *T. viride* show a greater biomass deposition in low-nutrient domains more distant from the point of inoculation. This suggests that there was a greater utilization of resources on such sites by 'older', or larger mycelia, albeit in the younger regions of the mycelium, i.e. those nearer the periphery. This may arise as a result of translocation of resources from nutrient-rich domains to augment growth on the impoverished domains. Alternatively, mycelial degeneration may be occurring on the older domains, and is more apparent in the less-dense zones overlaying low-nutrient sites. A fuller understanding of colony morphogenesis in such systems will only be achieved when it is possible to obtain detailed temporal as well as spatial maps.

Mathematical models

Foraging behaviour and coherence in mycelial form

Mathematical theories of the growth and development of fungi predominantly model the system either at the hyphal, or at the colony scale and few studies attempt to link the two levels of description. This is a consequence of (a) little data being available since most experiments are conducted in homogeneous, nutritionally rich environments and (b) an over-emphasis on the analogy between fungal and bacterial colonies, and therefore a failure to appreciate the significance of the coupling between hyphal compartments exemplified by the connectedness of the fungal mycelium.

In an attempt to address these issues, the LSMM technique was used to produce images of colonies grown on cellophane over nutrient gradients as described above (Fig. 3.2). These were digitized to quantify the distribution of individual hyphae across entire colonies. Far from being random, the results of the analyses showed that spatial distributions of hyphae were heterogeneous across all scales and could be approximated using fractals (Crawford *et al.*, 1993; Fig. 3.1). A considerable body of literature now attests to the fractal structure of mycelia (Ritz & Crawford, 1990; Obert, Pfeifer & Sernetz, 1990; Liddell & Hansen, 1993; Bolton & Boddy, 1993; Jones, Lonergan & Mainwaring, 1993; Matsuura & Miyazima, 1992; Mihail *et al.*, 1994, 1995; Hitchcock, Glasbey & Ritz, 1996). Broadly speaking, fractals are geometric shapes whose structural characteristics are repeated across all scales. Thus the structure at one point is correlated with the structure at any other.

Clearly fungal colonies are not fractals according to this strict definition, since only structure between hyphal and colony scales can be considered. However, in the same way that we approximate the surface of a table as two-dimensional between atomic scales and the dimension of the table, we can approximate a mycelium as a fractal between hyphal and colony scales.

In order to quantify the hyphal distribution, the LSMM images were analysed by calculating the moments of the mass distribution in the following way. If $P(m, L)$ is the probability that a square box of side L centred on a pixel belonging to an image of the mycelium contains m such pixels, then

$$\langle m^q(L) \rangle = \sum_m P(m, L) m^q \tag{1}$$

where q is an integer, and $\langle m^q(L) \rangle$ denotes the expectation of the mass raised to the power q, in a box of length L. The expectation was calculated using boxes centred on all pixels belonging to the mycelium image, and over a predefined range in length, L. A full description of the statistical properties of the hyphal distribution is contained in the $\langle m^q(L) \rangle$, and in particular the form of the dependence on L describes the scaling relation between hyphal and colony scales. Following the analysis of mycelia grown over nutritionally poor, homogeneous media, it was found that the dependence of $\langle m^q(L) \rangle$ on box-length, L, was a power-law of the form,

$$\langle m^q(L) \rangle \propto L^{d_q} \tag{2}$$

where d_q is a constant. If the hyphal distribution is homogeneous, then $d_q = 2$ for all values of q, reflecting the usual (Euclidean) dependence of mass or area on radius. If, however, $d_q < 2$, then the hyphal distribution is non-homogeneous across all measured scales. Such power-law relations between mass and radius L are a consequence of a fractal distribution.

Each d_q is then a fractal dimension describing the distribution, and the set of values corresponding to each q represent the (multifractal) spectrum of dimensions required to fully characterize the heterogeneous distribution. Formally, all the d_q are required to fully characterize the distribution; however, in some fractals (self-similar fractals) all the d_q are equal, and so a single value, e.g. d_1 suffices. In other cases, a considerable amount of information can be obtained from the value of d_1 corresponding to the first moment of the mass distribution. The 'mass' (more precisely the number of image pixels corresponding to hyphae) inside radius L can be written simply as,

$$m(L) = AL^{d_1} \tag{3}$$

where A is a constant and d_1 is the mass fractal dimension. In what follows, the subscript is omitted for convenience, and d denotes the mass fractal dimension as defined in equation (1) for $q = 1$.

The observation that a mycelium can be approximated by a fractal implies that the distribution of hyphae at one point is correlated with the distribution at another. Thus the mycelium grows coherently, implying that a form of 'communication' exists between distal parts of the colony. This striking conclusion is at odds with the current understanding of fungal growth (see below), yet such communication is known to exist between members of bacterial colonies (Budrene & Berg, 1995), and is exemplified in the slime mould (*Dictyostelium discoideum*). In these cases, the communication is related to substrate levels. In the case of fungi, the fractal distribution of hyphae can also be related to substrate. In experiments conducted at different substrate levels it was found that the fractal dimension increased with increasing nutritional quality of the substrate (Ritz & Crawford, 1990). This leads to an interpretation in terms of a foraging strategy through an understanding of the consequences of fractal structure for the distribution of hyphae. If R is a characteristic diameter of the region occupied by a colony of mass M, then equation (3) can be rewritten,

$$R \propto M^{1/d} \tag{4}$$

From equation (4) it is clear that a colony will cover a large region for a given investment in mass when d is low. Conversely, when d is high, that same mass will be concentrated in a small region of space. Since high values of d are associated with high levels of nutrition, this greater concentration of mass is consistent with an 'exploitative' strategy whereby the surface area of mycelia is maximized inside a region of space to enhance uptake. In low nutrition, the low values of d indicate an 'explorative' strategy that results in a large region being covered for small investment in biomass. Fractal structure can therefore be interpreted as arising as a consequence of a compromise between explorative and exploitative growth strategies (Ritz & Crawford, 1990).

Theoretical models of hyphal growth and mycelial form

Most theoretical models describe either hyphal or colony scale processes, reflecting the same bias in experimental studies alluded to in the previous section. However, some models attempt to predict colony scale parameters through a mechanistic description of hyphal-scale dynamics. The most important of these are described and compared below.

Peripheral growth zone models

A particularly important work aimed at explaining colony growth in terms of vesicle dynamics, hyphal growth and branching is that of Trinci (1971). A major consequence of this work was the emergence of the peripheral growth zone hypothesis which has been influential in shaping views on fungal morphogenesis. An explanation for the constant radial growth rate of colonies was sought in terms of the specific growth rate, under the assumption that biomass production is limited to independent hyphal units at the colony margin – the peripheral growth zone. Crucially, this implies that the hyphal distribution is governed by mechanisms operating locally at the hyphal apex, and precludes the possibility of large-scale coherence. The model of Prosser & Trinci (1979) has largely come to be accepted as providing confirming evidence in support of the peripheral growth zone hypothesis. Hyphal growth form is assumed to be a consequence of vesicle flow dynamics and septation. The rate of tip extension is assumed to be proportional to the rate of absorption of vesicles there, and septa are produced when the apical compartment reaches an assumed critical length. The central thesis, however, linking the model to the peripheral growth zone hypothesis, is that only that part of the hypha immediately below the hyphal tip influences tip growth. The model is based on detailed hypotheses relating to vesicle production rates, vesicle transport, branching and spore germination. It predicts (i) rate of hyphal extension, (ii) rate in increase of number of branches, (iii) increase in total hyphal length and (iv) the mean hyphal extension rate (a quantity derived from the last two). Agreement between theory and measurements is impressive but there are many free parameters in the model and much of the observed agreement can be attributed to these. Furthermore, since growth and branching of hyphal tips are assumed to occur independently of one another, the model cannot account for observed coherence in the mycelium.

The Spitzenkörper hypothesis

The hypothesis that hyphal growth and branching is controlled locally by conditions in the hyphal tip region was given further emphasis in the model of Bartnicki–Garcia (1990). A hypothesized *Spitzenkörper* or vesicle supply centre is a point source of vesicles which supplies wall precursors to the adjacent tip, hence determining its shape and growth rate and direction. Branching is hypothesized to occur via the creation of some new vesicle supply centre some way down the hypha, which generates a new tip. The Prosser and Trinci model was invoked to govern the position and frequency of these new centres. Such a model therefore fails for the same reasons as the Prosser and Trinci model and the peripheral growth zone hypothesis. It also replaces an unknown mechanism of branch initiation and tip dynamics by an unknown mechanism of Spitzenkörper formation and movement.

A model of tip dynamics

Edelstein (1982) and Edelstein & Segel (1983) presented a model of colony growth which circumvents the detailed molecular mechanisms, and concentrates on the dynamics of tip production and loss, and on the feedback between colony growth and nutrient depletion. Their model includes the effects of tip death, anastomosis, tip bifurcation, lateral branching and hyphal death on the mass and tip production dynamics. A phenomenological expression is used to relate tip production to hyphal and tip densities, which takes into account these tip–hyphal interactions. Branches are assumed to be created at a constant rate and hyphae assumed to extend at a constant linear rate. As such, colony morphology is entirely governed by spatially localized interactions between tips and hyphae, and by tip and hyphal death rates. Again, therefore, the model will fail to account for the observed spatial correlation resulting from fractal structure. However, through the feedback between nutrient uptake and internal metabolite depletion through growth, the model does attempt to reproduce hyphal density rings. Nutrient uptake occurs in proportion to hyphal density at low nutrient concentrations and saturates to some maximum rate at high concentrations. In order to induce hyphal rings it was assumed that branching occurred at some enhanced rate above a certain level of internal metabolite concentration. It was also assumed that tip extension was inhibited below some critical threshold metabolite concentration. In principle it was therefore possible that a build-up of metabolite could 'switch' on enhanced branching as it rose above the critical threshold. This would lead to a local increase in hyphal density, which would eventually result in dilution of metabolite and a consequential drop below the critical threshold for enhanced branching. In practice, however, the assumptions of the model leading to uptake being proportional to hyphal density mean that it is not possible to induce hyphal oscillations. An additional sink of metabolite is required and the strength of this sink must be a non-linear function of metabolite concentration. Even after this assumption, only very weak rings are reproduced by the model.

Stochastic models

A number of stochastic models for colony growth exist (Kotov & Reshetnikov, 1990; Obert, 1994). Stochastic models predict branch angle, inter-branch distance, growth rate, etc. by randomly choosing values from distribution functions determined from laboratory measurements of fungi grown under prescribed conditions. Being empirically based, these models are species specific, and by nature do not approach the question of the underlying mechanisms of colony form. At present, they are also unable to deal with the situation where the underlying nutrient is heterogeneously distributed.

Reaction-diffusion models

Reaction-diffusion models describe a system in terms of elements that interact and spread through space. Since the seminal paper by Turing (1952), it has been known that such systems are capable of spontaneous pattern formation under well-defined conditions. The patterns originate from non-linear reaction kinetics (e.g. autocatalysis, where the presence of one element, such as biomass, promotes the production of more biomass by uptake and synthesis of nutrients) and long-range interactions between the elements mediated by diffusion. To date, these are the only models of fungal growth capable of producing spatially coherent structures across a range in scales. Two examples of such models are examined below. The first model attempts to study colony-scale coherence in mycelia by invoking hyphal-level processes, and the second uses a model which averages out hyphal-level detail. Mathematical details are omitted here, but the interested reader is referred to the indicated references.

A reaction-diffusion model of hyphal growth and branching

The model of Regalado *et al.* (1996) is a development of a general model of network proliferation by Meinhardt (1982). Tip positioning and localization are governed by the interaction between two diffusible components: an autocatalytically produced activator of the growth process and an inhibitor that counteracts the production of the activator. As an example, the interaction between activator and inhibitor might represent the antagonistic regulation of cAMP by calcium (Reissig & Kinney, 1983). The formation of a new unit length of hypha is assumed to be completed when the concentration of the activator increases above a threshold value. Under quite general conditions, spatial localization of the activator results when the dispersal of activator is contained by the faster diffusing inhibitor. The concentration peak in activator increases locally above the threshold and produces a new unit length of hyphal tip. Tip extension arises from the interaction with an external substrate distribution. Uptake of substrate by the hyphae establishes gradients in substrate distribution. The activator peaks, representing hyphal tips, move up these gradients tracing out the distribution of hyphae in the growing colony. This two-way interaction between the activator/inhibitor kinetics and the spatial distribution in substrate leads to large-scale coherence in the spatial distribution of hyphae. As discussed above, this coherence implies 'communication' between distal parts of the colony and in this model the communication is mediated exogenously by diffusion-driven gradients in substrate concentration. The detailed response of the colony to these gradients depends on parameters which are representative of genetically controlled processes. Measurements made on the modelled hyphal distribution for a mycelium growing on a homogeneous substrate indicate that the model predictions are consistent with the fractal properties of real mycelia (Figs. 3.6, 3.7). The estimated fractal dimension was greater for a simulated colony growing on high substrate than one growing on low substrate. The equations were also solved for the case of a mycelium growing across a heterogeneous domain, mimicking the experimental system described by Crawford *et al.* (1993). Here again the estimated fractal dimension was higher in the high substrate regions than on the low, consistent with the results of the experiments (Fig. 3.7). The results of the model are consistent with the observed coherent behaviour of mycelia, and point to a two-way interaction between the developing hyphal distribution and the diffusible substrate as the origin of coherence. There are many inadequacies in such a simple model. For example, active internal translocation is not accounted for, and neither are the development of septa, which are important in sub-apical branch formation. However, it does illustrate how simple interactions can lead to complex and coherent patterns.

A reaction-diffusion model of colony form

A complementary model by Davidson *et al.* (1996) was derived to study the development of colony-scale coherent structures which arise as a consequence of spatial heterogeneity in substrate resulting from depletion by the mycelium. These structures include hyphal rings, irregular

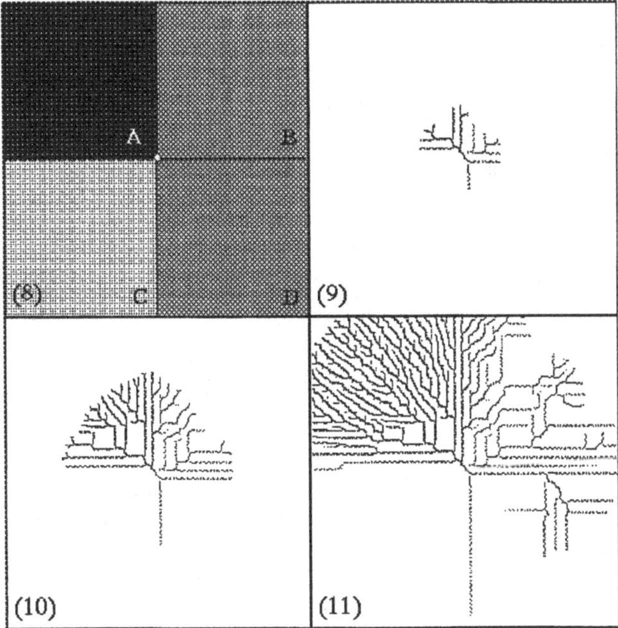

Fig. 3.6. Time sequence of the simulated branch pattern of a mycelium growing over a heterogeneous domain. Quadrant A has the highest nutrient level, quadrants B and D have the same intermediate level, and quadrant C has the lowest level. The reduced level of branching as lower level nutrient bases are encountered is clearly predicted, and accords with the results of experiments. Reprinted from Regalado *et al.* (1996).

hyphal fronts, and the global organizational forms which result from interactions between colonies. Furthermore, coherent temporal structures were also studied, including the cessation of propagation of the hyphal front. The model predicts the distribution of the activator of the growth process averaged over length-scales much larger than those corresponding to individual hyphae. Although hyphal-level details are not explicitly accounted for, the consequences of different network topologies are implicitly included in the model. In the absence of substrate replenishment (or when replenishment rate is very low), the model predicts radial expansion of the colony at a constant rate, mediated by a ring of active growth at the colony margin only. Whilst this is reminiscent of the peripheral growth zone hypothesis, the resulting structure arises as a consequence of non-local interaction between substrate and

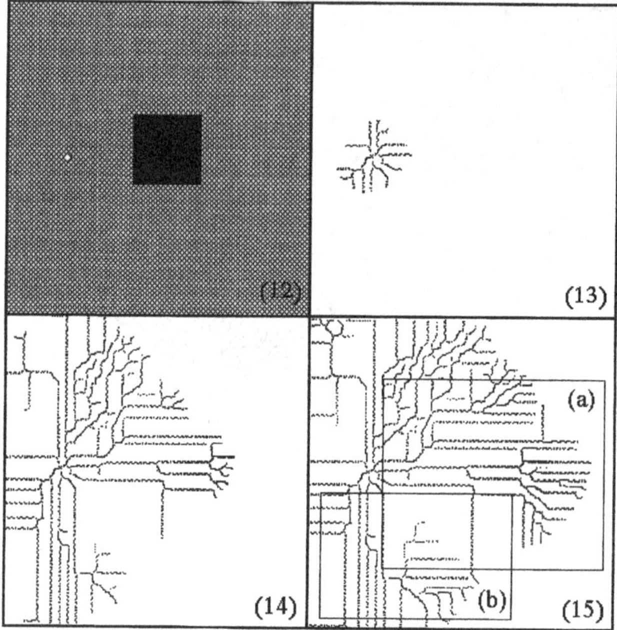

Fig. 3.7. Time sequence of the development of the branching pattern of a mycelium initially inoculated on the white spot in the top left picture some distance from a diffusible resource located in the black square. The fractal dimension of the structure contained in the box denoted (a) is 1.91, compared with the dimension in the box denoted (b), which is 1.86. This increase in the fractal dimension close to the resource is consistent with the experimental results of Crawford *et al.* (1993). Reprinted from Regalado *et al.* (1996).

the activator of growth mediated by diffusion. It is therefore an additional counter-example to the assumption that constant radial expansion must be driven by independent hyphal growth units. When the replenishment rate is larger, constant radial expansion still occurs, although important additional phenomena are predicted. In particular, coherent internal growth is generated behind the expanding hyphal front. If it is assumed that the diffusion of activator is predominantly radial, as would happen in colonies where anastomoses occur infrequently and hyphae are radially oriented on average, internal growth is confined to concentric rings which are fixed in space. These rings are commonly observed in mycelia grown in laboratory cultures (see Davidson *et al.*, 1996, for references). If the diffusion of the growth activator is increased in the model, after any particular stage in the development of structure outlined above, radial expansion ceases. Such an increase in the diffusion rate would occur if the underlying hyphal network developed in a manner which reduced resistance to internal transport, e.g. through increased anastomoses. If substrate levels and hence growth activation in the interior of the colony are then reduced by setting substrate replenishment to zero, the colony resumes expansion. However, growth activation is now confined to the colony margin. These results are consistent with the observed failure of colonies, or parts of colonies, to propagate. They are also consistent with observed autodegeneration of colony interiors, accompanied by the propagation of coherent ring structures, e.g. 'fairy-rings'. The predicted role of mycelial-induced substrate heterogeneity in this behaviour is clear.

Both these reaction-diffusion models acknowledge the importance of the two-way interaction between substrate distribution and mycelial form, in generating coherent structure either at the hyphal or colony scale. In particular, it is the generation of heterogeneity in the substrate distribution through depletion by the mycelium which plays a central role in generating large-scale coherent form. Heterogeneities superimposed on the substrate distribution have an additional influence on the macroscopic coherence of form.

It should be stressed that although these models reproduce many of the features of real mycelia which have otherwise gone unexplained, they are still very simple. An important omission is the lack of any account of active transport within the mycelium. Thus the models indicate an interaction between genetic factors and exogenous factors in governing mycelial form. By design, they exclude the possibility of endogenous factors such as translocation, and this means that the origin of highly organized distributive structures such as mycelial chords remains unknown.

Conclusions: some consequences of nutritional heterogeneity

Current knowledge of fungal growth responses to nutritional heterogeneity suggests a number of consequences from the perspective of the fungus and in relation to the functioning of the soil system which the organism inhabits. To date, the rationale for most of the research carried out has been in relation to a predominantly 'fungal' perspective; there is need for future research to focus on the latter, to understand the extent to which these growth responses actually have implications for soil ecosystem function.

Translocation

Translocational source/sink relationships within different mycelial zones are certainly influenced by the spatial organization of the substrata, but there appear to be no simple rules governing the direction of movement; different fungi respond in different ways. There is evidence for translocation of carbon and other nutrient elements from relatively low to high, and high to low regions, and even bi-directional simultaneous movement. From a fungal perspective, such movement may permit the augmentation of growth in zones lacking in certain elements (Olsson, 1995), or satisfy the high energetic demands of active mycelia (Cairney, 1992). From a soil perspective, this may result in an overall acceleration of nutrient cycling processes, but also modulation of transport patterns for elements which move slowly through the soil fabric, such as phosphorus. Mycelia may thus serve to disperse such elements from locally high concentrations, or reconcentrate them in spatially different zones.

Propagule synthesis

The site of propagule production within mycelia is also certainly influenced strongly by resource location, again with a variety of responses between species to nutritional heterogeneity. Formation of reproductive/survival structures in low nutrient domains (by e.g. *T. viride*, *R. solani*) may be localized responses to nutrient stress, and will most likely involve the translocation of nutrients to such domains, to enable the metabolically costly process of their production. Production on high nutrient domains may be more prevalent in those species that show low translocation ability, or it could be argued that such production is an effective strategy to synthesize such structures in the proximity of a resource providing substrate for such production, which is energetically more efficient since it does not involve translocation.

Soil structural dynamics

Hyphal binding of soil aggregates is an acknowledged biological component of soil structural dynamics (Tisdall, 1994). The general growth response of higher hyphal densities in high nutrient domains may thus result in stronger binding by this mechanism in nutrient 'hotspots'. The spatial extent of such binding may also be influenced by the distance beyond the resource where such hyphal proliferation occurs: for instance, *T. viride* appears to show very localized proliferation, whilst other species, such as *R. solani*, sometimes show a more uniform hyphal density even where nutrients are patchy (Fig. 3.6).

Fungal community dynamics

The combative ability of mycelia is influenced by the degree of nutrition available to the mycelium (Stahl & Christensen, 1992; Toyota *et al.*, 1996). This phenomenon is bound to influence the community dynamics of fungal populations, and will thus be modulated by the spatial distribution of resources. Little experimental work has been carried out in this arena, and designs such as the tessellated agar tile system clearly have great potential application in such studies.

References

Bartnicki-Garcia, S. (1990). Role of vesicles in apical growth and a new mathematical model of hyphal morphogenesis. In *Tip Growth in Plant and Fungal Cells*, ed. I. B. Heath, pp. 211–32. London: Academic Press.

Boddy, L. (1993). Saprotrophic cord-forming fungi: warfare strategies and other ecological aspects. *Mycological Research*, **97**, 641–55.

Bolton, R. G. & Boddy, L. (1993). Characterisation of the spatial aspects of foraging mycelial cord systems using fractal geometry. *Mycological Research*, **97**, 762–8.

Bryson, V. & Szybalski, W. (1952). Microbial selection. *Science*, **116**, 43–51.

Budrene, E. O. & Berg, H. C. (1995). Dynamics of formation of symmetric patterns by chemotactic bacteria. *Nature*, **376**, 49–53.

Cairney, J. W. G. (1992). Translocation of solutes in ectomycorrhizal and saprotrophic rhizomorphs. *Mycological Research*, **96**, 135–41.

Cairney, J. W. G. & Burke, R. M. (1996). Physiological heterogeneity within fungal mycelia: an important concept for a functional understanding of the ectomycorrhizal symbiosis. *New Phytologist*, **134**, 685–95.

Christensen, M. (1989). A view of fungal ecology. *Mycologia*, **81**, 1–19.

Crawford, J. W., Ritz, K. & Young, I. M. (1993). Quantification of fungal morphology, gaseous transport and microbial dynamics in soil: an integrated framework utilising fractal geometry. *Geoderma*, **56**, 157–72.

Davidson, F. A., Sleeman, B. D., Rayner, A. D. M., Crawford, J. W. & Ritz, K. (1996). Context-dependent macroscopic patterns in growing and interacting mycelial networks. *Proceedings of the Royal Society of London*, Series B, **263**, 873–80.

Dowson, C. G., Rayner, A. D. M. & Boddy, L. (1986). Outgrowth patterns of mycelial cord-forming basidiomycetes from and between woody resource units in soil. *Journal of General Microbiology*, **132**, 203–11.

Dowson, C. G., Rayner, A. D. M. & Boddy, L. (1988). Foraging patterns of *Phallus impudicus, Phanerochaete learis* and *Steccherinum fimbriatum* between discontinuous resource units in soil. *FEMS Microbiology Ecology*, **53**, 291–8.

Dowson, C. G., Springham, P., Rayner, A. D. M. & Boddy, L. (1989). Resource relationships of foraging mycelial systems of *Phanerochaete velutina* and *Hypholoma fasciculare* in soil. *New Phytologist*, **111**, 501–9.

Edelstein, L. (1982). The propagation of fungal colonies: a model for tissue growth. *Journal of Theoretical Biology*, **98**, 679–701.

Edelstein, L. & Segel L. A. (1983). Growth and metabolism in mycelial fungi. *Journal of Theoretical Biology*, **104**, 187–210.

Hitchcock, D., Glasbey, C. A. & Ritz, K. (1996). Image analysis of space-filling by networks: application to a fungal mycelium. *Biotechnology Techniques*, **10**, 205–10.

Howard, A. J. (1978). Translocation in fungi. *Transactions of the British Mycological Society*, **70**, 265–9.

Jennings, D. H. (1987). Translocation of solutes in fungi. *Biological Reviews*, **62**, 215–43.

Jones, C. L., Lonergan, G. T. & Mainwaring, D. E. (1993). Mycelial fragment size distribution: an analysis based on fractal geometry. *Applied Microbiology and Biotechnology*, **39**, 242–9.

Kotov, V. & Reshetnikov, S. V. (1990). A stochastic model for early mycelial growth. *Mycological Research*, **94**, 577–86.

Liddell, C. M. & Hansen, D. (1993). Visualising complex biological interactions in the soil ecosystem. *Journal of Visualisation and Computer Animation*, **4**, 3–12.

Marshall, T. J. & Holmes, J. W. (1988). *Soil Physics*, 2nd edn. Cambridge: Cambridge University Press.

Matsuura, S. & Miyazima, S. (1992). Self-affine fractal growth front of *Aspergillus oryzae*. *Physica A*, **191**, 30–4.

Meinhardt, H. (1982). *Models of Biological Pattern Formation*. London: Academic Press.

Mihail, J. D., Obert, M., Bruhn, J. N. & Taylor, S. J. (1995). Fractal geometry of diffuse mycelia and rhizomorphs of *Armillaria* species. *Mycological Research*, **99**, 81–8.

Mihail, J. D., Obert, M., Taylor, S. J. & Bruhn, J. N. (1994). The fractal dimension of young colonies of *Macrophomina phaseolina* produced from microsclerotia. *Mycologia*, **86**, 350–6.

Obert, M. (1994). Microbial growth patterns: fractal and kinetic characteristics of patterns generated by a computer model to simulate fungal growth. *Fractals*, **1**, 354–74.

Obert, M., Pfeifer, P. & Sernetz, M. (1990). Microbial growth patterns described by fractal geometry. *Journal of Bacteriology*, **172**, 1180–5.

Olsson, S. (1994). Uptake of glucose and phosphorus by growing colonies of *Fusarium oxysporum* as quantified by image analysis. *Experimental Mycology*, **18**, 33–47.

Olsson, S. (1995). Mycelial density profiles of fungi on heterogeneous media and their interpretation in terms of nutrient reallocation patterns. *Mycological Research*, **99**, 143–53.

Olsson, S. & Jennings, D. H. (1991*a*). A glass fiber filter technique for studying nutrient uptake by fungi: the technique used on colonies grown on nutrient gradients of carbon and phosphorus. *Experimental Mycology*, **15**, 292–301.

Olsson, S. & Jennings, D. H. (1991*b*). Evidence for diffusion being the mechanism of translocation in the hyphae of three moulds. *Experimental Mycology*, **15**, 302–9.

Parkinson, S. M., Wainwright, M. & Killham, K. (1989). Observations on oligotrophic growth of fungi on silica gel. *Mycological Research*, **93**, 529–34.

Prosser, J. I. & Trinci, A. P. J. (1979). A model for hyphal growth and branching. *Journal of General Microbiology*, **111**, 153–64.

Rayner, A. D. M. (1991). The challenge of the individualistic mycelium. *Mycologia*, **83**, 48–71.

Rayner, A. D. M. (1994). Pattern-generating processes in fungal communities. In *Beyond the Biomass: Compositional and Functional Analysis of Soil Microbial Communities*, ed. K. Ritz, J. Dighton & K. E. Giller, pp. 247–58. Chichester, UK: John Wiley.

Rayner, A. D. M. & Boddy, L. (1988). *Fungal Decomposition of Wood.* Chichester: John Wiley.

Read, N. D., Kellock, L. J., Collins, T. J. & Gundlach, A. M. (1997). Role of topographic sensing for infection-structure differentiation in cereal rust fungi. *Planta*, **202**, 163–70.

Regalado, C. M., Crawford, J. W., Ritz, K. & Sleeman, B. D. (1996). The origins of spatial heterogeneity in vegetative mycelia: a reaction-diffusion model. *Mycological Research*, **100**, 1473–80.

Reissig, J. L. & Kinney, S. G. (1983). Calcium as a branching signal in *Neurospora crassa. Journal of Bacteriology*, **154**, 1397–402.

Ritz, K. (1995). Growth responses of some soil fungi to spatially heterogeneous nutrients. *FEMS Microbiology Ecology*, **16**, 269–80.

Ritz, K. & Crawford, J. W. (1990). Quantification of the fractal nature of colonies of *Trichoderma viride. Mycological Research*, **94**, 1138–42.

Ritz, K., Millar, S. M. & Crawford, J. W. (1996). Detailed visualisation of hyphal distribution in fungal mycelia growing in heterogeneous nutritional environments. *Journal of Microbiological Methods*, **25**, 23–8.

Saunders, W., Frenk, C., Rowan-Robinson, M., Efstathiou, G., Lawrence, A., Kaiser, N., Ellis, R., Crawford, J. W., Xia, X-Y. & Parry, I. (1991). The density field of the local Universe. *Nature*, **349**, 32–8.

Smith, M. L., Bruhn, J. N. & Anderson, J. B. (1992). The fungus *Armillaria bulbosa* is among the largest and oldest living organisms. *Nature*, **356**, 428–31.

Stahl, P. D. & Christensen, M. (1992). *In vitro* mycelial interactions among members of a soil micro-fungal community. *Soil Biology and Biochemistry*, **24**, 309–16.

Thomson, W. (1984). Distribution, development and functioning of mycelial cord systems of decomposer basidiomycetes of the deciduous woodland floor. In *The Ecology and Physiology of the Fungal Mycelium*, ed. D. H. Jennings & A. D. M. Rayner, pp. 185–214. Cambridge: Cambridge University Press.

Tisdall, J. M. (1994). Possible role of soil microorganisms in aggregation in soils. *Plant and Soil*, **159**, 115–21.

Toyota, K., Ritz, K. & Young, I. M. (1996). Microbiological factors affecting the colonisation of soil aggregates by *Fusarium oxysporum* F. sp. *raphani. Soil Biology and Biochemistry*, **28**, 1513–21.

Tribe, H. T. & Mabadeje, S. A. (1972). Growth of moulds on media prepared without organic nutrients. *Transactions of the British Mycological Society*, **58**, 127–37.

Trinci, A. P. J. (1971). Influence of the width of the peripheral growth zone on the radial growth rate of fungal colonies on solid media. *Journal of General Microbiology*, **67**, 325–44.

Turing, A. (1952). The chemical basis of morphogenesis. *Philosophical Transactions of the Royal Society, London*, **B237**, 37–72.

Wainwright, M. (1993). Oligotrophic growth of fungi – stress or natural state? In *Stress Tolerance of Fungi*, ed. D. H. Jennings, pp. 127–44. London: Academic Press.

Wells, J. M. & Boddy, L. (1990). Wood decay, and phosphorus and fungal biomass allocation in mycelial cord systems. *New Phytologist*, **116**, 285–95.

Wells, J. M., Boddy, L. & Evans, R. (1995). Carbon translocation in mycelial cord systems of *Phanerochaete velutina* (DC.: Pers.) Parmasto. *New Phytologist*, **129**, 467–76.

Wells, J. M., Hughes, C. & Boddy, L. (1990). The fate of soil-derived phosphorus in mycelial cord systems of *Phanerochaete velutina* and *Phallus impudicus*. *New Phytologist*, **114**, 595–606.

Wimpenny, J. W. T. & Waters, P. (1984). Growth of microorganisms in gel-stabilised 2-dimensional gradient systems. *Journal of General Microbiology*, **130**, 2921–6.

Wimpenny, J. W. T. & Waters, P. (1987). The use of gel-stabilised gradient plates to map the responses of microorganisms to three or four environmental factors simultaneously. *FEMS Microbiology Letters*, **40**, 263–8.

4

Circadian rhythms in filamentous fungi

M. RAMSDALE

Time is nature's way of keeping everything from happening at once
Richard Feynman (*National Geographic Magazine*, 1991)

Introduction

This review focuses upon circadian rhythmicity in the fungi; in particular circadian rhythmicity in one of the best-studied model systems, the *Neurospora* clock. Consideration will also be given to the limited number of true circadian rhythms that have been detected in other filamentous fungi. Many fungi display non-circadian rhythms (Ingold, 1971; Lysek, 1978, 1984), which are not discussed in detail. Suffice to say that the output rhythms of most fungi probably reflect direct changes in the environment, triggering immediate developmental or physiological responses. Reactions to light are the most prevalent, although temperature also causes rhythmicity. Non-circadian rhythms may also arise from self-sustaining metabolic cycles. Examples of the latter may include the formation of concentric zones or archimedean spirals of conidia by *Nectria cinnabarina* (Bourret *et al.*, 1969), the sporulation of *Leptosphaeria* controlled by oscillations in the asparagine–pyruvate pathway (Jerebzoff & Jerebzoff-Quintin, 1982) and the 'hormonal' triggering of hyphal growth rhythms in *Ascobolus immersus* and *Podospora anserina* (Chevaugeon & Nguyen Van, 1969).

Fungi are useful model systems for the study of circadian rhythmicity for a number of reasons. They often have fast generation times, a long history of classical genetics, and now also molecular genetics. Many developmental and biochemical mutants are available that allow interactions between the clock, and input/output pathways, to be dissected. Fungi are easily manipulated in the laboratory and a record of their developmental activities is retained in their spatial trajectory, making output rhythms easy to monitor. Lastly, fungi display an enormous degree of variation which should allow important ecological and evolutionary comparisons of clock functions to be made.

Recurrent changes in the quality, intensity and duration of physical parameters such as light, temperature, pressure, humidity, salinity, and resource availability are particularly important to organisms that inhabit heterogeneous environments. Clocks represent useful devices to keep track of these changes so that the events in an organism's life can be appropriately orchestrated.

The ability to track time using biological clocks is a ubiquitous feature of eukaryotes (reviewed by Edmunds, 1988) and has also been reported in some prokaryotes (Sweeney & Borgese, 1989). The activities gated by biological clocks are extremely diverse, with over 200 overt metabolic rhythms being detectable in man (Moore-Ede, Sulzmann & Fuller, 1982). Rhythms observed in other eukaryotes involve sexual and asexual reproduction, cell proliferation and death, bioluminescence, photosynthesis, hormonal signalling, neuronal firing, and activity/rest cycles.

A complete description of the components of a biological clock is eagerly awaited. Until such time, it may be useful to view a clock as a biological sensory apparatus (Lakin-Thomas, Coté & Brody, 1990), transducing information from the environment (input), processing it through a central oscillator (clock) and then gating actions (output) – Eskin (1979). The input and output elements of clocks are probably interlinked with the central oscillator by complex regulatory networks with both positive and negative feedback components, but the simplified linearized view will be retained for clarity.

To be adaptive, a biological clock must not only track time and the events that occur within its domain, but it must also be able to anticipate them. Some of the most robust variations that occur in the environment occur on a daily basis. Clocks that monitor events in a 24 hour timeframe are termed circadian (*circa* – about, *diem* – a day). Other clocks exist that track recurrent features of the environment on shorter (ultradian), or longer (infradian) timescales, but they are not considered here.

The key properties of a circadian clock are: (1) An intrinsic period approximating to 24 h. (2) Responsiveness to recurrent environmental stimuli (zeitgebers), such as light or temperature, which reset (phase-shift), or entrain the clock on a daily basis. (3) Compensation, i.e. a constant period under a wide range of environmental conditions. For example, circadian clocks must run with the same speed at different temperatures such that the period displays a Q_{10} of about one. (4) The clock must be under genetic control. Rhythms driven by circadian clocks can be differentiated from non-circadian rhythms (with a 24 hour period) if the output rhythm is self-sustaining under constant, or 'free-running' conditions.

To allow comparisons between organisms running on different intrinsic periods the concept of circadian time has been introduced. This normalizes landmark events within a hypothetical 24 hour framework to the actual (sidereal) time that it takes for the organism to complete one full cycle under free-running conditions. By convention CT0 = dawn, and CT12 = dusk. Light–dark cycles are often used in experimental studies of biological clocks; correspondingly DD = constant darkness, LL = constant light and 12:12 LD is a light–dark cycle of 12 hours light followed by 12 hours darkness.

Circadian rhythmicity in *Neurospora crassa*

Numerous reviews have been written about the rapidly developing field of *Neurospora* circadian clock biology (Feldman, Gardner & Denison, 1979; Feldman, 1982; Feldman & Dunlap, 1983; Lakin-Thomas *et al.*, 1990; Dunlap, 1993, 1996; Nakashima & Onai, 1996; Bell-Pedersen, Garceau & Loros, 1997). The outline presented here covers the most salient features and highlights some of the more recent findings, in particular the generation of a molecular model based upon the cyclic behaviour of a component of the circadian oscillator, *frq*. However, it must be appreciated that many observations derived from classical genetic studies, as well as biochemical, physiological and developmental investigations, still remain to be adequately accommodated by the *frq* paradigm.

A conidiation rhythm in *Neurospora* was first described by Brandt (1953) in a proline auxotroph, *patch*. The rhythm was recognized as circadian by Pittendrigh *et al.* (1959) since it met some of the key criteria outlined earlier; *patch* displayed a consistent free-running period, was temperature compensated and was phase-shifted by light. Sussman, Lowry & Durkee (1964) identified another mutant, *clock*, which exhibited a hyphal branching rhythm with a period of 24 h, but this rhythm was highly sensitive to medium composition and temperature and could not be entrained to LD cycles. Later, Sargent, Briggs & Woodward (1966) described rhythmic conidiation in *timex*, which was subsequently found to be a double mutant carrying both the invertaseless (*inv*) and band (*bd*) mutations – (Sargent & Woodward, 1969). Periodic conidiation of wild-type strains was reported by Sargent & Woodward (1969) – although only when fresh air was passed over the colonies. Rhythmic conidiation of wild-type strains in culture is typically suppressed by the accumulation of CO_2 (Sargent & Kaltenborn, 1972). Band (*bd*) is now the strain most intensively studied by *Neurospora* chronobiologists because it is insensitive to CO_2. It is not known whether rhythmicity is expressed under field conditions, a matter that clearly deserves clarification.

The circadian rhythm of *Neurospora bd* mutants on solid agar medium is expressed as alternating zones of intense macroconidiation (band), and sparse mycelial zones (interbands) with few conidia (Fig. 4.1). The rhythm persists (free-runs) in constant darkness with an intrinsic period of 21–22 h. The clock has a cellular origin and runs in all regions of a fungal colony at the same time, even after a colony has differentiated. Slight phase differences can be detected between young marginal hyphae and older regions of a colony where the period of the rhythm appears to be a little longer (Dharmananda & Feldman, 1979).

Many factors influence conidiation in *Neurospora*, with perhaps the most important stimuli being light and temperature. Both light and temperature can set the phase of the conidiation rhythm, though neither has any significant effect upon period length. Constant light damps the rhythm so that colonies sporulate more or less continuously. Nutritional status may also alter the rhythm in some mutants.

Phase-shifting the *Neurospora* clock

Light responses and temperature effects

Circadian clocks may show time-of-day specific responses to environmental inputs or perturbations, which advance or delay the phase of the clock. A typical phase-response curve to light for *Neurospora* shows advances in the late subjective night/early subjective morning, and delays in the late subjective evening/early subjective night (Sargent & Briggs, 1967). Responses to other stimuli may deviate from this typical pattern, potentially giving useful information about clock mechanisms and components.

Neurospora can be entrained to a precise 24 h period (cf. intrinsic period of 21–22 h) by 'natural' 12:12 LD cycles (Fig. 4.2) and even to the very short photoperiod of 5 minutes light once per 24 h (Gardner, unpublished, in Edmunds 1988). The action spectrum for the phase-shifting response resembles that for blue-light-absorbing photoreceptors such as flavins and carotenoids (Sargent *et al.*, 1966 ; Sargent & Briggs, 1967).

Fig. 4.1. *Neurospora crassa* (*csp-1;bd*) entrained to a 12:12 LD cycle displaying alternating regions of high (band) and low (interband) macroconidiation. Colony grown at 22°C on Maltose–Arginine–Vogel's salts medium containing 0.002% [w/v] sodium dodecyl sulphate and 0.15 M KCl.

Albino-mutants (*al*-2 and *al*-3) blocked in carotenoid biosynthesis, show normal constant light-induced damping of the conidiation rhythm (Sargent & Briggs, 1967), suggesting that carotenoids may not be involved. Munoz & Butler (1975) concluded that the light-induced effects upon the clock were brought about by the flavoprotein-mediated reduction of a *b*-type cytochrome because *poky* mutants were less sensitive to light. Two riboflavin auxotrophs, *rib*-1 and *rib*-2, also display a reduced sensitivity to light for both constant light damping and phase-shifting (Paietta & Sargent, 1981), but three light-insensitive mutants (*lis*-1, *lis*-2

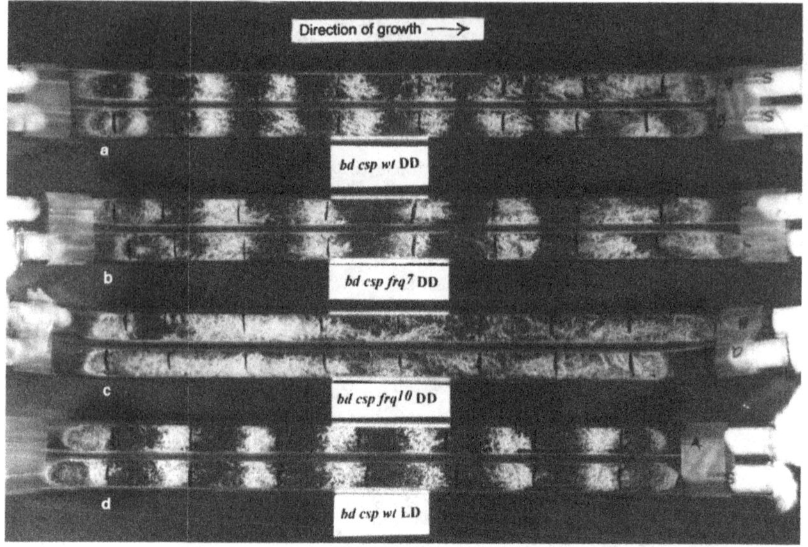

Fig. 4.2. Wild-type and clock mutants of *Neuropora crassa* in a *csp-1; bd* background grown in race-tubes. Daily growth marks are indicated on the tube to give time reference points. (a) Wild-type strain in constant darkness (period = 21 h); (b) frq^7 mutant with a long period in constant darkness; (c) frq^{10} null mutant exhibiting arrhythmicity; (d) wild-type strain entrained to a 12:12 LD cycle. Strains grown at 22°C on Maltose–Arginine–Vogel's salts medium.

and *lis*-3), which appear to have normal complements of flavin and cytochrome-*b*, do not exhibit constant light suppression (Paietta & Sargent, 1983). Nitrate reductase (a molybdo-flavoprotein containing a *b*-type cytochrome) does not play any role in providing an input to the circadian clock since three (*nit*) mutants exhibit normal clock photoresponses (Paietta & Sargent, 1982). Although the identity of the blue-light receptor remains to be resolved, recent work has identified two components of light signal transduction, white collar-1 and white collar-2 (Ballario *et al.*, 1996 ; Linden & Macino, 1997), which interact with both light perception and the clock (Crosthwaite, Dunlap & Loros, 1997).

Light-induced phase shifts are both temperature sensitive and pH sensitive. Higher amplitude responses are elicited at low temperatures (25°C) when compared to high temperatures (34°C) – Nakashima & Feldman, (1980). The mycelium is also more sensitive to light at pH 5.7 than at pH 6.7 (Nakashima, 1982*a*,*b*; Nakashima & Fujimura, 1982). Whether these changes are due to changes in light perception itself, alterations in the signal tranduction pathway that connects the clock and the input pathway, or changes in the amplitude of the oscillator is not clear. Nakashima & Feldman (1980) proposed that some component of the photoreceptor or the clock is heat labile, making the mycelium less responsive to light. As an alternative they also proposed that the amount of photoreceptor at higher temperatures may decrease so that light exerted a smaller effect on the clock. A third possibility is that at higher temperatures the amounts (but not amplitude) of an oscillating clock component might increase, so that the input from the photoreceptor was proportionally reduced. If both input and clock components increased at higher temperatures, their interaction would be temperature compensated.

Temperature compensation of period length is a canonical feature of circadian clocks. Whereas the growth rate of the band mutant is strongly influenced by temperature (Q_{10} = 2.35 in the range 16°C–35°C), the length of the free-running period at 18°C–25°C remains fairly constant (Q_{10} = 0.95) and only increases slightly at the upper temperature range of 25°C–35°C (Q_{10}=1.21) – Sargent *et al.* (1966). Sudden temperature steps (increases or decreases) advance or delay the phase of the rhythm. Temperature pulses also produce advances or delays, which can be predicted by the algebraic sum of the phase responses elicited by the component temperature steps (Pavlidis, Zimmerman & Osborn, 1968).

Outputs from the clock

Biochemical activities

Rhythms that accompany conidiogenesis, but which may not be independently driven by the clock, have been detected in CO_2 production (Woodward & Sargent, 1973), citrate synthase, glucose-6-phosphate dehydrogenase, glyceraldehyde-phosphate dehydrogenase, isocitrate lyase, NADase and phosphogluconate dehydrogenase (Hochberg & Sargent, 1974). Spatial variations in the levels of oxidized and reduced pyridine nucleotides (NAD^+, $NADP^+$, NADH and NADPH) have also been reported (Brody & Harris, 1973), but since the redox ratio remains constant over the circadian cycle they are probably not clock controlled either (Dieckmann, 1980 in Lakin-Thomas *et al.*, 1990).

Rhythms of overall adenylate energy charge occur at the growing front of *bd* strains, with maxima at points destined to become band centres (Delmer & Brody, 1975). Concentrations of ADP and ATP remain constant, whereas that of AMP oscillates between 0.5 and 6.0 μmol/g RDW.

The rhythm cannot be due to conidiation *per se*, since it also occurs in liquid cultures (Schulz, Pilatus & Rensing, 1985). Furthermore, the rhythm can be phase-shifted by light and damped in constant light. AMP oscillations might have been generated if mitochondrial oxidative phosphorylation was cyclic and partially uncoupled.

Both wild-type and *bd* strains exhibit rhythms of RNA and DNA content at the growth front which are in phase with the conidiation rhythm (Martens & Sargent, 1974). Nucleic acid oscillations were separated from the conidiation process in aconidial fluffy (*fl*) mutants, with peak abundance running out of phase with that in *bd* and wild-type strains.

Heat shock proteins display a circadian rhythm of sensitivity to induction, with maximal sensitivity occuring at CT12 (Cornelius & Rensing, 1986). The adaptive significance of such a change in sensitivity is clear since under normal environmental conditions, CT12 represents the phase at which the ambient temperature normally drops.

Circadian rhythms in fatty acid metabolism have been detected in *Neurospora*. Roeder, Sargent & Brody (1982) demonstrated oscillations in the mole percentages of linoleic (18:2) and linolenic (18:3) unsaturated fatty acids, which fluctuated 180° out of phase with one another. Such rhythms were detected in wild-type (bd^+) strains too, suggesting that the rhythm might be a distinct output of the clock, or even an essential feature of the clock machinery. Several models of circadian time-keeping are based upon changes in the rates of metabolic reactions resulting from altered membrane potentials and fluidity (e.g. Njus, Sulzman & Hastings, 1974), so these findings are particularly important.

Oscillations in diacylglycerol (DAG) occur within fully differentiated band regions, and at the growing margin of *Neurospora* colonies (M. Ramsdale & P.L. Lakin-Thomas, unpublished). Maximal levels occur between CT0 and CT3, just after the peak commitment to conidiation. Fluctuations of DAG may be related to the overall changes in lipid metabolism that accompany conidiation, or they may play an inductive role in the expression of the rhythm, altering rates of Golgi vesicle formation (Martin, 1997), or activating protein kinase C.

Isolation of clock-controlled genes

Subtractive hybridization of morning- and evening-specific mRNA populations, and differential screening of time-of-day-specific cDNA libraries has led to the identification of eight clock-controlled genes (Loros,

Denome & Dunlap, 1989; Bell-Pedersen *et al.*, 1996b). Two morning-specific genes, *ccg*-1 and *ccg*-2, which were abundantly expressed during conidiation, were found to be analogous to the glucose-repressed gene, *grg*-1 (McNally & Free, 1988) and the hydrophobin encoding gene, *eas* or *bli*-1 (Selitrennikoff, 1976; Sommer *et al.*, 1989). These genes were cloned and their clock regulation verified by finding that the period of the mRNA abundance rhythm matched the period of the conidiation rhythm in both wild-type strains and in a long-period mutant (frq^7) – (Loros *et al.*, 1989). The clock itself is unaffected in *ccg*-1 and *ccg*-2 null mutants, which suggests that *ccg*-1(*grg*-1) and *ccg*-2 (*eas*) are true output genes (Bell-Pedersen, Dunlap & Loros, 1992; Lindegren, 1994).

Molecular dissection of the *ccg*-2(*eas*) promoter has revealed a positive-activating-clock element (ACE) that is distinct from the other elements regulating the expression of this developmentally important gene (Bell-Pedersen, Dunlap & Loros, 1996*a*). Furthermore, this regulatory element is in itself sufficient to confer rhythmicity on a naïve reporter gene. Three independent pathways for the regulation of this gene can therefore be identified, one driven by the clock, one regulated by light and one through developmental/nutritional feedback (Arpaia *et al.*, 1993).

A further six *ccg*s identified by Bell-Pedersen *et al.* (1996*b*) encode mRNAs which also peak during conidiation. The identities of two of these genes are known; *ccg*-7 codes for glyceraldehyde-3-phosphate dehydrogenase, and *ccg*-12 is allelic to the stress-induced copper metallothionein gene, *cmt*. Of the remaining clock-controlled genes, *al*-3(c) encodes one of the three geranylgeranyl pyrophosphate synthase transcripts involved with carotenoid biosynthesis (Arpaia, Carattoli & Macino, 1995*b*), *con*-6, *con*-10 are conidiation-specific genes (Lauter & Yanofsky, 1993), and no information has been published about the identity of *ccg*-4.

Although the precise function of *ccg*-1 remains unclear, it represents over 8% of all mRNA in a morning-specific cDNA library (Bell-Pedersen *et al.*, 1996*b*), and can be induced by heat shock (Lindegren, 1994; Garceau, 1996 in Bell-Pedersen, Garceau & Loros, 1997). The cyclic behaviour of two genes involved with stress responses (*ccg*-1 and *ccg*-12) may be particularly significant since this confers a selective advantage to the *Neurospora* clock, which is not based upon the timing of conidiation. Clock-controlled genes, such as *ccg*-7, *ccg*-8 and *ccg*-12, are neither developmentally nor light-induced, further suggesting that the clock controls output pathways other than conidiation (Bell-Pedersen *et al.*, 1996*b*).

Although conidiation is linked to the oscillator mechanism, several observations indicate that this linkage is one-way (Lakin-Thomas *et al.*, 1990). Biochemical rhythms persist in the absence of conidiation, the phase of conidiation is retained under conditions non-permissive for its expression (e.g. liquid cultures) and morphological mutations affecting conidiation do not change the underlying rhythms (Brody & Martins, 1973).

Biochemical and genetic dissection of the *Neurospora* clock

The availability of metabolic mutants of *Neurospora* and a wide range of metabolic inhibitors has allowed a detailed investigation of the components of the circadian clock. Studies of this kind have revealed that there may be a great deal of cross-talk between different components of the clock and that the distinction between input, oscillator and output pathways may be somewhat blurred.

Electrophysiological studies

Plasma membrane electrochemical gradients may play an important role in the response of *Neurospora* to light, but probably not in the determination of phase. For example, inhibitors of H^+-translocating plasma membrane ATPase, such as diethylstilbestrol and *N,N*-dicyclohexylcarbodiimide, reduce the ability of light pulses to produce a phase-shift in cultures grown at pH 6.7 (but not at pH 5.7) – Nakashima (1982*a,b*). In contrast, specific inhibitors of mitochondrial ATPases do not affect the light response, but can bring about phase shifts (Nakashima, 1982*a*). Diethylstilbestrol and its related analogues DIE and HEX also produce phase shifts; however, HEX only partially inhibits the activity of isolated plasma membrane ATPases and DIE not at all, leading to the conclusion that the phase-shifting did not arise from disrupted proton gradients (Nakashima, 1982*b*).

Sato, Kondo & Miyoshi (1985) reported circadian rhythms of K^+ and Na^+ content at the margin of developing *Neurospora* colonies. Pulses of nystatin or the K^+ ionophore valinomycin produce large phase delays in the late subjective night, but only small advances in the early subjective day. Furthermore, such responses are attenuated in ergosterol-deficient mutants, such as *erg*-3 and *erg*-1 (ergosterol is the primary target of nystatin) – Koyama in Feldman & Dunlap, (1983). No correlation between circadian periodicity and intracellular pH has been found in *Neurospora* (Johnson, 1983; Pilatus and Techel, 1991) perhaps indicating that phase-dependent extrusion of protons at the hyphal tip is not necessary for rhythmic growth and differentiation (cf. Lysek, 1984).

Nakashima (1984*b*) found that a variety of respiratory inhibitors phase shifted the clock during the subjective day, but not during the subjective night. The effects could be attributed to a direct inhibition of clock functions, or to the uncoupling of respiration from the clock at some phases but not others. The magnitude of phase shifts were proportional to the reduction in respiratory activity and ATP content for azide and *m*-chlorophenylhydrazone; but not for antimycin A and cyanide. A failure to detect a correlation between ATP content and the magnitude of phase-shifts further strengthens the suggestion that adenylate energy charge is an output of the clock rather than an oscillator component.

Cell signalling components

Calcium controls many intracellular events in eukaryotes (Berridge, 1993) making it a prime target for circadian clock studies. Long (3 h) pulses of a calcium ionophore, A23187 (μM), induce phase-shifts in Ca^{2+} free media (Nakashima, 1984*a*), which can be inhibited by the addition of 0.1 mM $CaCl_2$ (but not by $MgCl_2$). Calcium at concentrations lower than 0.3 mM at CT10 produce phase delays, but advances above 10 mM. Furthermore, rhythmicity, entrainability and period are all affected in calcium dependent (*Ca4* and *Ca23*) mutants of *Neurospora* (Nakashima, 1985; Ohnishi, Cornelius & Nakashima, 1992), clearly implicating calcium in clock functions.

Calmodulin antagonists such as chlorpromazine, trifluoperazine, W7 and W12 also induce phase-shifts in *Neurospora*, whereas imipramine and alprenolol have virtually no effect (Nakashima 1985, 1986; Sadakane & Nakashima, 1996). The results correlate well with the effectiveness of the inhibitors in reducing the calmodulin-induced activation of phosphodiesterase. The ineffectiveness of chlorinated analogues of W7 and W12 implies that the phase-shifting is specific to the inhibition of calmodulin, and not to inhibition of other targets (Nakashima, 1985). Moreover, levels of calmodulin did not vary over time, which leads to the conclusion that calmodulin antagonists bring about phase-shifts as a result of their activity towards calmodulin-dependent reactions rather than calmodulin itself. Although calmodulin regulates the phosphorylation of proteins via its effects upon protein kinases and phosphatases, specific inhibitors of protein kinase C and calcium/calmodulin-dependent protein kinase II failed to inhibit the light-induced phase shifting of *Neurospora* (Sadakane & Nakashima, 1996). Two mutants, *cpz*-1 and *cpz*-2, which are hypersensitive to certain calmodulin antagonists, could provide excellent opportunities to explore further the relationship between time-keeping and protein phosphorylation (Suzuki, Katagiri & Nakashima, 1996).

Mutants and inhibitors affecting cyclic AMP metabolism have a variety of effects upon the *Neurospora* conidiation rhythm. Tonic increases in intracellular cAMP slow the clock down, whereas tonic decreases have no effect. Inhibition of phosphodiesterase by aminophylline, theophylline or caffeine brings about an increase in the endogenous level of cAMP and causes a dose-dependent increase in period (Feldman, 1975). Quinidine, an inhibitor of adenylate cyclase, has virtually no effect upon the period (Feldman & Dunlap, 1983), but does produce time-of-day specific phase shifts (Perlman, Nakashima & Feldman, 1981). Time of day specific advances or delays are also produced when cAMP levels are transiently increased.

After exposure to short light pulses, cAMP levels in *Neurospora* rise, but then fall after only 10 minutes – the subsequent levels dropping below their start values (Kallies, Gebauer & Rensing, 1996) . In the same time period cGMP levels fall, whereas $InsP_3$ levels remain constant (see also Lakin-Thomas, 1993; Prior, Robson & Trinci, 1994). An analogous response occurs in *Trichoderma viride*, where it has been proposed that light induces a transient increase in membrane potential, which activates H^+-ATPases. A rise in ATP then triggers the production of cAMP by adenylate cyclase (Farkǎs, Sulová & Lehotský, 1985 ; Gresík, Nadezda & Farkaš, 1988). The subsequent loss of cAMP might be due to the a stimulation of phosphodiesterase by light (Sokolovsky & Kritsky, 1985).

Not all information regarding cAMP and the *Neurospora* clock is so consistent. Partial revertants of *crisp*-1, with reduced endogenous levels of adenylate cyclase (Feldman *et al.*, 1979) and the phosphodiesterase mutants *cpd*-1/*cpd*-2 (Hasunuma, 1984*a,b*; Hasunuma & Shinohara, 1985) display normal periods and light entrainment. Furthermore, rhythms of cAMP have been detected in both *bd* strains (Techel *et al.*, 1990) and the phosphodiesterase mutants (Hasunuma, 1984*a,b*; Hasunuma & Shinohara, 1985). Cyclic AMP and cGMP-dependant protein kinases may therefore affect the clock when inhibited, but are probably not essential parts of its machinery (Techel *et al.*, 1990). Perhaps the effects of high cAMP titres might be attributed to clock-independent repression of conidiation. Little is known of the function of cGMP in fungi (Gadd, 1994), but it is clear that further investigation is required.

Role of fatty acids and membranes in clock control

Neurospora strains with known metabolic defects and altered clock phenotypes have been used to reveal additional intricacies of the clock (see Lakin-Thomas *et al.*, 1990; Dunlap, 1993). Mutants of amino acid metabolism (*arg*-13, *cys*-4 and *cys*-12), (ergosterol biosynthesis) *phe*-1 and glycerol utilization (*glp*-3) have altered clocks. In addition, three maternally inherited mutations ([*mi*-2], [*mi*-3], and [*mi*-5]), four cytochrome-deficient mutants (*cya*-5, *cyb*-2, *cyb*-3 and *cyt*-4), the oligomycin-resistant (*oli*) mutant, and several fatty acid metabolism mutants exhibit clock defects.

The *cel* mutant of *Neurospora* requires saturated fatty acids for normal growth and rhythmicity, and displays a long period/slow growth when supplemented with unsaturated fatty acids or short-chain fatty acids (Brody & Martins, 1979; Lakin-Thomas *et al.*, 1990). A choline biosynthesis mutant, *chol*-1, also displays clock defects (Lakin-Thomas, 1996). On low choline medium, the mutant grows slowly and the period is long. Moreover, long-period cultures are sensitive to light, can be entrained to long-period light–dark cycles, and show damping of the rhythm in constant light indicative of a true clock mutant.

Studies of double mutant strains constructed between *cel* or *chol-1* and various *frq* strains reveal that both are epistatic to *frq*. When supplemented appropriately, double mutants display the period of the fatty-acid mutant parent rather than the conditional rhythmicity of the *frq* parent (P.L. Lakin-Thomas, unpublished). For example, on high choline medium, when the *chol-1* defect is not expressed, the phenotype is the same as the *frq* parent. However, on low choline medium, the strains behave like the *chol-1* parent displaying long periods.

Both *cel* and *chol-1* have altered lipid compositions and exhibit defective temperature compensation. The adaptation of membrane composition to changes in ambient temperature has long been a candidate mechanism for temperature compensation of circadian oscillators (Njus *et al.*, 1974; Coté, Lakin-Thomas & Brody, 1996; Lakin-Thomas, Brody & Coté, 1997). The *cel* and *chol-1* strains therefore provide some of the strongest evidence in favour of the membrane models.

The importance of translation

Cycloheximide induces large phase shifts in *Neurospora*, with maximal sensitivity occurring in the middle of the subjective day (CT5). No phase shifting is seen at CT19, the middle of the subjective night (Nakashima, Perlman & Feldman, 1981*a*). Although inhibition of protein synthesis was the same in both night and day phases the effects of the inhibitor can be directly attributed to the inhibition of protein synthesis because the cycloheximide resistant mutants, *cyh*-1 and *cyh*-2, cannot be phase shifted (Nakashima, Perlman & Feldman, 1981*b*). Such data could indicate that protein synthesis at specific times of the day is necessary for the operation of the circadian clock (Nakashima, *et al.*, 1981*a*). However, this interpretation needs to be viewed with caution since one of the canonical clock mutants, frq^{7} is cycloheximide insensitive with respect to phase shifting, but is sensitive with respect to 80 S ribosomal protein synthesis (Feldman & Dunlap, 1983). Furthermore, Sadakane & Nakashima (1996) could find no effect of another protein synthesis inhibitor (5-azacytidine) on phase in wild-type strains.

Canonical clock mutants

A variety of canonical clock mutants have been identified in *Neurospora* with periods significantly different from the wild-type. Seven of these map to discrete loci; *frq* (frequency), *chr* (chrono), *prd-1*, *prd-2*, *prd-3*, *prd-4* (period), and *cla-1* (clock-affecting) – (Feldman & Hoyle, 1976; Feldman & Atkinson, 1978; Feldman *et al.*, 1979; Gardner & Feldman, 1980, 1981; Feldman, 1982; Loros & Feldman, 1986 Loros, Richman & Feldman, 1986; Brody, Willert & Chuman, 1988).

Most studies of clock genes have concentrated upon the *frequency* locus. Eight independent mutations of *frq* have been isolated, representing an allelic series which maps to the right arm of chromosome VII. Periods exhibited by these mutants range from 16.5 h to 29.0 h. Most alleles of *frq* display incomplete dominance when compared with the wild-type frq^{+} allele, but frq^{9} is a recessive null mutant (Feldman & Hoyle, 1973, 1976; Gardner & Feldman, 1980). A second recessive *frq*-allele (frq^{10}) has been synthesized using molecular genetic approaches by replacing the native *frq* gene with a dominant selectable marker (Aronson, Johnson & Dunlap, 1994*a*).

Some of the *frq* alleles have altered temperature compensation properties. Whereas the short-period *frq* mutants ($frq^{1,2,4,6}$) exhibit near-normal temperature compensation up to 30°C ($Q_{10} = 1$); the longer period mutants (*frq*^{3,7,8}) are much more temperature sensitive. *Frq*^{9} shows a complete loss of temperature compensation, with a period ranging from 34 h at 18°C to 16 h at 34°C. Furthermore, the period of *frq*^{9} is also affected by the composition of the growth medium, becoming completely arrhythmic under some circumstances (Loros & Feldman, 1986; Loros *et al.*, 1986). Conditional arrhythmicity of the null alleles implies that *frq* is not required for rhythmicity *per se*, but may be required for its stability and the effective coupling of the clock to the environment, weakening some claims that *frq* is a state-variable of the central oscillator (Lakin-Thomas, 1998; cf. Dunlap, 1996)

Differences in the phase response curves of *frq* alleles to light indicate that their periods are not altered proportionally throughout the circadian cycle, rather the alterations relate to events that occur within the early subjective night (Feldman, 1982 Nakashima, 1985). Any model of circadian rhythmicity including *frq* must be able to explain all of its properties, and also the interaction of *frq* with other loci such as *wc*-1, *wc*-2, *cel* and *chol*-1.

Molecular analysis of clock genes

The *frq* gene was cloned using a chromosome walk strategy starting at a closely linked gene, *oli* (McClung, Fox & Dunlap, 1989). *Frq* containing cosmids were identified on the basis of their ability to rescue a circadian phenotype when transformed into a null *frq*-9 background. The *frq* locus was located within an 8 kb region of the chromosome encoding two genes of about 1.5 kb and 5 kb respectively (Dunlap, 1993). Transcripts from both genes were polyadenylated, but subsequent genetic analyses revealed that the smaller gene, overlying the *frq* promoter, lacked clock function (Dunlap, 1996). An antisense transcript has also been detected with no known function at present. The longer gene encodes two major transcripts of 4 and 4.5 kb length, each containing 1.5 kb untranslated regions.

Two FRQ translational products of 890 and 989 amino acids arise from a choice of ATGs at the start of the *frq* ORF (Garceau *et al.*, 1997). Dominant landmarks on the putative protein include a mixed repeat of threonine/glycine and serine/glycine residues surrounded by potential glycosylation sites, a helix-turn-helix DNA binding motif; an RKKRK nuclear localization signal; an acid tail region characteristic of transcription factors; several serine/threonine phosphorylation sites and a number of PEST sequences associated with proteins that show rapid turnover (Dunlap, 1993 Lewis & Feldman, 1993; Aronson *et al.*, 1994*a*).

Frq homologues have also been cloned from eleven other ascomycetes (Merrow & Dunlap, 1994; Lewis & Feldman, 1996). In all species examined the *frq* homologues exhibit the dominant features of the *Neurospora frq* gene. When the *Sordaria frq* homologue was placed under the control of a *frq* promoter, the chimaeric transgene rescued rhythmicity in frq^{10} null strains (Merrow & Dunlap, 1994; Lewis *et al.*, 1996) even though *Sordaria* itself does not conidiate rhythmically.

Cloning of the wild-type *frq* locus made it possible to examine the defects in *frq* mutant alleles – providing insight into how changes in period and temperature compensation might arise. All alleles produced *frq* mRNA in amounts similar to that of the wild-type, but all differed from the wild-type by single nucleotides (Aronson *et al.*, 1994a). Frq^{9} bears a single base-pair deletion at amino acid codon 674, resulting in a nonsense transcript which encodes a truncated FRQ protein. Mutant alleles $frq^{2,4,6}$ are identical, with Ala to Thr substitution at codon 895. Mutations with long periods have Glu 364 to Lys (frq^{3}) and Gly 459 to Asp ($frq^{7,8}$) substitutions. Such changes may alter the charge of the proteins, their side-chain bulk, and potential phosphorylation sites perhaps influencing the time of entry of the protein into the nucleus, the stability of its association with other proteins and its rate of degradation.

Regulation of frq *transcription*

Analysis of the *frq* gene and its regulation has given insight into the way a cellular oscillator might be assembled (Aronson *et al.*, 1994*a,b*; Garceau *et al.*, 1997; Liu *et al.*, 1997). Aronson *et al.* (1994*b*) propose that the oscillator in *Neurospora* requires an autoregulatory feedback cycle.

Evidence for the autoregulatory feedback model comes from several critical observations: (1) The period of *frq* mRNA and FRQ protein abundance is characteristic of the strain examined (Aronson *et al.*, 1994b; Garceau *et al.*, 1997). (2) *Frq* overexpression abolishes the conidiation rhythm (Aronson *et al.*, 1994*b*). (3) There is a lag phase between peak mRNA and peak protein levels. (4) When there is no functional FRQ gene product, *frq* transcription is high. (5) Non-rhythmic expression of *frq* (from an inducible promoter) in a frq^{+} strain abolishes the conidiation rhythm and the rhythmicity of the endogenous *frq* gene (Aronson *et al.*, 1994*b*). (6) No level of non-cyclic frq^{+} mRNA expression can rescue rhythmicity in frq^{9} null mutants (Bell-Pedersen *et al.*, 1997).

The kinetics and feedback between the expression of *frq*, FRQ and the output from the clock can all be monitored over time (Aronson *et al.*, 1994*a*; Crosthwaite, Dunlap & Loros, 1995; Garceau *et al.*, 1997; Merrow, Garceau & Dunlap, 1997), giving rise to a temporally discretized model for the feedback cycle.

Frq ***oscillator model***

Frq mRNA and FRQ protein are low at CT0, but *frq* mRNA levels begin to rise, reaching a peak at CT2-6. Following a 2–4 h lag, FRQ protein appears, reaching a maximum at around CT8. Both products of the *frq* gene are translated and the proteins formed become partially phosphorylated. Phosphorylation continues in a more or less time-of-day specific manner producing a number of FRQ isoforms (Garceau *et al.*, 1997). FRQ then enters the nucleus and represses its own transcription, and probably that of a number of clock-controlled genes (Garceau, 1996 in Bell-Pedersen *et al.*, 1997). As *frq* levels fall, FRQ protein levels also decline. The decline in FRQ arises from both a reduction in its rate of production, and possibly from its enhanced degradation (due to its high degree of phosphorylation – Karin & Hunter, 1995). FRQ levels remain low from CT12 onwards, the de-repression of *frq* taking about 14–18 h for completion.

In order to produce persistent rhythmicity, Dunlap (1996) proposed that such a loop must be boosted by a positive acting element, implying that *frq* transcription must be turned on by another (clock) element (Merrow *et al.*, 1997; Aronson *et al.*, 1994*b*). Recent studies have revealed that white collar-1 (*wc*-1) and white collar-2 (*wc*-2), which regulate all of the known light responses in *Neurospora* (Harding & Shropshire, 1980), may not only link the *frq* rhythm to environmental cycles (light detection), but may provide the necessary positive feedback under free-running conditions (Crosthwaite *et al.*, 1997).

This suggests a slightly different model for *frq* output. If FRQ negatively regulated *wc*-2 rather than itself, and *wc*-2 was an inducible, oscillating positive regulator of *frq* transcription, the *frq* mRNA/FRQ cycle would still be maintained. Such a model would also explain why: (1) light increases *frq* mRNA, (2) low levels of *frq* mRNA are produced by *wc* mutants, (3) there is no effect of light on *frq* mRNA in *wc* mutants, (4) *frq* mRNA expression is high in *frq* (FRQ) null mutants. Furthermore, if *frq*/FRQ negatively regulated clock-controlled genes there would also be explanations for the constitutive conidiation of *frq* null mutants, high *ccg*-2 (*eas*) expression in *frq*[10] null mutants (Lakin-Thomas, personal communication), the generation of arrhythmicity in the *frq* inducible system, and constant conidiation of *wc* mutants. If *wc* is found to be cyclically expressed there will be little to distinguish the WC proteins as input components, and the FRQ proteins as oscillator components.

Frq expression may be closely linked to temperature responses since growth temperature dictates the relative abundance of the two *frq* mRNA translation products (Garceau *et al.*, 1997; Liu *et al.*, 1997). At high temperatures (30°C) the long product is the dominant species, whereas at low temperatures (18°C) the short product is dominant. Furthermore if the first ATG codon is disrupted, rhythmicity is lost at 30°C, but when the third ATG is disrupted rhythmicity is lost at 18°C. This effectively sets physiological limits to the temperature range over which rhythms are expressed and also indicates how temperature might reset the clock. Such findings demonstrate that temperature compensation of period length is probably not a simple homeostatic process that holds *frq* at a constant level over a range of temperatures, rather it is a dynamic process that involves changes in the amount of the clock components.

Models of morphological patterning

All of the data described so far have concerned the activities of putative clock components at the cellular level. This is reasonable if the aim is to understand how time is monitored *per se*. On the other hand, the clearest manifestation of the clock in *Neurospora* is a macroscopically visible differentiation pattern – a fungal colony. If the clock is simply giving cues to developmental processes, we may wonder why developmental determination signals are only registered at the colony margin. Since growth also continues behind the margin, a simple question that needs to be addressed is: why don't interbands 'fill' in? Sadly there is no immediate answer, though models have been proposed that might account for certain aspects of spatial patterning.

When mycelia of *Neurospora* with different phases interact, several possible outcomes might be expected (Deutsch, Dress & Rensing, 1993). (1) No mutual influence of interacting mycelia; (2) averaging of phases; (3) dominance of one phase over another; (4) expression of both phases; or (5) expression of neither phase, i.e. constant conidiation or no conidiation. When the appropriate experiments were performed, outcomes, 2, 4 and 5 were observed. Deutsch *et al.* (1993) modelled these outcomes with cellular automata, leading them to suggest that concentric ring formation could arise from the synchronous oscillation of a population of endogenous, temporally regulated, coupled activator–inhibitor mechanisms.

The cellular automaton simulations predicted the formation of spiral patterns as well as concentric rings. For spiral pattern formation to occur, initial phase discontinuities must be maintained and not damped. Since *Neurospora* has not yet been shown to produce spirals, this could imply that some form of mechanical synchronization of hyphal tips occurs at the margin, that 'phasing substances' transmit information from one part of the colony to another, or that the model is inappropriate.

There is no strong evidence in favour of the occurrence of 'phasing substances' in *Neurospora*. On the contrary, light pulses given to one part of a colony do not cause phase shifts in distal regions (Winfree & Twaddle, 1981) and phase differences between young marginal hyphae and older regions of the colony are not homogenized (Dharmananda & Feldman, 1979). However, unpublished experiments by Woodward (reported by Feldman & Dunlap, 1983), revealed that heterokaryons generated between strains entrained to different LD cycles occasionally produced mycelia that simultaneously expressed the phase of both parents. Furthermore, some phases were more dominant than others, leading to the suggestion that a phasing substance might be entraining the clock components.

The models of Deutsch *et al.* indicated how a single phase would predominate if some hyphae (of one phase) developed more quickly than others (with a different phase). This would generate a mycelium with a single phase at the margin through mechanical synchronization. Circadian changes in growth rate have been detected in *Neurospora*, with hyphae in late band/early interband phase extending more slowly than hyphae in late interband/early band phase (Sargent *et al.*, 1966; Lakin-Thomas, personal communication).

Two additional mathematical models have indicated how concentric zonation might develop in fungal mycelia (Edelstein & Segel, 1983; Davidson *et al.*, 1996), although neither specifically addresses the issue of circadian zonation. Whereas the Edelstein & Segel (1983) model represented microscopic events and then extrapolated them to macroscopic appearance (very much akin to the cellular automaton approach already described), the Davidson *et al.* (1996) model takes a more systemic approach, viewing the fungal mycelium as an integrated entity (see Chapter 3).

Davidson *et al.* (1996) proposed that large-scale heterogeneous distributions of biomass could be explained in terms of the ability of fungi to distribute (and re-distribute) resources, alter internal resistances, compartmentalize metabolic activities and degenerate. The model treats the mycelium as a pooled system of variable resistances that bound the components necessary for the reaction-diffusion driven kinetics of a developmental activator. Model parameters were augmented with appropriate data in order to model mycelial pattern formation realistically. The model predicted the formation of both fixed and travelling waves of activator substances across a colony, uneven biomass densities, stop–start growth and the formation of aversion zones between paired mycelia.

What these models tell us about how the mycelium interprets a persistent, oscillatory differentiation signal is unclear, but certainly warrants further attention. This issue is not addressed at all by the automaton model (Deutsch *et al.*, 1993) since it only simulates decisions regarding differentiation at the margin. Inhibition/nutrient deprivation behind the growing front would probably be required to prevent the filling in of interbands. The formation of a fixed mycelial distribution pattern in the reaction-diffusion model does not require the additional elements such as growth inhibitors. The two types of model could be compared if inhibitory elements (or nutritional deprivation) could be experimentally demonstrated/ruled out as requirements for the stable expression of zoned patterns.

The production of concentric rings or spirals by fungi, rather than any of the many other conceivable patterns, might be due to the tendency of self-organizing systems to 'direct' pattern formation towards one of a limited number of 'morphogenetic attractors' – Goodwin (1995), Rayner *et al.* (Chapter 1 this volume). Clearly further work (both theoretical and experimental) is required to ascertain the possible involvement of external/internal inhibitors, reaction-diffusion kinetics, and genotype–environment interactions.

Circadian rhythms in other fungi

There are many reports of fungi that display alternating zones of aerial and appressed mycelium, conidial and aconidial mycelium, or pigmented and non-pigmented mycelium (e.g. Kafi & Tarr, 1963; Lysek, 1978, 1984). However, the number of studies that have investigated the origins of such rhythmicity is comparatively small, and only rarely is there good evidence for a circadian basis.

In most studies where the involvement of a biological clock has been implicated, circadian rhythmicity was not the focus of the study, rather an incidental aside. As a result, the data available to a reviewer are often insufficiently detailed for definite conclusions to be reached. In many cases the evidence collected is neither systematic, nor obtained under appropriately rigorous conditions. Frequently there has been a failure to appreciate the sheer sensitivity of fungi to light, which subsequently casts doubts upon some conclusions. For example, if the dim-light of a torch allows the observer to see a culture (Ingold & Cox, 1955), it may be sufficient to induce a phase shift; similarly, placing a culture in a cupboard may not exclude external light cues (see Carpenter, 1949). Notwithstanding such misgivings, a number of fungi (Table 4.1) do display rhythms that free-run under constant conditions, are light entrainable and display periods close to 24 h.

Although little is known of the biochemistry or genetics of circadian rhythmicity in fungi other than *Neurospora*, studies focusing upon the identification of rhythms and the conditions under which they occur may provide much needed insight into the organization of biological clocks, the range of developmental processes that are driven by them, and their adaptive significance.

Circadian rhythmicity in fungi is not just a feature of multicellular organization. *Saccharomyces cerevisiae* exhibits a circadian-gated rhythm of cell division and amino acid uptake (Edmunds *et al.*, 1979). Similarly, cell division and sensitivity of stationary phase cultures to heat shock is driven by a clock in *Schizosaccharomyces pombe* (Kippert, 1989; Kippert *et al.*, 1991). In both organisms the rhythms are temperature compensated and can be entrained to light–dark cycles. The immense amount of information available about the yeast genome may soon open up a wealth of opportunities for chronobiologists to dissect further the molecular basis of circadian clocks.

The known output rhythms of filamentous fungi (other than *Neurospora*) can without exception be classified as patterns of spore production (coremia, pycnidia, conidia, sclerotia, etc.), or discharge (zygospores, ascospores and basidiospores). Spore development is somewhat conspicuous, which has probably led to its over-representation/investigation. Biochemical and morphological aspects of development, such as aerial-submerged, or fast-effuse transitions may also be rhythmic, but are clearly more difficult to detect.

In most cases fungal rhythms can be entrained to alternating (natural or artificial) light–dark cycles – some with peak activities that occur

Table 4.1. *Circadian rhythms in fungi*

Organism	Output	Timing	FRP (h)	Entrain	TC	FRC	Transients	References
Zygomycetes								
Pilobolus sphaerosporus	Sporangiospore discharge	Diurnal	24	LD or Temperature	Yes	LL or DD	Yes 1 or 2	Uebelmesser (1954); Bruce *et al.* (1960)
Pilobolus crystallinus	Sporangiospore discharge	Diurnal	n.d.	n.d.	n.d.	LL or DD	n.d.	Uebelmesser (1954)
Cokeromyces spp.	Sporangiospore discharge	?	?	?	?	LL or DD	?	Poitras (1954)
Ascomycetes								
Neurospora crassa	Ascospore discharge	Nocturnal (?)	?	LD	n.d.	DD	n.d.	Lakin-Thomas *et al.* (1990)
	Conidiation	Nocturnal/ Diurnal	21 h	LD Temperature	Yes Q_{10} = 1	DD only	No	Various
Sclerotinia fructigena	Conidiation	Diurnal	22–24	LD	Yes Q_{10} = 1.031	DD only	n.d.	Jensen & Lysek (1983)
Apiosordaria verruculosa	Ascospore discharge	Diurnal	21.5	LD	n.d.	DD only	n.d.	Ingold & Marshall (1963)
Sordaria fimicola	Ascospore discharge	Diurnal	24	LD	n.d.	LL or DD	Yes	Ingold & Marshall (1963); Austin (1968)
Sordaria macrospora	Ascospore discharge	?	?	?	?	?	?	Walkey & Harvey (1967)
Nectria cinnabarina	Ascospore discharge	Diurnal	n.d.	LD	n.d.	LL only	n.d.	Ingold (1933)
Daldinia concentrica	Ascospore discharge	Nocturnal	22	LD	n.d.	DD only	No	Ingold (1946); Ingold & Cox (1955)
Hypoxylon coccineum	Ascospore discharge	Diurnal	22.3	LD	n.d.	DD only	n.d.	Ingold (1933)
Hypoxylon fuscum	Ascospore discharge	Nocturnal	n.d.	LD	n.d.	DD only	n.d.	Ingold (1933)
Lopadostoma turgidum	Ascospore discharge	Nocturnal	n.d.	n.d.	n.d.	n.d.	n.d.	Walkey & Harvey (1967)
Melanomma spp.	Ascospore discharge	Nocturnal	n.d.	n.d.	n.d.	n.d.	n.d.	Ingold (1933)
Hyphomycetes								
Penicillium diversum	Conidiation	n.d.	24	No zeitgeber	Yes Q_{10} = 1.15	LL or DD	n.d.	Bourret *et al.* (1969)
Helminthosporium rostrum	Conidiation	n.d.	24	LD	n.d.	DD only	n.d.	Kafi & Tarr (1963)
Verticillium albo-atrum	Conidiation	?	?	?	?	LL or DD	?	Chaudhuri (1923)
Phaeoisariopsis personata	Conidial discharge	Nocturnal	n.d.	LD	n.d.	DD only	n.d.	Butler *et al.* (1995)
Basidiomycetes								
Thanetophorus cucumeris	Basidiospore discharge	Nocturnal	27	LD	n.d.	DD	n.d.	Carpenter (1949)
Sphaerobolus stellatus	Basidiospore discharge	Nocturnal	24	LD	n.d.	LL or DD (slight damp)	Yes	Freiderichsen & Engel (1960); Engel & Friederichsen (1964)
Yeasts								
Candida utilis	Growth	?	?	LD	?	?	?	Wille (1974)
Saccharomyces	Mitosis	Nocturnal	26	LD	n.d.	DD	n.d.	Edmunds *et al.* (1979)
	Amino acid transport	Nocturnal	22	LD	n.d.	DD	n.d.	Edmunds *et al.* (1979)
S. pombe	Mitosis	Nocturnal		LD	n.d.	DD	n.d.	Kippert *et al.* (1991)
	Heat sensitivity	Nocturnal	27	LD	Yes	DD	n.d.	Kippert (1989)

FRP, free running period; TC, temperature compensated; FRC, free running conditions; L, light; D, dark; n.d., no data.

during the day (diurnal) and some that occur at night (nocturnal). In *Pilobolus*, entrainment has also been demonstrated to temperature cycles (Uebelmesser, 1954; Bruce *et al.*, 1960). Temperature compensation has only been examined in four filamentous fungi, and in each case it was found to occur (Uebelmesser, 1954; Bruce *et al.*, 1960; Bourret *et al.*, 1969; Jensen & Lysek, 1983). These are perhaps the strongest indications that circadian rhythms in fungi might be more common than previously suspected.

Jensen & Lysek (1983) found that 92% of strains ($n = 196$) of *Sclerotinia fructigena* developed rhythmically in light–dark cycles and 30, 15%, exhibited rhythmicity under free-running conditions. Four strains demonstrated good circadian properties including temperature compensation in the range 22°C–32°C ($Q_{10} = 1.031$). Strain specific differences in the endogenous period length were detected at 22°C, indicating a possible genetic component to the timing system in *Sclerotinia*.

The conidiation rhythm of *Penicillium diversum* is of particular interest since the rhythm persists in both constant light and constant darkness, and neither light nor temperature changes reset the clock. This could indicate the involvement of non-photic environmental cues/zeitgebers. An investigation of changes in nutrient availability might be particularly productive in this regard. Nutritional inputs to a biological clock have recently been detected in a marine diatom, *Gonyaulax*, which is sensitive to variations in the levels of nitrate (Roenneberg & Mittag, 1996).

Periodic discharge of ascospores has been detected from numerous ascomycetes (see Ingold, 1971 for a review) – including *Neurospora* (Lakin-Thomas *et al.*, 1990). Ascospores from stromata of *Daldinia concentrica* are maximally ejected at night in natural light–dark cycles, but free-run with a period of about 22 h in constant conditions. Although the design of the experiments might be questioned (torch-lights at night!) the rhythm did persist for about 12 d in constant darkness, and then gradually lost synchrony. The rhythm could then be restored by exposing the stromata to a 12:12 LD cycle. Moreover, the rhythm could also be entrained to short-day (6:6 LD), but not to long-day (24:24 LD) cycles. The total number of spores produced in comparable 24 h time frames were considerably lower than in 12:12 LD cycles (see later discussion). On transfer to constant darkness the intrinsic 22 h rhythm was immediately restored, with no evidence of transient behaviour.

The retention of a phase set by an entrainment cycle for several cycles after transfer of an organism to free-running conditions, or the display of two phases after a light/temperature pulse is termed a transient.

Transients are an intriguing feature of the circadian clock in a number of eukaryotes (particularly mammals), and also appear to occur in some filamentous fungi, most notably *Pilobolus sphaerosporus* (Uebelmesser, 1954; Bruce *et al.*, 1960) and *Sphaerobolus stellatus* (Friederichsen & Engel, 1960). The occurrence of a transient could be taken to indicate that a population of synchronous cells gradually desynchronized upon release into free-running conditions. Alternatively, a single complex limit-cycle oscillator, with three or more state-variables, when perturbed could produce a colony with a period set by a bifurcation series – a colony entrained to a 12 h period (6:6 LD) would then reset to a 'virtual' 24 h period – fortuitously close to its 'real' intrinsic period. A biologically more plausible explanation is that rhythmicity is controlled by more than one oscillator; a master clock oscillator, and a slave output oscillator. Weak coupling between the two might then generate a delay between the clock being reset and the rhythm responding.

More unusual transients, with a complete suppression of output for several days after entry into constant conditions, have also been seen in *Sordaria fimicola* (Austin, 1968) and *Sph. stellatus* (Engel & Friederichsen, 1964). It is unclear if these result from a direct effect upon the clock, or if darkness temporarily suppresses the output in 'diurnal' species. Inhibition of spore discharge in the dark has been described in a number of fungi including *Apiosordaria verruculosa,* but it has not been detected in *S. fimicola* (Ingold & Marshall, 1963). In *S. fimicola*, peaks of ascospore discharge that eventually occur are still in phase with parallel cultures maintained in a LD cycle (Austin, 1968), which indicates that the clock was probably still running and that constant darkness had no effect upon it. The situation in *Sph. stellatus* is even less clear, with no correlation between the timing of the onset of basidiospore discharge in constant darkness and the previous LD cycle. Furthermore, the final output rhythm also displayed a much reduced amplitude.

The timing of spore production in some fungi may be expected to have a large impact upon their fitness. As noted previously for *D. concentrica*, total spore productivity was reduced in short-day (6:6 LD) and long-day (24:24 LD) cycles when compared to the 'natural' 12:12 LD cycle. A re-examination of the data provided by Friederichsen & Engel (1960) on basidiospore discharge in *Sph. stellatus* reinforces this finding. Maximal spore productivity (per 24 h) occurred when fruiting cultures were maintained in a 12:12 LD cycle. If maximal spore output = 100%, then the total output decreased as follows: 4:20 (43%), 7.5 min:24 h (31%), 6:6 (28%), 2:22 (27%) and 24:24 (22%). Some fungi therefore appear to be strongly dependent upon a natural light–dark cycle and their fitness upon an ability to predict the most appropriate time to liberate spores. Notwithstanding all of the other physiological processes (including biochemical activities and growth phases) which may be more efficiently carried out at specific times of the day, such a fitness advantage is in itself probably sufficient for many fungi to require and maintain a biological clock.

Acknowledgements

The author wishes to acknowledge the support and encouragement provided by Dr Pat Lakin-Thomas in the preparation of this review.

References

Aronson, B. D., Johnson, K. A., Lui, Q. & Dunlap, J. C. (1992). Molecular analysis of the *Neurospora* clock. *Chronobiology International*, **9**, 231–9.

Aronson, B. D., Johnson, K. A. & Dunlap, J. C. (1994*a*). The circadian clock locus *frequency*: a single ORF defines period length and temperature compensation. *Proceedings of the National Academy of Sciences, USA*, **91**, 7683–7.

Aronson, B. D., Johnson, K. A., Loros, J. J. & Dunlap, J. C. (1994*b*). Negative feedback defining a circadian clock: autoregulation in the clock gene *frequency*. *Science*, **263**, 1578–84.

Arpaia, G., Carattoli, A. & Macino, G. (1995*b*). Light and development regulate the expression of the *albino*-3 gene in *Neurospora crassa*. *Developmental Biology*, **170**, 626–35.

Arpaia, G., Loros, J. J., Dunlap, J. C., Morelli, G. & Macino, G. (1993). The interplay of light and the circadian clock: independent dual regulation of clock-controlled gene *ccg*-2(*eas*). *Plant Physiology*, **102**, 1299–305.

Austin, B. (1968). An endogenous rhythm of spore discharge in *Sordaria fimicola*. *Annals of Botany*, **32**, 261–78.

Ballario P., Vittorioso, P., Magrelli, A., Talora, C., Cabibbo, A. & Macino, G. (1996). White collar-1, a central regulator of the blue light responses in *Neurospora*, is a zinc finger protein. *EMBO Journal*, **15**, 1650–57.

Bell-Pedersen, D., Dunlap, J. C. & Loros, J. J. (1992). The *Neurospora* circadian clock controlled gene, *ccg*-2, is allelic to *eas* and encodes a fungal hydrophobin required for the formation of the conidial rodlet layer. *Genes and Development*, **6**, 2382–94.

Bell-Pedersen, D., Dunlap, J. C. & Loros, J. J. (1996*a*). Distinct *cis*-acting elements mediate clock, light and developmental regulation of the *Neurospora crassa eas* (*ccg*-2) gene. *Molecular and Cellular Biology*, **16**, 513–21.

Bell-Pedersen, D., Garceau, N. & Loros, J. J. (1997). Circadian rhythms in fungi. *Journal of Genetics*, **75**, 387–401.

Bell-Pedersen, D., Shinohara, M., Loros, J. J. & Dunlap, J. C. (1996*b*). Clock-controlled genes isolated from *Neurospora crassa* are late-night specific to early-morning specific. *Proceedings of the National Academy of Sciences, USA*, **93**, 13096–101.

Berridge, M. J. (1993). Inositol triphosphate and calcium signalling. *Nature*, **312**, 315–21.

Bourret, J. A., Lincoln, R. G. & Carpenter, B. H. (1969). Fungal endogenous rhythms expressed by spiral figures. *Science*, **166**, 763–4.

Brandt, W. H. (1953). Zonation in a prolineless strain of *Neurospora. Mycologia*, **45**, 194–208.

Brody, S. & Harris, S. (1973). Circadian rhythms in *Neurospora*: spatial differences in pyrimidine nucleotide levels. *Science*, **180**, 498–500.

Brody, S. & Martins, S. A. (1973). Circadian rhythms in *Neurospora crassa*: effects of unsaturated fatty acids. *Journal of Bacteriology*, **137**, 912–15.

Brody, S. & Martins, S. A. (1979). Effects of morphological and auxotrophic mutations on the circadian rhythm of *Neurospora crassa. Genetics*, **74**, s31–s40.

Brody, S., Willert, K. & Chuman, L. (1988). Circadian rhythms in *Neurospora crassa*: the effects of mutations at the *ufa* and *cla-1* loci. *Genome*, **30** (supplement), 299–309.

Bruce, V. G., Weight, F. & Pittendrigh, C. S. (1960). Resetting the spoıulation rhythm in *Pilobolus* with short light flashes of high intensity. *Science*, **131**, 728–30.

Butler, D. R., Wadia, K. D. R. & Reddy, R. K. (1995). Effects of humidity, leaf wetness, temperature and light on conidial production by *Phaeoisariopsis personata* on groundnut. *Plant Pathology*, **44**, 662–74.

Carpenter, J. B. (1949). Production and discharge of basidiospores of *Pellicularia filamentosa* on *Hevea* rubber. *Phytopathology*, **39**, 980–5.

Chaudhuri, H. (1923). A study of the growth in culture of *Verticillium albo-atrum*, B. et Br. *Annals of Botany*, London, **37**, 519–39.

Chevaugeon, J. & Nguyen Van, H. (1969). Internal determinism of hyphal growth rhythms. *Transactions of the British Mycological Society*, **53**, 1–14.

Cornelius, G. & Rensing, L. (1986). Circadian rhythm of heat shock protein synthesis of *Neurospora crassa. European Journal of Cell Biology*, **40**, 130–2.

Coté, G. G., Lakin-Thomas, P. L. & Brody, S. (1996). Membrane lipids and circadian rhythms in *Neurospora crassa*. In *Membranes and Circadian Rhythms*, ed. Th. Vanden Driessche, pp. 13–46. Berlin: Springer.

Crosthwaite, S. K., Dunlap, J. C. & Loros, J. J. (1997). *Neurospora wc-1* and *wc-2*: transcription, photoresponses, and the origins of circadian rhythmicity. *Science*, **276**, 763–9.

Crosthwaite, S. K., Loros, J. J. & Dunlap, J. C. (1995). Light-induced resetting of a circadian clock is mediated by a rapid increase in *frequency* transcript. *Cell*, **81**, 1003–12.

Davidson, F. A., Sleeman, B. D., Rayner, A. D. M., Crawford, J. W. & Ritz, K. (1996). Context-dependent macroscopic patterns in growing and interacting mycelial networks. *Proceedings of the Royal Society of London, Series B*, **263**, 873–80.

Delmer, D. P. & Brody, S. (1975). Circadian rhythms in *Neurospora crassa*: oscillations in the level of an adenine nucleotide. *Journal of Bacteriology*, **121**, 548–53.

Deutsch, A., Dress, A. & Rensing, L. (1993). Formation of morphological differentiation patterns in the ascomycete *Neurospora crassa. Mechanisms of Development*, **44**, 17–31.

Dharmananda, S. & Feldman, J. F. (1979). Spatial distribution of circadian clock phase in aging cultures of *Neurospora crassa*. *Plant Physiology*, **63**, 1049–54.

Dieckmann, C. L. (1980). Circadian rhythms in *Neurospora crassa:* A biochemical and genetic study of the involvement of mitochondrial metabolism in periodicity. PhD Thesis, University of California, San Diego.

Dunlap, J. C. (1993). Genetic analysis of circadian clocks. *Annual Review of Physiology*, **55**, 683–728.

Dunlap, J. C. (1996). Genetic and molecular analysis of circadian rhythms. *Annual Review of Genetics*, **30**, 579–601.

Edelstein, L. & Segel, L. A. (1983). Growth and metabolism in mycelial fungi. *Journal of Theoretical Biology*, **104**, 187–210.

Edmunds, L. N. (1988). *Cellular and Molecular Bases of Biological Clocks*. New York: Springer-Verlag.

Edmunds, L. N., Apter, R. I., Rosenthal, P.-J. Shen, W.-K. & Woodward, J. R. (1979). Light effects in yeast: persisting oscillations in cell division activity and amino acid transport in cultures of *Saccharomyces cerevisiae* entrained by light–dark cycles. *Photochemistry and Photobiology*, **30**, 595–601.

Engel, H. & Friederichsen, I. (1964). Der abschuss der sporangiolen von *Sphaerobolus stellatus* (Thode) Pers. in kontinuierlicher dunkelheit. *Planta*, **61**, 361–70.

Eskin, A. (1979). Identification and physiology of circadian pacemakers. *Federation Proceedings*, **38**, 2573–9.

Farkaš, V., Sulová, Z. & Lehotský, J. (1985). Effects of light on the concentration of adenine nucleotides in *Trichoderma viride*. *Journal of General Microbiology*, **131**, 317–320.

Feldman, J. F. (1975). Circadian periodicity in *Neurospora*: alteration by inhibitors of cyclic AMP phosphodiesterase. *Science*, **190**, 789–90.

Feldman, J. F. (1982). Genetic approaches to circadian clocks. *Annual Review of Plant Physiology*, **33**, 583–608.

Feldman, J. F. & Atkinson, C. A. (1978). Genetic and physiological characteristics of a slow-growing circadian clock mutant of *Neurospora crassa*. *Genetics*, **88**, 255–65.

Feldman, J. F. & Dunlap, J. C. (1983). *Neurospora crassa*: a unique system for studying circadian rhythms. *Photochemical and Photobiological Reviews*, **7**, 319–68.

Feldman, J. F., Gardner, G. & Denison, R. (1979). Genetic analysis of the circadian clock of *Neurospora*. In *Biological Rhythms and their Central Mechanism* , ed. M. Suda, O. Hayaishi & H. Nakagawa, pp. 57–66. Amsterdam: Elsevier/North Holland Biomedical Press.

Feldman, J. F. & Hoyle, M. N. (1973). Isolation of circadian clock mutants of *Neurospora crassa*. *Genetics*, **75**, 605–13.

Feldman, J. F. & Hoyle, M. N. (1976). Complementation analysis of linked circadian clock mutants in *Neurospora*. *Genetics*, **82**, 9–17.

Friederichsen, I. & Engel, H. (1960). Der abschussrhythmus der fruchtkörper von *Sphaerobolus stellatus* (Thode) Pers. *Planta*, **55**, 313–26.

Gadd, G. M. (1994). Signal transduction in fungi. In *The Growing Fungus*, ed. N. A. R. Gow & G. M. Gadd, pp. 183–210. London: Chapman & Hall.

Garceau, N. Y. (1996). Molecular and genetic studies on the *frq* and *ccg-1* loci of Neurospora. PhD Thesis, Dartmouth Medical School, Hanover, USA.

Garceau, N. Y., Liu, Y., Loros, J. J. & Dunlap, J. C. (1997). Alternative initiation of translation and time-specific phosphorylation yield multiple forms of the essential clock protein FREQUENCY. *Cell*, **89**, 469–76.

Gardner, G. F. & Feldman, J. F. (1980). The *frq* locus in *Neurospora crassa*: a key element in circadian clock organization. *Genetics*, **96**, 877–86.

Gardner, G. F. & Feldman, J. F. (1981). Temperature compensation of circadian period length in clock mutants of *Neurospora crassa*. *Plant Physiology*, **68**, 1244–8.

Gooch, V. D., Wehseler, R. A. & Gross, C. G. (1994). Temperature effects on the resetting of the phase of the *Neurospora* circadian rhythm. *Journal of Biological Rhythms*, **9**, 83–94.

Goodwin, B. (1995). *How the Leopard Changed its Spots*. London: Weidenfeld & Nicolson.

Gresík, M., Nadežda, V. & Farkaš, V. (1988). Membrane potential, ATP and cyclic AMP changes induced by light in *Trichoderma viride*. *Experimental Mycology*, **12**, 295–301.

Harding, R. W. & Shropshire, W. (1980). Photocontrol of carotenoid biosynthesis. *Annual Review of Plant Physiology*, **31**, 217–38.

Hasunuma, K. (1984*a*). Rhythmic conidiation and light sensitivity of mutants in orthophosphate repressible cyclic phosphodiesterase in *Neurospora crassa*. *Proceedings of the Japan Academy, Series B*, **60**, 260–4.

Hasunuma, K. (1984*b*). Circadian rhythm in *Neurospora* includes oscillation of cyclic 3′,5′-AMP level. *Proceedings of the Japanese Academy, Series B*, **60**, 377–80.

Hasunuma, K. & Shinohara, Y. (1985). Characterization of *cpd-1* and *cpd-2* mutants which affect the activity of orthophosphate regulated cyclic phosphodiesterase in *Neurospora*. *Current Genetics*, **10**, 197–203.

Hochberg, M. L. & Sargent, M. L. (1974). Rhythms of enzyme activity associated with circadian conidiation in *Neurospora crassa*. *Journal of Bacteriology*, **120**, 1164–75.

Ingold, C. T. (1933). Spore discharge in the ascomycetes. I. Pyrenomycetes. *New Phytologist*, **32**, 175–96.

Ingold, C. T. (1946). Spore discharge in *Daldinia concentrica*. *Transactions of the British Mycological Society*, **34**, 43–51.

Ingold, C. T. (1971). Periodicity. In *Fungal Spores*, pp. 215–38. London: Oxford University Press.

Ingold, C. T. & Cox, V. J. (1955). Periodicity of spore discharge in *Daldinia*. *Annals of Botany*, **19**, 201–9.

Ingold, C. T. & Marshall, B. (1963). Further observations on light and spore discharge in certain pyrenomycetes. *Annals of Botany*, **27**, 481–91.

Jensen, C. & Lysek, G. (1983). Differences in the mycelial growth rhythms in a population of *Sclerotinia fructigena* (Pers.) Schröter. *Experientia*, **39**, 1401–2.

Jerebzoff, S. & Jerebzoff-Quintin, S. (1982). Rhythmic regulation of the asparagine-pyruvate pathway controls the *Leptosphaeria* sporulation rhythm. Abstract: *First European Congress of Cell Biology, Paris*. Number 74. *Biology of the Cell*, **45**, 69.

Johnson, C. H. (1983). Changes in intracellular pH are not correlated with the circadian rhythm of *Neurospora*. *Plant Physiology*, **72**, 129–33.

Kafi, A. & Tarr, S. A. J. (1963). Concentric zonation in colonies of *Helminthosporium* spp. with special reference to *H. rostratum*. *Transactions of the British Mycological Society*, **46**, 549–59.

Kallies, A., Gebauer, G. and Rensing, L. (1996). Light effects on cyclic nucleotide levels and phase shifting of the circadian clock in *Neurospora crassa*. *Photochemistry and Photobiology*, **63**, 336–43.

Karin, M. & Hunter, T. (1995). Transcriptional control by protein phosphorylation: signal transmission from the cell surface to the nucleus. *Current Biology*, **5**, 747–57.

Kippert, F. (1989). Circadian control of heat tolerance in stationary phase cultures of *Schizosaccharomyces pombe*. *Archives of Microbiology*, **151**, 177–9.

Kippert, F., Ninnemann, H. & Engelmann, W. (1991). Photosynchronization of the circadian clock of *Schizosaccharomyces pombe* – mitochondrial cytochrome-B is an essential component. *Current Genetics*, **19**, 103–7.

Lakin-Thomas, P. L. (1993). Evidence against a direct role for inositol metabolism in the circadian oscillator and the blue-light signal transduction pathway in *Neurospora crassa*. *Biochemical Journal*, **292**, 813–18.

Lakin-Thomas, P. L. (1996). Effects of choline depletion on the circadian rhythm in *Neurospora crassa*. *Biological Rhythms Research*, **27**, 12–30.

Lakin-Thomas, P. (1998). Circadian rhythmicity in *Neurospora crassa*. In *Biological Rhythms and Photoperiodism in Plants*, ed. P. Lumsden & A. Millar, pp. 119–34. Oxford: BIOS Scientific publishers.

Lakin-Thomas, P. L., Brody, S. & Coté, G. G. (1997). Temperature compensation and membrane composition in *Neurospora crassa*. *Chronobiology International*, **14**, 445–54.

Lakin-Thomas, P. L., Coté, G. G. and Brody, S. (1990). Circadian rhythms in *Neurospora crassa*: biochemistry and genetics. *Critical Reviews in Microbiology*, 17, 365–416.

Lauter, F. R. & Yanofsky, C. (1993). Day/night and circadian rhythm control of *con* gene expression in *Neurospora*. *Proceedings of the National Academy of Sciences, USA*, **90**, 8249–53.

Lewis, M. T. & Feldman, J. F. (1993). The putative *frq* clock protein of *Neurospora crassa* contains sequence elements that suggest a nuclear transcriptional regulatory role. *Protein Sequence Data Analysis*, **5**, 315–23.

Lewis, M. T. & Feldman, J. F. (1996). Evolution of the *frequency* (*frq*) clock locus in ascomycete fungi. *Molecular Biology and Evolution*, **13**, 1233–41.

Lindegren, K. M. (1994). Characterization of *ccg-1*, a clock-controlled gene of *Neurospora crassa*. PhD Thesis, Dartmouth Medical School, Hanover, USA.

Linden, H. & Macino, G. (1997). White collar-2, a partner in blue-light signal transduction, controlling expression of light-regulated genes in *Neurospora crassa*. *EMBO Journal*, **16**, 98–109.

Liu Y., Garceau, N. Y., Loros, J. J. & Dunlap, J. C. (1997). Thermally regulated translational control of FRQ mediates aspects of temperature responses in the *Neurospora* circadian clock. *Cell*, **89**, 477–86.

Loros, J. (1995). The molecular basis of the *Neurospora* clock. *Seminars in the Neurosciences*, **7**, 3–13.

Loros, J. J., Denome, S. A. & Dunlap, J. C. (1989). Molecular cloning of genes under control of the circadian clock in *Neurospora*. *Science*, **243**, 385–8.

Loros, J. J. & Feldman, J. F. (1986). Loss of temperature compensation of circadian period length in the *frq*-9 mutant of *Neurospora crassa*. *Journal of Biological Rhythms*, **1**, 187–98.

Loros, J. J., Richman, A. & Feldman, J. F. (1986). A recessive circadian clock mutant at the *frq* locus in *Neurospora crassa. Genetics*, **114**, 1095–110.

Lysek, G. (1978). Circadian rhythms. In *The Filamentous Fungi, Volume 3, Developmental Mycology*, ed. J. E. Smith & D. R. Berry, pp. 376–8. London: Edward Arnold.

Lysek, G. (1984). Physiology and ecology of rhythmic growth and sporulation. In *The Ecology and Physiology of the Fungal Mycelium*, ed. D. H. Jennings & A. D. M. Rayner, pp. 323–42. Cambridge: Cambridge University Press.

Martens, C. L. & Sargent, M. L. (1974). Conidiation rhythms of nucleic acid metabolism in *Neurospora crassa. Journal of Bacteriology*, **117**, 1210–15.

Martin, T. F. J. (1997). Greasing the Golgi budding machine. *Nature*, **387**, 21–2.

Mattern, D. L., Forman, L. R. & Brody, S. (1982). Circadian rhythms in *Neurospora crassa*: a mutation affecting temperature compensation. *Proceedings of the National Academy of Sciences, USA*, **79**, 825–9.

McClung, C. R., Fox, B. A. & Dunlap, J. C. (1989). The *Neurospora* clock gene frequency shares a sequence element with the *Drosophila* clock gene *period. Nature*, **339**, 558–62.

McNally, M. T. & Free, S. J. (1988). Isolation and characterization of a *Neurospora* glucose repressible gene. *Current Genetics*, **14**, 545–51.

Merrow, M. W. & Dunlap, J. C. (1994). Intergeneric complementation of a circadian rhythmicity defect: phylogenetic conservation of structure and function of the clock gene *frequency. EMBO Journal*, **13**, 2257–66.

Merrow, M. W., Garceau N. Y. & Dunlap, J. C. (1997). Dissection of a circadian oscillation into discrete domains. *Proceedings of the National Academy of Sciences, USA*, **94**, 3877–82.

Moore-Ede, M. C., Sulzmann, F. M. & Fuller, C. A. (1982). The clocks that time us: physiology of the circadian timing system. Cambridge, MA: Harvard University Press.

Munoz, V. & Butler, W. L. (1975). Photoreceptor pigment for blue light in *Neurospora crassa. Plant Physiology*, **55**, 421–6.

Nakashima, H. (1982*a*). Effects of membrane ATPase inhibitors on light induced phase shifting of the circadian clock in *Neurospora crassa. Plant Physiology*, **69**, 619–23.

Nakashima, H. (1982*b*). Phase shifting of the circadian clock by diethylstilbestrol and related compounds in *Neurospora crassa. Plant Physiology*, **70**, 982–6.

Nakashima, H. (1984*a*). Calcium inhibits phase shifting of the circadian conidiation rhythm of *Neurospora crassa* by the calcium ionophore A23187. *Plant Physiology*, **74**, 268–71.

Nakashima, H. (1984*b*). Effects of respiratory inhibitors on respiration, ATP contents, and the circadian conidiation rhythm of *Neurospora crassa. Plant Physiology*, **76**, 612–14.

Nakashima, H. (1985). Biochemical and genetic aspects of the conidiation rhythm in *Neurospora crassa*: phase-shifting by metabolic inhibitors. In *Circadian Clocks and Zeitgebers*, ed. T. Hiroshige & K. Honma, pp. 35–44. Sapporo, Japan: Hokkaido University Press.

Nakashima, H. (1986). Phase shifting of the circadian conidiation rhythm in *Neurospora crassa* by calmodulin antagonists. *Journal of Biological Rhythms*, **1**, 163–9.

Nakashima, H. & Feldman, J. F. (1980). Temperature sensitivity of light-induced phase shifting of the circadian clock of *Neurospora. Photochemistry and Photobiology*, **32**, 247–52.

Nakashima, H. & Fujimura, Y. (1982). Light-induced phase-shifting of the circadian clock in *Neurospora* requires ammonium salts at high pH. *Planta*, **155**, 431–6.

Nakashima, H. & Onai, K. (1996). The circadian conidiation rhythm in *Neurospora crassa. Seminars in Cell and Developmental Biology*, **7**, 765–74.

Nakashima, H., Perlman, J. & Feldman, J. F. (1981*a*). Cycloheximide-induced phase shifting of the circadian clock of *Neurospora. American Journal of Physiology*, **241**, r31–5.

Nakashima, H., Perlman, J. & Feldman, J. F. (1981*b*). Genetic evidence that protein synthesis is required for the circadian clock of *Neurospora. Science*, **212**, 361–2.

Njus, D., Sulzman, F. M. & Hastings, J. W. (1974). Membrane model for the circadian clock. *Nature*, **248**, 116–19.

Ohnishi, T., Cornelius, G. & Nakashima, H. (1992). A mutant of *Neurospora crassa* that has a long lag phase in low calcium medium. *Journal of General Microbiology*, **138**, 1573–8.

Paietta, J. & Sargent, M. L. (1981). Photoreception in *Neurospora crassa:* correlation of reduced light sensitivity with flavin deficiency. *Proceedings of the National Academy of Sciences, USA*, **78**, 5573–7.

Paietta, J. & Sargent, M. L. (1982). Blue light response in nitrate reductase mutants of *Neurospora crassa. Photochemistry and Photobiology*, **35**, 853–5.

Paietta, J. & Sargent, M. L. (1983). Isolation and characterization of light-insensitive mutants of *Neurospora crassa. Genetics*, **104**, 11–21.

Pavlidis, T., Zimmerman, W. F. & Osborn, J. (1968). A mathematical model for temperature effects on circadian rhythms. *Journal of Theoretical Biology*, **18,** 210–21.

Perlman, J., Nakashima, H. & Feldman, J. F. (1981). Assay and characteristics of circadian rhythmicity in liquid cultures of *Neurospora crassa. Plant Physiology*, **67**, 404–7.

Pilatus, U. & Techel, D. (1991). ^{31}P-NMR-studies on intracellular pH and metabolite concentrations in relation to the circadian rhythm, temperature and nutrition in *Neurospora crassa. Biochimica et Biophysica Acta*, **1091**, 349–55.

Pittendrigh, C. S., Bruce, V. G., Rosenweig, N. S. & Rubin, M. L. (1959). A biological clock in *Neurospora. Nature*, **184**, 169–70.

Poitras, A, W. (1954). Response of *Cokeromyces recurvatus* (Mucorales) to certain wavelengths of the visible spectrum. *Proceedings of the Paris Academy of Sciences*, **28**, 52–8.

Prior, S. L., Robson, G. D. & Trinci, A. P. J. (1994). Phosphoinositide turnover does not mediate the effects of light or choline, or the relief of derepression of glucose metabolism in filamentous fungi. *Mycological Research*, **98**, 291–4.

Roeder, P. E., Sargent, M. L. & Brody, S. (1982). Circadian rhythms in *Neurospora crassa*: oscillations in fatty acids. *Biochemistry*, **21**, 4909–16.

Roenneberg, T. & Mittag, M. (1996). The circadian program of algae. *Seminars in Cell and Developmental Biology*, **7**, 753–63.

Sadakane, Y. & Nakashima, H. (1996). Light-induced phase shifting of the circadian conidiation rhythm is inhibited by calmodulin antagonists in *Neurospora crassa. Journal of Biological Rhythms*, **11**, 234–40.

Sargent, M. L. & Briggs, W. R. (1967). The effects of light on a circadian rhythm of conidiation in *Neurospora. Plant Physiology*, **42**, 1504–10.

Sargent, M. L., Briggs, W. R. & Woodward, D. O. (1966). The circadian nature of a rhythm expressed by an invertaseless strain of *Neurospora crassa. Plant Physiology*, **41**, 1343–9.

Sargent, M. L. & Kaltenborn, S. H. (1972). Effects of medium composition and carbon dioxide on circadian conidiation in *Neurospora. Plant Physiology*, **50**, 171–5.

Sargent, M. L. & Woodward, D. O. (1969). Genetic determinants of circadian rhythmicity in *Neurospora. Journal of Bacteriology*, **97**, 861–6.

Sato, R., Kondo, T. & Miyoshi, Y. (1985). Circadian rhythms of potassium and sodium contents in the growing front of *Neurospora crassa. Plant Cell Physiology*, **26**, 447–53.

Schulz, R., Pilatus, U. & Rensing, L. (1985). On the role of energy metabolism in *Neurospora* circadian clock function. *Chronobiology International*, **2**, 223–33.

Selitrennikoff, C. P. (1976). *Easily-wettable*, a new mutant. *Neurospora Newsletter*, **23**, 23.

Sokolovsky, V. S. & Kritsky, M. S. (1985). Photoregulation of cyclic AMP phosphodiesterase activity of *Neurospora crassa. Biochemistry (Russia)*, **282**, 1017–20.

Sommer, T, Chamber, J. A. A., Eberle, J., Lauter, F. R. & Russo, V. E. (1989). Fast light-regulated genes of *Neurospora crassa. Nucleic Acids Research*, **17**, 5713–22.

Sulzman, F. M., Ellman, D., Fuller, C. A., Moore-Ede, M. C. & Wassmer, G. (1984). *Neurospora* rhythms in space: a re-examination of the endogenous–exogenous question. *Science*, **225,** 232–4.

Sussman, A. S., Lowry, R. J. & Durkee, T. (1964). Morphology and genetics of a periodic colonial mutant of *Neurospora crassa. American Journal of Botany*, **51**, 243–52.

Suzuki, S., Katagiri, S. & Nakashima, H. (1996). Mutants with altered sensitivity to a calmodulin antagonist affect the circadian clock in *Neurospora crassa. Genetics*, **143**, 1175–80.

Sweeney, B. M. & Borgese, M. B. (1989). A circadian rhythm in cell division in a prokaryote, the cyanobacterium *Synechococcus* WH7803. *Journal of Phycology*, **25**, 183–6.

Techel, D., Gebauer, G., Kohler, W., Braumann, T., Jastorff, B. & Rensing, L. (1990). On the role of Ca^{2+}-calmodulin-dependent and cAMP-dependent protein-phosphorylation in the circadian rhythm of *Neuropora crassa. Journal of Comparative Physiology, Series B*, **159**, 695–706.

Uebelmesser, E. R. (1954). Über den endonomen Tagesrhythmus der Sporangienträgerbildung von *Pilobolus. Archive für Mikrobiologie*, **20**, 1–33.

Walkey, D. G. A. & Harvey, R. (1967). Spore discharge rhythms in Pyrenomycetes. II. Endogenous and exogenous rhythms of discharge. *Transactions of the British Mycological Society*, **50**, 229–40.

Wille, J. J. (1974). Light entrained circadian oscillations of growth rate in the yeast *Candida utilis*. In *Chronobiology*, ed. L. E. Scheving, F. Halberg, & J. E. Pauly, pp. 72–7. Tokyo: Igaku Shoin.

Winfree, A. T. & Twaddle, G. M. (1981). The *Neurospora* mycelium as a two-dimensional continuum of coupled circadian clocks. In *Mathematical Biology*, ed. T. A. Burton, pp. 237–49. New York: Pergamon Press.

Woodward, D. O. & Sargent, M. L. (1973) Circadian rhythms in *Neurospora*. In *Behavior of Micro-Organisms*, ed. A. Perez-Miravete, pp. 282–96. London: Plenum Press.

5

Growth, branching and enzyme production by filamentous fungi in submerged culture

A. P. J. TRINCI, S. BOCKING, R. J. SWIFT,
J. M. WITHERS, G. D. ROBSON AND M. G. WIEBE

Introduction

This chapter considers (a) the generation of highly branched (colonial) mutants during prolonged fermentation of the Quorn® myco-protein fungus, *Fusarium graminearum* A3/5, (b) the influence of hyphal branch frequency on the production of extracellular enzymes by *Aspergillus oryzae*, and (c) the stability of *Aspergillus niger gla* A transformants in prolonged continuous flow cultures. We believe that the results obtained with these three species can be extended to most, if not all, filamentous fungi.

Highly branched mutants do not have a selective advantage because of their morphology

When filamentous fungi or streptomycetes are grown in prolonged, continuous culture, it is common for the relatively sparsely branched parental strain to be supplanted by a relatively highly branched mutant (Fig. 5.1). Such mutants are called colonial because in Petri dish culture they form dense colonies that expand in radius more slowly than parental colonies. Selection of colonial mutants has been observed in continuous cultures of *Byssochlamys nivea*, *Paecilomyces variotii*, *Paecilomyces puntonii*, *Gliocladium virens*, *Trichoderma viride* (Forss *et al*., 1974), *Penicillium chrysogenum* (Righelato, 1976), *F. graminearum* (Solomons & Scammell, 1976) and *Acremonium chrysogenum* (A. Trilli, personal communication).

When the Quorn® myco-protein fungus, *F. graminearum* A3/5, was grown in glucose-limited chemostat culture at a dilution rate of 0.19 h^{-1} (doubling time of 3.65 h), colonial mutants were first detected about 107 generations (*c*. 389 h) after the onset of continuous flow (Fig.

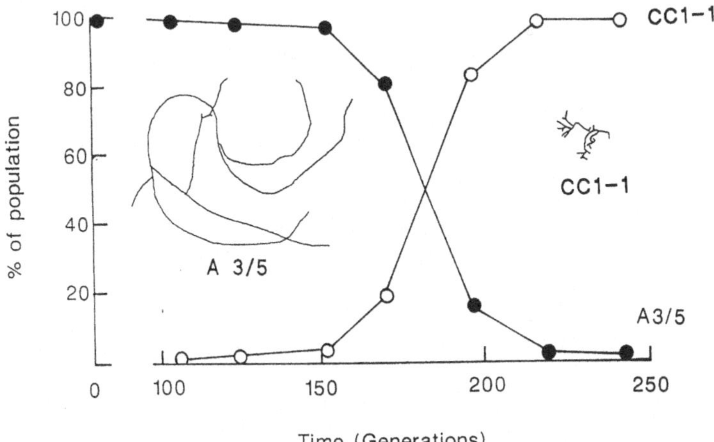

Fig. 5.1. Population of a highly branched (colonial) mutant CC1-1 (expressed as a percentage of the total population) generated during a glucose-limited, continuous flow culture of *F. graminearum* A3/5 grown at 25°C on modified Vogel's medium (Vogel, 1956) containing 3 g glucose l^{-1} at a dilution rate of 0.19 h^{-1} (pH of 5.8 and a stirrer speed of 1400 rpm). The morphology of representative mycelia of A3/5 and CC1-1 is also shown.

5.1; Trinci, 1994). Wiebe *et al.* (1991) isolated 20 morphological (colonial) mutants from two chemostat cultures; all were more highly branched (hyphal growth unit values from 14 to 174 μm) than A3/5 (hyphal growth unit of 232 μm) and produced colonies that expanded in radius more slowly (colony radial growth rate, K_r, values ranging from 18 to 105 μm h^{-1}) than A3/5 (K_r 135 μm h^{-1}). Given the recessive nature of the mutations (Wiebe *et al.*, 1991), the colonial phenotype of mutants can only be expressed once sufficient mutant nuclei have become separated from parental nuclei. This kind of separation may follow sporulation (macroconidia are formed from uninucleate phialides) or mycelial fragmentation (Wiebe *et al.,* 1996).

The above results led to speculation about the reasons why highly branched mycelia have a selection advantage over sparsely branched mycelia in continuous flow cultures. The outcome of competition between the sparsely branched parental strain (A3/5) and a highly branched colonial mutant (CC1-1) under glucose- and sulphate-limitation in chemostat culture provided a crucial insight into this question (Fig. 5.2). For both nutrient limitations, CC1-1 retained its highly branched phenotype but behaved as a selectively advantageous mutant

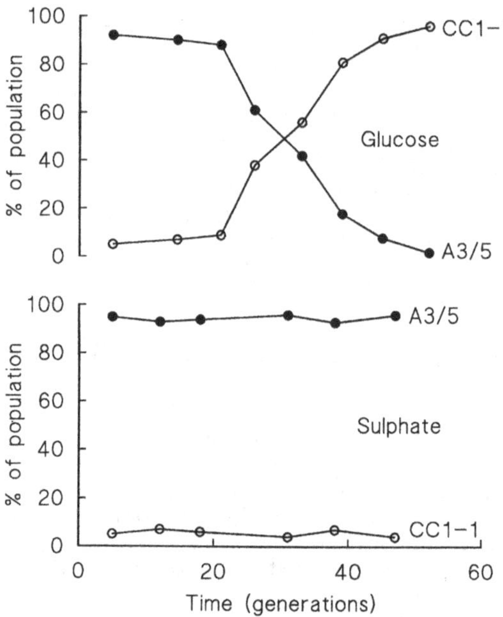

Fig. 5.2. Competition between *Fusarium graminearum* A3/5 and a highly branched (colonial) mutant CC1-1 in glucose-limited and sulphate-limited chemostat cultures grown at 25°C, pH 5.8 and a stirrer speed of 1400 rpm at a dilution rate of 0.19 h^{-1}.

under glucose-limitation but as a neutral mutant under sulphate-limitation. Thus, the highly branched phenotype does not confer a selection advantage on the mutant. Instead, the mutation alters glucose metabolism and it is this change which confers a selective advantage on CC1-1 when it is grown in mixed culture with A3/5 in glucose-limited chemostat culture (Wiebe *et al.,* 1992); thus, the highly branched phenotype is a pleiotropic effect of the CC1-1 mutation. Significantly, mutants of *Neurospora crassa* with altered activities of glucose-6-phosphate dehydrogenase (Brody & Tatum, 1966) and phosphoglucomutase (Brody & Tatum, 1967) also have highly branched phenotypes, as does *N. crassa* mycelium treated with compounds (paramorphogens like L-sorbose) which affect membrane or wall biosynthesis (Trinci, Wiebe and Robson, 1994). Finally, a putative protein kinase mutant of *N. crassa* forms a highly branched mycelium (Yarden *et al.,* 1992). These observations suggest that highly branched (colonial) phenotypes are the pleiotropic effects of mutations (or paramorphogens) whose primary effects are on the activities of enzymes involved in carbon metabolism, membrane biosynthesis, wall biosynthesis, etc. Indeed we estimate that probably up to about 5% of all random mutations have colonial phenotypes. The knowledge gained from the results of the experiment shown in Fig. 5.2 facilitated the development of strategies to prevent or delay the appearance of colonial mutants during industrial production of biomass of *F. graminearum* A3/5 for Quorn® myco-protein production. It is hoped that the successful exploitation of these strategies will increase the cost-efficiency of the process (Trinci, 1994).

Increasing hyphal tip number per unit biomass does not result in an increase in the yield of extracellular enzymes

Following the observation of Wösten *et al.* (1991) that glucomylase is secreted (by a process known as 'bulk-flow') at the hyphal tips of mycelia of *A. niger* (Wessels, 1993), there has been speculation about whether or not the productivity of industrial enzyme fermentations can be increased by using highly branched fungal strains, i.e. by increasing the number of hyphal tips per unit fungal biomass. Figure 5.3 shows batch growth of *Aspergillus oryzae* and a more highly branched (colonial) mutant derived from this strain: the two strains had the same specific growth rate and approximately the same rate of α-amylase accumulation. A similar result was obtained with *A. oryzae* which, following treatment with echinocandin, was four times more highly branched than untreated control mycelia. However, some highly branched mutants of *A. oryzae* produced enzymes more rapidly than the sparsely branched parental strain during exponential growth in batch culture.

Of course, enzyme secretion may not be the factor limiting enzyme productivity in the strains shown in Fig. 5.3 and if this is so increasing branch frequency (and hence the number of points for enzyme secretion) will not increase enzyme productivity. Further, the percentage of the wall of a mycelium which is 'extensible' may not be changed by altering branch frequency since the hyphae of highly branched mycelia will have shorter extension zones than those of sparsely branched mycelia. Thus, the spatial distribution of 'extensible' wall in a mycelium may be changed in colonial mutants, but not its relative amount; if this is the case there is no reason to believe that enzyme secretion would occur at a faster rate in a highly branched mycelium than in a sparsely branched mycelium.

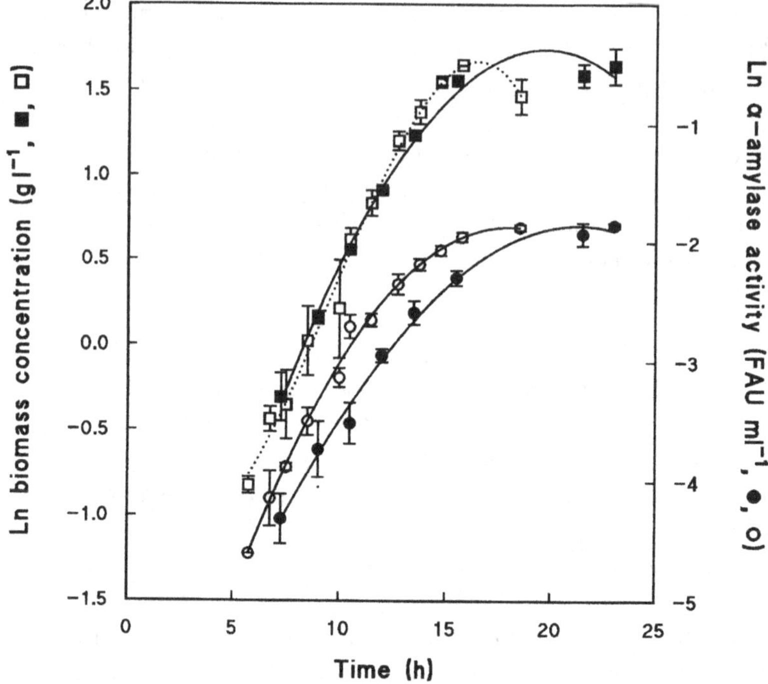

Fig. 5.3. Growth (■, □) and α-amylase production (●, ○) of a parental strain (IFO4177) of *A. oryzae* (■, ●) and of a more highly branched morphological mutant (HNP12) (□, ○) grown at 30°C in batch culture on Vogel's medium containing 10 g glucose l^{-1}. The parental strain and the more highly branched mutant had hyphal growth unit lengths of 101 ± 4 and 53 ± 2 μm respectively (mean of 5 replicates ± SEM).

Increasing hyphal tip number per unit biomass causes a decrease in culture viscosity

Although increasing hyphal tip number per unit biomass does not increase enzyme yield, it does have an appreciable effect on culture rheology. Figure 5.4 shows that for morphological mutants of *A. oryzae* culture viscosity (torque) increases with increase in hyphal growth unit length. Culture viscosity is important in industrial fermentations because oxygen solubility decreases with an increase in culture viscosity, and consequently steps have to be taken when culturing filamentous fungi to ensure that oxygen supply does not become a growth-limiting factor. Thus, it may be possible to increase the cost-efficiency of some industrial fungal fermentations by using highly branched strains to reduce culture viscosity and hence increase oxygen supply.

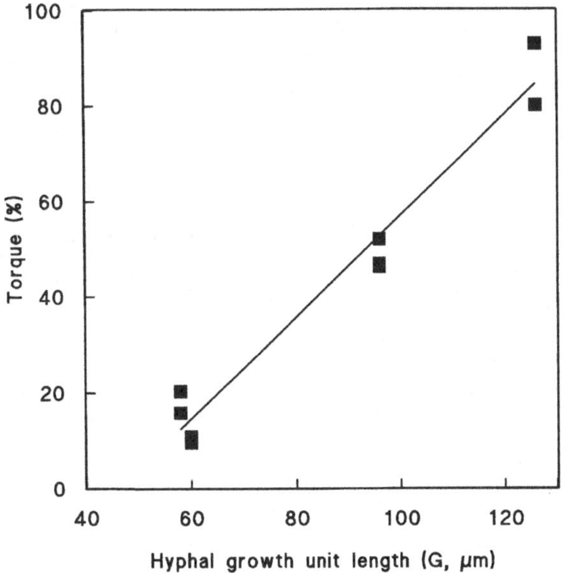

Fig. 5.4. Effect of hyphal branch frequency on the torque of biomass (4 g dry wt l^{-1}) of cultures of various strains of *A. oryzae* ($r^2 = 0.97$).

Production of growth-associated enzymes in continuous flow culture

A continuous flow culture operated at a high dilution rate is the most cost-effective way of producing a growth-associated product (Pirt, 1975); these culture systems include chemostats in which an organism can be grown at a value just below its maximum rate of growth (μ_{max}) and pH auxostats (Simpson *et al.*, 1995) in which an organism can be grown at μ_{max}. Therefore, to maximize productivity, growth-associated products (including at least some recombinant proteins) should be produced in continuous flow culture systems. However, although *F. graminearum* A3/5 is grown by Marlow Foods Ltd in a 155 m^3 continuous flow culture (a glucostat), and although the industrial potential of growing fungi in continuous flow systems to produce recombinant proteins is enormous, there is currently little information about the production of recombinant proteins by filamentous fungi in such cultures. Recombinant protein productivity of such systems is most meaningfully described in terms of specific production rate (q_p, g product [g biomass]$^{-1}$ h^{-1}). Continuous culture at a constant dilution rate is the simplest experimental system for determining q_p; for such systems, q_p is equal to $D \cdot C_p \cdot C_x^{-1}$ in which D is dilution rate, C_p is product concentration and C_x is biomass concentration.

The *A. niger* B1 transformant used here originated from strain N402, a *cspA1* (conferring short conidiophores) derivative of ATCC 9029 and has been described by Verdoes *et al.* (1993, 1994). Transformation was based on a cosmid vector (42.8 kb) carrying four copies of the *A. niger glaA* (glucoamylase) gene and a single copy of the *amdS* (acetamidase) gene of *A. nidulans* as a selectable marker. Transformant N402 [pAB6-10]B1 (subsequently referred to as B1) carries an additional 20 copies of *glaA*.

Glucoamylase (GAM) production by A. niger *B1 in batch and chemostat culture*

A batch culture of *A. niger* B1 grown in an Infors IFS 100 fermenter (1.5 l volume of culture; pH 5.5 ± 0.2; aeration 0.7 l air [l culture]$^{-1}$ min^{-1}) on 5 g maltodextrin l^{-1} produced 320 ± 8 mg GAM l^{-1} (mean ± SEM), giving a specific production rate of 5.6 ± 0.6 mg GAM [g dry weight]$^{-1}$ h^{-1} (Table 5.1). Specific growth rate (dilution rate) had no significant ($P > 0.05$) effect on the concentration of GAM in maltodextrin-limited chemostat cultures (Fig. 5.5a) but did have a significant ($P < 0.05$) effect on the specific production rate (Fig. 5.5b). Table 5.1 compares the productivity of the batch and chemostat culture systems; it shows that since GAM is a growth-related product (i.e. productivity increases with specific growth rate), a continuous flow culture operated at a high dilution rate will be a more efficient system for GAM production than batch or fed-batch cultures. GAM production in glucose-limited chemostat cultures of *A. niger* B1 was only slightly less than in maltodextrin-limited cultures (Table 5.1).

Stability of transformant A. niger *B1 in prolonged cultures*

Table 5.1 shows the potential economical benefit of producing a growth-associated product (GAM, a recombinant protein) in a continuous flow culture operated at a high dilution rate. However, the extent of this benefit will be critically dependent on the stability of the transformed strain used in the production process (Trilli, 1977). Although industrial fermentations involve a substantial number of generations, even for high-value pharmaceutical proteins produced in reactors with a volume of 0.1–5 m^3, few studies have been made of the long-term stability of transformants of filamentous fungi (Dunn-Coleman *et al.*, 1994; Withers *et al.*,

Table 5.1. *Concentrations of GAM and specific GAM productivities for* A. niger *B1 grown on 5 g maltodextrin l^{-1} Vogel's medium in batch and in chemostat culture at a dilution rate of 0.13 h^{-1} (pH 5.4; 30°C; 0.7 l air [l culture]$^{-1}$ min^{-1}; 1000 rpm)*

Type of culture	GAM concentration (C_p, mg l^{-1})	Specific production rate of GAM (q_p, mg GAM [g biomass]$^{-1}$ h^{-1})
Batch culture grown on maltodextrin	320 ± 8	5.6
Maltodextrin-limited chemostat culture	373 ± 9	16.0
Glucose-limited chemostat culture	303 ± 12	12.0

Mean ± S.E.M. (5 replicates)

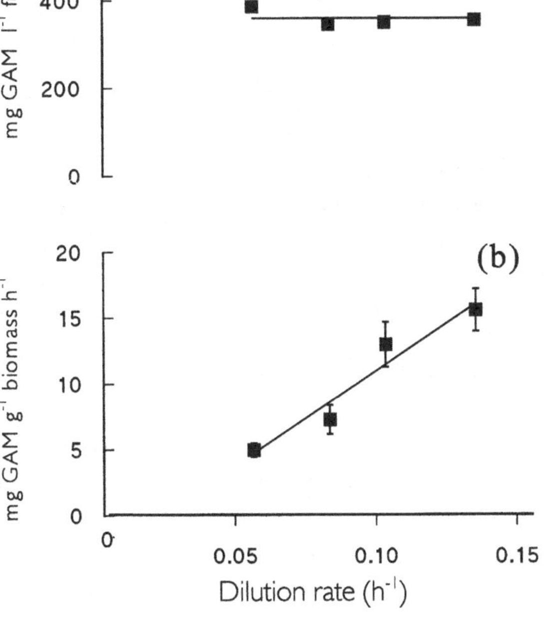

Fig. 5.5. Effect of dilution rate on GAM production by *A. niger* B1 in a maltodextrin-limited chemostat culture (pH 5.5 ± 0.1; 30.0 ± 0.2°C; 0.7 l air [l culture]$^{-1}$ min^{-1}; 1000 rpm). (a) Concentration of GAM in the culture filtrate; (b) specific production rate of GAM. Error bars represent mean of 5 replicates ± SEM.

1995). Fed-batch fermentations are generally characterized by an initial phase of rapid growth followed by a period of deceleration during which production of the recombinant protein may be induced (Hensing *et al.*, 1995). However, instability in product formation will be enhanced if recombinant product formation is delayed until the final stages of the process. A continuous flow fermentation (chemostat or turbidostat) is more efficient than a fed-batch culture for the production of microbial biomass or a biomass-associated product (Pirt, 1975; Trinci, 1992) but such cultures demand even greater strain stability for product formation.

Importantly, numerous studies have shown that, in the absence of selective agents, plasmid carriage reduces the competitive fitness of microorganisms (Goodwin & Slater, 1979; Helling, Kinney & Adams, 1981; Lenski, Simpson & Nguyen, 1994) and plasmid-free segregants tend to overgrow plasmid-bearing strains (Cooper, Brown & Caulcott, 1987; Lenski & Nguyen, 1988). The selective advantage of one strain compared to another can be determined by calculation of the selection coefficient, which is based on the proportion of each colony type present in viable counts (Dykhuizen & Hartl, 1981). Selection coefficient (s) is defined as:

$$s = \frac{\ln\left[\frac{p_{(t)}}{q_{(t)}}\right] - \ln\left[\frac{p_{(o)}}{q_{(o)}}\right]}{t}$$

where $p_{(t)}$ is the concentration of the mutant at time t, $q_{(t)}$ *is the concentration of the parental strain at time t,* and $p_{(o)}$ and $q_{(o)}$ are the initial concentrations of each strain.

In bacteria and *S. cerevisiae*, two main factors have been found to influence plasmid stability in non-selective culture conditions: the segregation frequency of the plasmid and the effect of the plasmid's presence on growth rate (Goodwin & Slater, 1979). Segregation frequency has been shown to be dependent on various factors including dilution rate, nature of the growth-limiting substrate, and temperature (Roth, Noack & Geuther, 1985; Roth *et al.*, 1994; O'Kennedy, Houghton & Patching, 1995) as well as the host strain and the plasmid itself (Kumar *et al.*, 1991; Leonhardt & Alonso, 1991; D'Angio *et al.*, 1994; Roth *et al.*, 1994). In contrast to the self-replicating plasmids commonly used in the transformation of bacteria and *S. cerevisiae*, however, integrative plasmids are used to introduce extrachromasomal DNA into filamentous fungi. Although loss of the plasmid DNA as a result of unequal segregation during mitosis in filamentous fungi is therefore unlikely, deletions and substitutions in both native and foreign DNA sequences may occur (Adams *et al.*, 1992; Gilbert *et al.*, 1994; Hensing *et al.*, 1995).

Figure 5.6 shows that the concentration of GAM (303 ± 3 mg l^{-1}) in *A. niger* B1 grown in a glucose-limited chemostat culture at a dilution rate of 0.14 h^{-1} (doubling time, 4.95 h) remained constant for 948 h (191 generations), having a mean specific production rate of 18.2 mg GAM [g biomass]$^{-1}$ h^{-1}. After 213 h (43 generations) the original B1 strain, which produced densely sporing black colonies, was completely displaced by a morphological mutant which produced less densely sporing, brown colonies. However, no change in GAM production was associated with the appearance of this mutant. Although other morphological mutants (forming fawn and white colonies, with few conidia on agar-solidified Vogel's medium) were occasionally observed in samples from the culture, the brown mutant continued to comprise over 90% of the population throughout the final 735 h of the fermentation. An isolate of the brown mutant (B1-E) was obtained at the end of the fermentation.

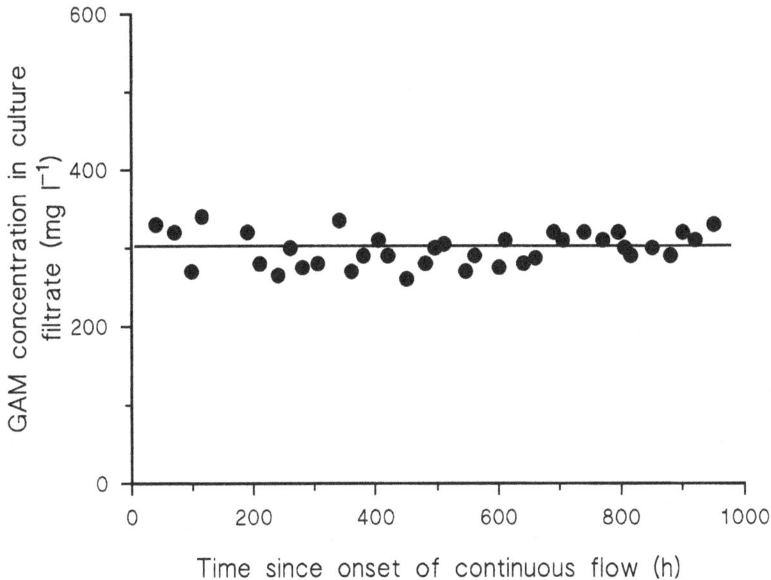

Fig. 5.6. GAM concentration in glucose-limited (5 g l^{-1}) chemostat cultures of *A. niger* B1 cultures grown on Vogel's medium at a dilution rate of 0.13 h^{-1}, 30.0 ± 0.2°C, pH 5.5 ± 0.1, aeration 0.6 l air [l culture]$^{-1}$ min^{-1}, agitation 1000 rpm). GAM assays were performed in quadruplicate for each sample. Error bars represent mean ± SEM.

Figure 5.7 shows the lack of stability of GAM production in cultures of *A. niger* B1 grown at a dilution rate of 0.14 h^{-1} for 354 h (71 generations) in a glucose-limited chemostat culture grown on Vogel's modified medium enriched with 5 g mycopeptone l^{-1}; all nutrients other than glucose were present in excess in these cultures and an increase in biomass concentration was observed when glucose was added to batch cultures grown on the same medium. Table 5.2 compares the production of biomass and GAM in mycopeptone-enriched and non-enriched medium during the first 130 h of chemostat culture and shows that GAM concentration and the specific production rate of GAM were increased by 64% and 37% respectively by the mycopeptone enrichment. However, after 136 h, GAM concentration in the mycopeptone-enriched culture decreased to 178 ± 14 mg l^{-1} following which no further significant change in the GAM concentration was observed (Fig. 5.6); this corresponded to a specific production rate of 9.1 mg GAM [g biomass]$^{-1}$ h^{-1}. The decline in GAM concentration was associated with the accumulation in the

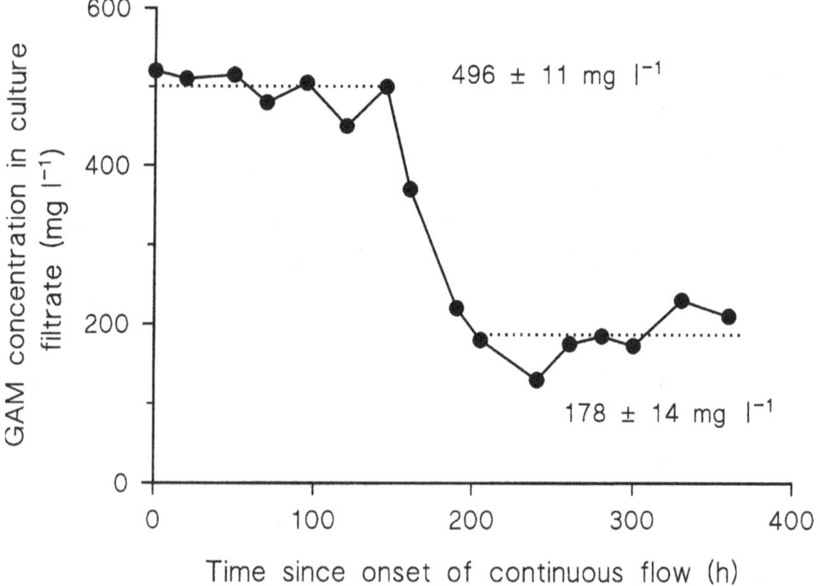

Fig. 5.7. GAM concentration in the culture filtrate of a glucose-limited (5 g l^{-1}) chemostat culture of *A. niger* B1 grown on modified Vogel's medium containing 5 g mycopeptone l^{-1} (cultures were grown at a dilution rate of 0.13 h^{-1}, 30.0 ± 0.2°C, pH 5.5 ± 0.1, aeration 0.6 l air [l culture]$^{-1}$ min^{-1}, agitation 1000 rpm).

Table 5.2. *GAM production for the first 130 h of the growth of* A. niger *B1 in glucose-limited and mycopeptone-supplemented glucose-limited chemostat culture at a dilution rate of 0.13 h^{-1} (pH 5.5; 30°C; 0.7 l air [l culture]$^{-1}$ min^{-1}; 1000 rpm)*

Parameter of GAM production	Units of measurement	Modified Vogel's medium containing: 5 g glucose l^{-1}	Modified Vogel's medium containing: 5 g glucose and 5 g mycopeptone l^{-1}	% Increase
Biomass	g l^{-1}	2.95 ± 0.2*	3.95 ± 0.2	34
GAM concentration (C_p)	mg l^{-1}	303 ± 12	496 ± 10	64
Specific production rate of GAM (q_p)	mg GAM [g biomass]$^{-1}$ h^{-1}	12.0 ± 0.4	16.4 ± 1.6	37

*Mean ± S.E.M. (5 replicates)

population of a brown mutant (B1-M) that was less densely sporulating than B1.

Selective advantage of strains B1-E and B1-M relative to N402 and B1

There was no selective advantage for either B1 or N402 when these strains were grown together in chemostat culture (Fig. 5.8). However, as shown in Table 5.3, each evolved strain (B1-E and B1-M) isolated from prolonged chemostat cultures had a selective advantage over the parental strain (B1); this selective advantage was largest under the cultural conditions in which the mutant had been originally isolated, i.e. glucose limitation for B1-E, and mycopeptone-enriched, glucose-limitation for B1-M. However, all evolved strains retained their selective advantage in the other conditions tested, including fructose limitation, which does not support GAM production. Further competition studies between B1-M and N402 (which produced similar levels of GAM to B1) demonstrated that B1-M also had a selective advantage over N402, although the selection coefficients were lower than for B1-M over B1.

glaA *gene copy number in fermenter-selected strains*

To analyse B1-E and B1-M for the presence and expression of the *gla*A gene copies the isolates were plated on acetamide/acrylamide and hygro-

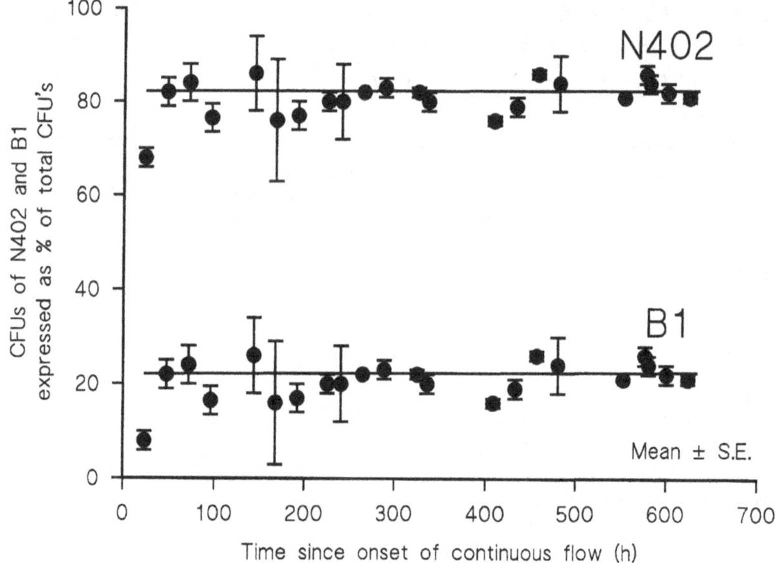

Fig. 5.8. Mixed culture of *A. niger* N402 and B1 grown in glucose-limited (5 g l^{-1}) chemostat cultures at a dilution rate of 0.13 h^{-1}, 30.0 ± 0.2°C, pH 5.5 ± 0.1, aeration 0.6 l air [l culture]$^{-1}$ min^{-1}, agitation 1000 rpm.

mycin B plates (Verdoes *et al.*, 1993, 1994) to verify the transformed phenotype. From this analysis it was clear that the B1-E had retained the transformed phenotype, whereas B1-M showed reduced growth on acrylamide plates, indicating reduced *amdS* expression and thus loss of introduced gene copies. This was further confirmed by Southern analysis using a *glaA*-specific probe (and the single-copy *gpdA* control probe). It was clear from these analyses that B1-M had lost most of the introduced *glaA* gene copies, while B1-E retained all or most of its *glaA* gene copies. Thus, although the specific production rate of GAM was significantly increased by the addition of mycopeptone to the medium, GAM production was unstable and decreased to less than half its previous level after 136 h (27 generations) of chemostat cultivation. B1-M, isolated from the end of this fermentation, produced significantly less GAM in mycopeptone-enriched, glucose-limited culture than B1. Rapid loss of productivity has been observed in cultures of *Streptomyces* sp. grown in a complex medium (Fazeli, Core & Baumberg, 1995) and more rapid plasmid loss has been observed from *S. cerevisiae* in complex than in defined medium

Table 5.3. *Selection coefficients (*s, *a measure of differences in specific growth rates during mixed culture in chemostat cultures; Dykhuizen & Hartl, 1981) for strains B1, B1-E, and B1-M relative to N402 and B1 in carbon-limited chemostat culture at a dilution rate of 0.13 h^{-1} (pH 5.5; 0.6 to 1.0 l air [l culture]$^{-1}$ min^{-1}; 30°C; 1000 rpm).*

Competition between		Carbon source limiting growth rate[a]	GAM produced?	Selection coefficient s for strain A (h^{-1})
Strain A	Strain B			
B1	N402	Glucose	Yes	0.000[b]
B1-E	B1	Glucose	Yes	0.058
B1-E	B1	Fructose	No	0.038
B1-M	N402	Glucose	Yes	0.055
B1-M	N402	Glucose + mycopeptone	Yes	0.077
B1-M	B1	Glucose	Yes	0.070
B1-M	B1	Fructose	No	0.069
B1-M	B1	Glucose + mycopeptone	Yes	0.144

[a]Glucose or fructose (5 g l^{-1}) were used as the carbon source. In addition, some cultures were enriched with mycopeptone (5 g l^{-1}). All cultures were inoculated with mycelial suspensions of the two strains used in the competition, to give initial concentrations of 5% to 50% of one of the strains and 50% to 95% of the other.
[b]i.e. B1 behaves as a neutral mutation.

(O'Kennedy *et al.*, 1995) but only when the culture was grown at a high dilution rate.

For stable integrated gene inserts, it is generally assumed that instability will primarily be the result of growth rate differences between the insert-bearing and insert-free cell (Devchand & Gwynne, 1991). Segregation instability (the unequal distribution of plasmids between daughter cells during cell division), a problem for non-integrated plasmids in bacteria, yeast and fungi (Roth *et al.*, 1994; Moreno *et al.*, 1994), is unlikely to occur, although structural instability (loss of heterologous DNA by deletions, insertions or rearrangements) may occur (Hensing *et al.*, 1995; Gilbert *et al.*, 1994; Numan, Venables & Wimpenny, 1991). As indicated by the measurement of selection coefficients for each of the evolved strains (B1-E and B1-M) relative to the parental strain (B1), growth rate differences probably accounted for the displacement of B1 in the long-term fermentations (Table 5.3). However, the growth rate differences between the original strains and the evolved strains cannot be explained on the basis of loss of gene copies and thus lower gene expression, since no copy loss was observed in B1-E. Furthermore, the selective advantages observed were similar whether or not glucoamylase expression occurred (Table 5.3). Thus, the selective advantages of these strains result from other alterations which affect their physiology.

B1-M had a selective advantage of 0.055 h^{-1} relative to N402 in glucose-limited chemostat culture and a selective advantage of 0.069 h^{-1} relative to B1 in fructose-limited culture (Table 5.3). Thus, in this case, part of the growth rate increase may be GAM-expression independent. However, B1-M had also lost copies of the *gla*A gene, and this may have contributed to its growth rate advantage. Nevertheless, it is more likely that in B1-M, as in B1-E, a physiological alteration resulted in the selective advantage and that by chance the strain also suffered gene copy loss.

Concluding remarks

When an organism is cultured under constant conditions (Novick & Szilard, 1950*a*; Monod, 1950) in a continuous flow culture, the population evolves and progressively becomes better adapted to its environment (Novick & Szilard, 1950*b*). This evolution involves mutations which confer selective advantages (such as an increase in μ_{max} or decrease in K_s) to the mutants, and consequently, the mutants eventually replace the wild type. Since the chemostat cultures of *F. graminearum* A3/5 and *A. niger* described above were grown at relatively high dilution rates, the advantageous mutants selected probably had increased μ_{max} values.

Provided that the production strain is stable in prolonged culture, a continuous flow culture (chemostat or turbidostat) operated at high dilution rate (Fig. 5.5) is a cost-efficient system for the production of a growth related product. As far as the latter is concerned, it is noteworthy that 155 m^3 glucostat cultures are used to produce biomass of *F. graminearum* for Quorn® myco-protein production (Trinci, 1994). However, engineering problems related to scaling-up and strain stability and loss of productivity during long-term cultivation would need to be addressed before continuous cultures could be used on an industrial scale for recombinant protein production.

For *A. oryzae,* the use of highly branched mutants failed to improve the enzyme production (Fig. 5.3) although they did decrease culture viscosity (Fig. 5.4). Finally, recognition of the pleiotropic nature of the highly branched (colonial) phenotype of mutants arising in *F. graminearum*

A3/5 fermentations was an important step in developing strategies to prevent/delay the appearance of these mutants (Trinci, 1994).

The studies clearly show the importance of mycelial morphology and strain stability in the productivity of industrial fermentations of filamentous fungi.

Acknowledgements

We thank the BBSRC and the EU for sponsoring some of the work reported in this chapter.

References

Adams, J., Puskas-Rozsa, S., Simlar, J. & Wilke, C. M. (1992). Adaptation and major chromosomal changes in populations of *Saccharomyces cerevisiae*. *Current Genetics*, **22**, 13–19.

Brody, S. & Tatum, E. C. (1966). The primary biochemical effect of a morphological mutation in *Neurospora crassa*. *Proceedings of the National Academy of Sciences, USA*, **56**, 1290–7.

Brody, S. & Tatum, E. C. (1967). Phosphoglucomutase mutants and morphological changes in *Neurospora crassa*. *Proceedings of the National Academy of Sciences, USA*, **58**, 923–30

Cooper, N. S., Brown, M. E. & Caulcott, C. A. (1987). A mathematical method for analysing plasmid stability in micro-organisms. *Journal of General Microbiology*, **133**, 1871–80.

D'Angio, C., Béal, C., Boquien, C-Y., Langella, P. & Corrieu, G. (1994). Plasmid stability in recombinant strains of *Lactococcus lactis* subsp. *lactis* during continuous culture. *FEMS Microbiology Letters*, **116**, 25–30.

Devchand, M. & Gwynne, D. I. (1991). Expression of heterologous proteins in *Aspergillus*. *Journal of Biotechnology*, **17**, 3–10.

Dunn-Coleman, N. S., Bodie, E. A., Carter, G. L. & Armstrong, G. L. (1994). Stability of recombinant strains under fermentation conditions. In *Applied Molecular Genetics of Filamentous Fungi*, ed. J. R. Kinghorn & G. Turner, pp. 152–174. Glasgow: Blackie & Son Ltd.

Dykhuizen, D. E. & Hartl, D. L. (1981). Evolution of competitive ability in *Escherichia coli*. *Evolution*, **35**, 581–94.

Fazeli, M. R., Cove, J. H. & Baumberg, S. (1995). Physiological factors affecting streptomycin production by *Streptomyces griseus* ATCC 12475 in batch and continuous culture. *FEMS Microbiology Letters*, **126**, 55–62.

Forss, K. G., Gadd, G. O., Lundell, R. O. & Williamson, H. W. (1974). Process for the manufacture of protein-containing substances for fodder, foodstuffs and technical application. U.S. Patent Office, Patent No. 3,09,614.

Gilbert, S. C., van Urk, H., Greenfield, A. J., McAvoy, M. J., Denton, K. A., Coghlan, D., Jones, G. D. & Mead, D. J. (1994). Increase in copy number of an integrated vector during continuous culture of *Hansenula polymorpha* expressing functional human haemoglobin. *Yeast*, **10**, 1569–80.

Goodwin, D. & Slater, J. H. (1979). The influence of the growth environment on the stability of a drug resistance plasmid in *Escherichia coli* K12. *Journal of General Microbiology*, **111**, 201–10.

Helling, R. B., Kinney, T. & Adams, J. (1981). The maintenance of plasmid-containing organisms in populations of *Escherichia coli. Journal of General Microbiology*, **123**, 129–41.

Hensing, M. C. M., Rouwenhorst, R. J., Heijnen, J. J., van Dijken, J. P. & Pronk, J. T. (1995). Physiological and technological aspects of large-scale heterologous-protein production with yeasts. *Antonie van Leeuwenhoek*, **67**, 261–79.

Kumar, P. K. R., Maschke, H-E., Friehs, K. & Schügerl, K. (1991). Strategies for improving plasmid stability in genetically modified bacteria in bioreactors. *Trends in Biotechnology*, **9**, 279–84.

Lenski, R. E. & Nguyen, T. T. (1988). Stability of recombinant DNA and its effects on fitness. *Trends in Ecology and Evolution*, **3**, S18–S20.

Lenski, R. E., Simpson, S. C. & Nguyen, T. T. (1994). Genetic analysis of a plasmid-encoded, host genotype-specific enhancement of bacterial fitness. *Journal of Bacteriology*, **176**, 3140–7.

Leonhardt, H. & Alonso, J. C. (1991). Parameters affecting plasmid stability in *Bacillus subtilis*. *Gene*, **103**, 107–11.

Monod, J. (1950). La technique de culture continue: théorie et applications. *Annales de l'Institut Pasteur* (Paris), **79,** 390–410.

Moreno, M. A., Pascual, C., Gibello, A., Ferrer, S., Bos, C. J., Debets, A. J. M. & Suárez, G. (1994). Transformation of *Aspergillus parasiticus* using autonomously replicating plasmids from *Aspergillus nidulans*. *FEMS Microbiology Letters*, **124**, 35–42.

Novick, A. & Szilard, L. (1950*a*). Description of the chemostat. *Science*, **112**, 715–16.

Novick, A. & Szilard, L. (1950*b*). Experiments with chemostat on spontaneous mutation of bacteria. *Proceedings of the National Academy of Sciences, USA*, **36**, 708–19.

Numan, Z., Venables, W. A. & Wimpenny, J. W. T. (1991). Competition between strains of *Escherichia coli* with and without plasmid RP4 during chemostat growth. *Canadian Journal of Microbiology*, **37**, 509–12.

O'Kennedy, R., Houghton, C. J. & Patching, J. W. (1995). Effects of growth environment on recombinant plasmid stability in *Saccharomyces cerevisiae* grown in continuous culture. *Applied Microbiology and Biotechnology*, **44**, 126–32.

Pirt, S. J. (1975). *Principles of Microbe and Cell Cultivation*. Blackwell: Oxford, UK.

Righelato, R. C. (1976). Selection of strains of *Penicillium chrysogenum* with reduced penicillin yields in continuous cultures. *Journal of Applied Chemistry and Biotechnology*, **26**, 153–9.

Roth, M., Hoffmeier, C., Geuther, R., Muth, G. & Wohlleben, W. (1994). Segregational stability of pSG5-derived vector plasmids in continuous cultures of *Streptomyces lividans* 66. *Biotechnology Letters*, **16**, 1225–30.

Roth, M., Noack, D. & Geuther, R. (1985). Maintenance of the recombinant plasmid pIJ2 in chemostat cultures of *Streptomyces lividans* 66 (pIJ2). *Journal of Basic Microbiology*, **25**, 265–71.

Simpson, D. R., Wiebe, M. G., Robson, G. D. & Trinci, A. P. J. (1995). Use of pH auxostats to grow filamentous fungi in continuous flow culture at maximum specific growth rate. *FEMS Microbiology Letters*, **126**, 151–8.

Solomons, G. L. & Scammell, G. W. (1976). Production of edible protein substances. United States Patent Office Patent No. 3,937,654.

Trilli, A. (1977). Prediction of costs in continuous fermentations. *Journal of Applied Chemistry and Biotechnology*, **27**, 251–9.

Trinci, A. P. J. (1992). Mycoprotein – a twenty year overnight success story. *Mycological Research*, **96**, 1–13.

Trinci, A. P. J. (1994). Evolution of the Quorn® myco-protein fungus, *Fusarium graminearum*. *Microbiology*, **140**, 2181–8.

Trinci, A. P. J., Wiebe, M. G. & Robson, G. D. (1994). The fungal mycelium as an integrated entity. In *The Mycota, Volume I, Growth and Differentiation and Sexuality*, ed. J. G. H. Wessels & F. Meinhardt, pp. 175–93. Berlin: Springer-Verlag.

Verdoes, J. C., Punt, P. J., Schrickx, J. M., van Verseveld, H. W., Stouthamer, A. H. & van den Hondel, C. A. M. J. J. (1993). Glucoamylase over expression in *Aspergillus niger*: molecular genetic analysis of strains containing multiple copies of the *glaA* gene. *Transgenic Research*, **2**, 84–92.

Verdoes, J. C., van Diepeningen, A. D., Punt, P. J., Debets, A. J. M., Stouthamer, A. H. & van den Hondel, C. A. M. J. J. (1994). Evaluation of molecular and genetic approaches to generate glucoamylase overproducing strains of *Aspergillus niger*. *Journal of Biotechnology*, **36**, 165–75.

Vogel, H. J. (1956). A convenient growth medium for *Neurospora* (Medium N). *Microbial Genetics Bulletin*, **243**, 112–19.

Wessels, J. (1993). Wall growth, protein excretion and morphogenesis in fungi. *New Phytologist*, **123**, 397–413.

Wiebe, M. G., Blakebrough, M. L., Craig, S. H., Robson, G. D. & Trinci, A. P. J. (1996). How do highly branched (colonial) mutants of *Fusarium graminearum* A3/5 arise during Quorn® myco-protein fermentations? *Microbiology*, **142**, 525–32.

Wiebe, M. G., Robson, G. D., Cunliffe, B., Trinci, A. P. J. & Oliver, S. G. (1992). Nutrient-dependent selection of morphological mutants of *Fusarium graminearum* A3/5 isolated from long term continuous flow cultures. *Biotechnology & Bioengineering*, **40**, 1181–9.

Wiebe, M. G., Trinci, A. P. J., Cunliffe, B., Robson, G. D. & Oliver, S. G. (1991). Appearance of morphological (colonial) mutants in glucose-limited, continuous flow cultures of *Fusarium graminearum* A3/5. *Mycological Research*, **95**, 1284–8.

Withers, J. M, Wiebe, M. G., Robson, G. D., Osborne, D., Turner, G. & Trinci, A. P. J. (1995). Stability of recombinant protein production by *Penicillium chrysogenum* in prolonged chemostat culture. *FEMS Microbiology Letters*, **133**, 245–51.

Wösten H. A. B., Moukha S. M., Sietsma J. H. & Wessels J. G. H. (1991). Localization of growth and secretion of proteins in *Aspergillus niger*. *Journal of General Microbiology*, **137**, 2017–23.

Yarden, O., Plamann, M., Ebbole, D. J. & Yanofsky, C. (1992). *cot-1*, a gene required for hyphal elongation in *Neurospora crassa*, encodes a protein kinase. *EMBO Journal*, **11**, 2159–66.

6

Metabolism and hyphal differentiation in large basidiomycete colonies

S. WATKINSON

Introduction

Mycelial fungi are adapted for the colonization of solid surfaces (Carlile, 1994). These are unlike aquatic environments because natural mixing processes are insufficient for effective colonization of solid substrates. A common colonization and dispersal method, typical of the majority of ascomycetes (but by no means restricted to them) is the production of very large numbers of spores, allowing the facile, but random, creation of new foci for mycelial exploitation of substrate.The disadvantage of this process is that the whole mycelium cannot be converted to spores when a particular mycelium has exhausted all the substrate in its range.

Within the basidiomycetes an alternative strategy has evolved, which is the formation of very large colonies that behave in a non-random manner with respect to colonization and, crucially, employ long-range translocation mechanisms to enable the re-use of scarce resources where new substrate is detected. This phenomenon is best described for some wood-degrading fungi of the forest floor and the dry-rot fungus *Serpula lacrymans*. In both cases wood substrate is dispersed in an otherwise nutrient-poor environment, requiring 'foraging'. This is an unusual process in microorganisms (and in plants), where growth or movement towards *detectable* centres of nutrient is the common pattern. Again, the responses of the wood-degrading fungi have to be different in that their substrate is non-diffusing and cannot be detected chemotactically. Thus, the surface of the environment is interrogated by hyphae radiating from a nutrient centre, until a new nutrient source is detected. This leads to a remarkable series of events – the formation of complex multihyphal strands/cords and the bulk transfer of nutrients from the exhausted nutrient focus to the newly found one.

Before considering these events in more detail, it is worth pointing out that there is a fairly obvious advantage of nutrient reuse by these fungi. The substrate they are adapted to exploit – dead plant remains – is a ready source of carbon and energy, but is very poor in nitrogen. Thus the reuse of nitrogenous materials at the expense of much energy for translocation and of both energy and carbon in cord formation can still have profound adaptive advantages.

In basidiomycetes the metabolism and development of sometimes extremely large colonies (Smith, Bruhn & Anderson, 1992) are coordinated with respect to their discontinuous nutrient supply (Rayner, Griffith & Ainsworth, 1994). This requires the flow through the system of both materials and information. The following discussion first considers the morphogenesis of a redistributive system, and its relation to nitrogen metabolism. It goes on to propose an experimental approach to the question of how long-distance signals might be transmitted through mycelium to coordinate morphogenesis in large systems, and focuses largely on *S. lacrymans* as a model.

Mycelial differentiation

Cord development and structure

The hyphal differentiation involved in the formation of a cord has been well described for several species (Butler 1957, 1958; Jennings & Watkinson, 1982; Cairney, Jennings & Veltkamp, 1989). In *S. lacrymans*, the hyphae of primary mycelium differentiate, with some becoming relatively wide and empty, lacking septa, and with walls containing neutral lipid material (Helsby, 1976). Development begins round a single hypha of this type when its branches adhere to it instead of diverging, and grow both basipetally and acropetally alongside it (Butler, 1957,1958). Further growth of these, adhesion of other hyphae which meet it during their growth, and massive production of extracellular matrix material, produces a visible strand several tens of microns thick and up to several metres in length. The structure seems exemplary of the 'hyphal knot' in which one initiating hypha becomes the organizing centre for others, which then differentiate under its influence (Moore, 1994). Very thick-walled wiry fibre hyphae develop, running longitudinally through the cord and arranged in a cylindrical band round its periphery. Throughout the cord there are longitudinal spaces, apparently formed by the lysis of the contents of a few wide hyphae which formed the original nucleus for the cord, but with the lumen of the spaces wider than that of any of the hyphae seen at an earlier stage. Narrower hyphae with cytoplasmic contents are scattered through the strand and arranged longitudinally within the non-cellular carbohydrate matrix and inside the peripheral ring of fibre hyphae (Jennings & Watkinson, 1982). The mature structure thus consists of a peripherally strengthened, spongy mass of matrix material with longitudinal spaces in which a number of widely spaced living hyphae are embedded.

The functions of cords

In *S. lacrymans* and woodland fungi, cords develop in mycelium that has colonised a new food source, and mycelium which has not found new sources dies away. Ultimately a mature mycelium consists mainly of cords. Patches of diverging separate hyphae may, however, develop, when the mycelium encounters new sources or sinks for nutrients (Donnelly & Boddy, 1997). While these patches of separate hyphae appear to be adapted for the uptake of nutrients such as phosphorus and nitrogen, hyphae in cords act mainly as conduits for translocation. Nutrients can be carried along cords to be sequestered in already colonized wood, as shown for phosphorus in *Phanerochaete velutina* (Wells, Hughes & Boddy, 1990; Wells & Boddy 1995). Water (Jennings, 1984), carbon (Brownlee & Jennings,1982*a*,*b*), and nitrogen (Arnebrant *et al.*, 1993; Watkinson, 1984*a*) can also travel in cords. The velocity of translocation of water and carbon compounds in *S. lacrymans* is too fast to be explained by diffusion. Bulk flow appears to occur, driven by water potential gradient as in plants (Jennings, 1987). However, movement of labelled, nonmetabolized amino acid can be towards or away from the direction of growth, and occur in the absence of a water potential gradient (Watkinson, 1984*a*), therefore other mechanisms must also be involved.

Apart from carrying translocated nutrients, cords may have other functions, for example to protect against desiccation or predation, and to provide electrical insulation for hyphae carrying electrical signals (see Olsson, Chapter 2 and below).

The pattern of cord development in natural environments

The formation of cords between food sources, and the occasional development of patches of diverging separate hyphae, produce the appearance of 'foraging' in the mature mycelium (Boddy, 1993; Dowson, Rayner & Boddy 1986, 1988*a*,*b*; Dowson, Springer & Rayner 1989). Foraging patterns differ in an adaptive manner between species. *P. velutina* forms long-range mycelial systems connecting large pieces of wood, and forms cords only when it encounters a large new food resource, while *Stropharia caerulea* is adapted to small food sources such as nettle stems and rhizomes in nutrient-enriched sites, and is more developmentally sensitive than *P. velutina* to small changes in nutrient supply (Donnelly & Boddy, 1997).

Cords also form to connect developing fruiting bodies to sources of supply. It is common to find *S. lacrymans* fruiting in the corners of ceilings, supplied by a wood source embedded in masonry several metres away (Hennebert, Boulenger & Balon, 1990; Jennings & Bravery, 1991; Watkinson, 1984*a*).

Cord induction as part of a process of resource reallocation

The mechanism by which cords are developed in a foraging mycelium is not understood. Microscopic observation of cord formation in *S. lacrymans* shows that colonization of a new food source is followed within days by increased branching of the hyphae which have connected to the new source (Watkinson, 1971; Fig. 6.1). A stimulus to branch appears to travel basipetally in the connecting hyphae. The new branches are directed towards the new food source and quickly connect to it. At the same time, two other responses are seen: the cords start to form round connecting hyphae, and there is lysis of hyphae which have not connected to the new source. Thus three separate responses are initiated by the encounter with fresh food. Together they comprise an adaptive response which is here referred to as the 'reallocation response'. While the immediate consequence of colonizing a fresh food source must be a gain in nutrients by the colonizing hyphae and translocation of nutrients in the mycelium, information also passes through the connecting hyphae, which signals the non-connecting hyphae to lyse, and the connecting hyphae to differentiate forming a cord. The long-distance transmission of signals is similar to (although not homologous with) a nervous system (see Chapter 2).

What stimulus passes from a nutrient source to bring about this hyphal differentiation? Experiments using synthetic media as food sources for *S. lacrymans* and *Penicillium claviforme* suggest that changes in intracellular levels of amino acids control developmental changes in hyphal extension and branching, and the initiation of multihyphal structures. These experiments are reviewed below, followed by a discussion of the processes of nitrogen metabolism in fungi which are affected by the

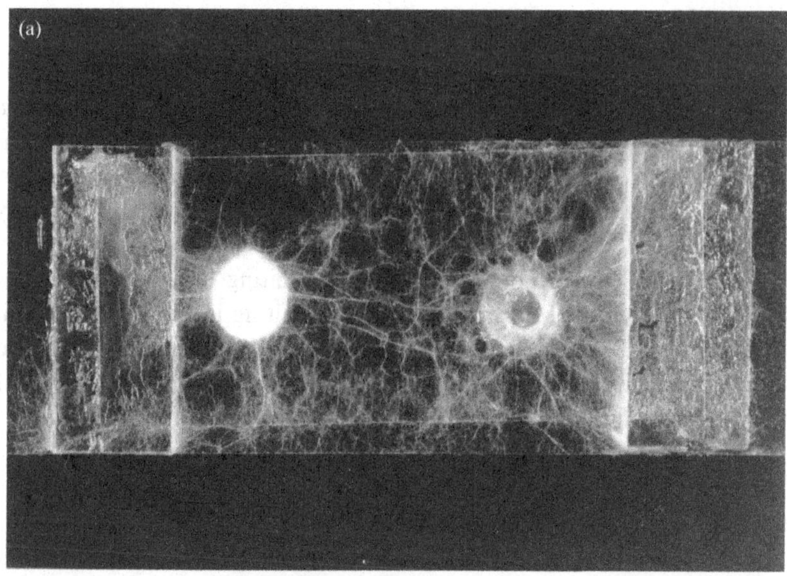

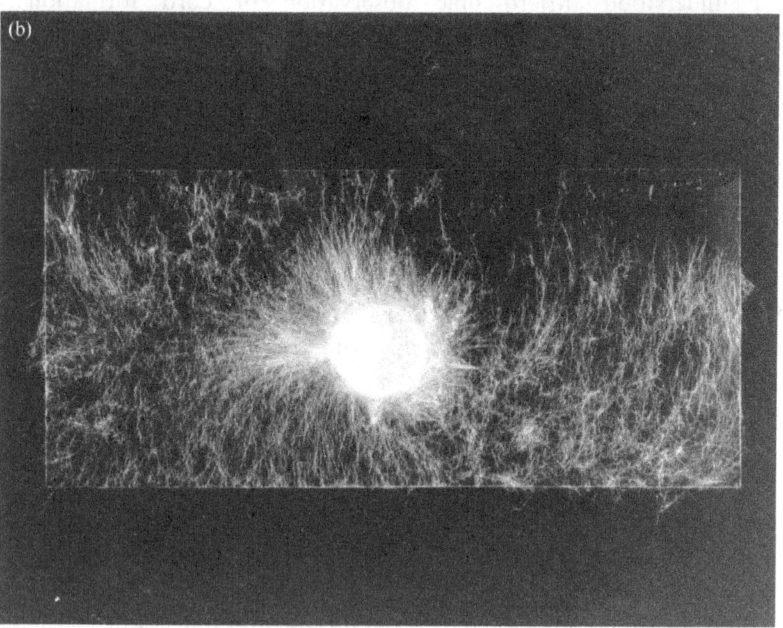

Fig. 6.1. (a) Growth of *S. lacrymans* from an agar inoculum disc on a glass microscope slide, with colonization of a fresh agar disc of the same nutrient medium. Cords have started to form, most abundantly in the bridging mycelium. Unconnected mycelium appears thinner. Some cords have formed at the sides of the slide where mycelium has begun to colonize the damp filter paper used to maintain high humidity for growth. (b) Growth from a single agar inoculum disc, showing absence of cord initiation in the absence of a second food source (Watkinson, 1971) (Photo: T. Pullen).

environment. A model is put forward for a role of nitrogen metabolism in the reallocation response of *S. lacrymans*.

Evidence that intracellular nitrogen status can control mycelial differentiation

In the reallocation response observed in a microcosm on glass slides, (as in Fig. 6.1), it is clear that the stimulus for hyphal differentiation is intracellular, and can be propagated intracellularly. There is no medium for signal molecules to travel through, apart from connecting hyphae.

By what mechanisms might hyphae sense and respond to intracellular nutrient concentration? Intracellular nitrogen has a morphogenetic effect in *S. lacrymans,* controlling the rate of mycelial extension from a resource with experimentally varied nitrogen content (Watkinson, Davison & Bramah, 1981). With sodium nitrate as sole nitrogen source in an agar disc from which hyphae were grown over Perspex, there was an optimum nitrogen content in the disc for the rate and final extent of mycelial growth. When the nitrogen source was sodium nitrate, abundant nitrogen in the nutrient source, or an insufficent supply, produced a more localized mycelium. With glutamine as sole nitrogen source, the control with no nitrogen added to the food source produced the fastest extension, with all additions of glutamine producing denser, more localized growth. Glutamine is a better repressor than sodium nitrate of catabolic pathways of nitrogen metabolism (see below).

The non-metabolized amino acid α-aminoisobutyric acid (AIB) (also called 2-methylalanine) is taken up by active transport and accumulated in mycelium (Ogilvie-Villa, DeBusk & DeBusk, 1981). It is not metabolized and is not incorporated into protein. It does, however, depress intracellular glutamate levels and inhibit incorporation of utilizable amino acids into protein in *S. lacrymans* (Watkinson, 1984*b*). It has profound and long-lasting effects on mycelial morphology. AIB applied to a wood nutrient source from which mycelium is growing inhibits extension just as excess glutamine does (Elliott & Watkinson, 1989). The effect of a single application persisted for several months (Fig. 6.2), making AIB a useful chemical for controlling the spread of dry rot (Dobson *et al.*, 1993). A further effect of AIB is its effect on increasing the frequency of branching (Elliot & Watkinson, 1989). AIB-treated mycelium develops a cushion-like form with a mound of hyphae, rather than extending widely. This suggests that intracellular amino acid levels might be sensed directly rather than via amino acid metabolism. AIB

Fig. 6.2. Inhibition of extension of mycelium of *S. lacrymans* mycelium from a wood block embedded in brickwork, by α-aminoisobutyric acid. (AIB). Mycelium was allowed to grow into the bricks from an infected wood block (w) embedded in the middle of the columns. When 2–3 courses of bricks had been infiltrated by the mycelium, column (a) was treated by drip-feeding 125 ml of aqueous AIB into the wood block for 24 h. Column (b) received 125 ml water. The photograph was taken four months after application, and shows the long-lasting effect of this non-metabolized amino acid on the growth form of the mycelium. Treated mycelium did not continue to extend into the bricks but developed a localized, cushion-like growth form (arrow). Water-treated mycelium extended the full length of the column. The experiment was conducted in a cellar maintained at close to 99% relative humidity (Dobson *et al.*, 1993).

accumulated in free amino acid pools may signal, falsely, that intracellular amino acid levels are high, and induce localized growth appropriate to exploiting a rich nutrient resource.

Nitrogen nutrition controls initiation of multihyphal structures such as coremia (Watkinson, 1975*b*) and sclerotia (Shapira *et al.*, 1989), as well as cords. The nature and concentration of nitrogen sources and the C:N ratio are critical for cord formation in *S. lacrymans* (Watkinson, 1975*b*, 1979). *P. claviforme* coremium initiation is quantitatively determined by the concentration and nature of amino acids in the medium, and comparison of the effect of glutamate with that of other amino acids equally well utilized for growth indicate a special morphogenetic role for glutamate (Watkinson, 1977, 1981).

Developmental responses by fungi to nitrogen nutrition have been analysed at the genetic level. Of the genes which act sequentially to control conidiation in *Aspergillus nidulans*, the final one in the sequence, *brlA* (bristle A) is activated by active growth or a high nutrient status of the mycelium (Adams *et al.*, 1992). Gimeno *et al.* (1992) described the induction of filamentous 'pseudohyphal' growth in diploid *Saccharomyces cerevisiae* by starvation or in the presence of proline as sole nitrogen source in the nutrient medium. The phenomenon was analysed by investigating the effects of mutations affecting amino acid uptake and the RAS signal transduction pathway. A mutant strain deficient in amino acid uptake showed enhancement of the pseudohyphal growth form. The response to nitrogen starvation was also increased in a mutant strain in which the mutation conferred constitutive activation of the RAS signal transduction pathway with consequent elevated intracellular cAMP levels, leading the authors to suggest that filamentous growth in response to nitrogen starvation is regulated through induction of the RAS pathway by nitrogen starvation. They interpreted this as a foraging response. The involvement of the RAS pathway indicates signal transduction via changes in the levels of intracellular cAMP. This pathway is known to be involved in fungal development (Ishikawa, 1989; Gadd, 1994). In *P. claviforme,* glutamate, which induces coremium initiation, was also shown to cause an elevation of cAMP, while serine, which is equally well utilized for growth but does not induce coremia, had less effect (S.C. Watkinson, unpublished observation).

Genes of major pathways are conserved in fungi and in eukaryotes generally (Blumer & Johnson, 1994; Trinci, 1995), so these observations on the control of gene expression in Ascomycetes are likely to be relevant to Basidiomycetes. It is clearly feasible for intracellular levels of nitrogen metabolites to regulate the expression of genes involved in development.

In fungi, intracellular levels of nutrients are strongly affected by the extracellular growth medium. Fungi have large expandable pools of amino acids. The composition of the amino acid pool and of mycelial protein was investigated in *S. lacrymans* to establish how the composition of intracellular pools varied with environmental supply.

The effects of nitrogen availability in the environment on intracellular nitrogen status in **S. lacrymans**

Like all fungi, *S. lacrymans* mycelium can contain widely varying levels of nitrogen. Nitrogen is concentrated from wood by the mycelium. Mycelium grown exclusively from wood which contained 1.5 mg g^{-1} or less total nitrogen was found to contain up to 37 mg g^{-1} nitrogen (Watkinson *et al.*, 1981). The amount and chemical form of nitrogen was compared in mycelium grown on each of two synthetic nutrient media containing ammonium sulphate as sole nitrogen source, either at nitrogen levels similar to those in seasoned wood or 100-fold greater. At the higher level, nitrogen presumably in excess of metabolic requirements was accumulated in mycelial protein and in soluble form in the mycelium (Venables & Watkinson, 1989*a*).

The free amino acid pool had a distinctive composition at high nitrogen levels and some amino acids were more affected than others by environmental nitrogen levels (Table 6.1). These included lysine and ornithine, basic amino acids that are known to be accumulated in vacuoles and act as nitrogen stores. The non-protein amino acid gamma-aminobutyric acid (GABA) was detected in the free amino acid pool of *S. lacrymans* grown on media containing very high levels of nitrogen. Glutamate/glutamine (the analysis did not distinguish them) was the most abundant amino acid in the pool at all times, but its concentration was less sensitive to supply than ornithine, lysine or GABA and it is presumed to be located in the cytosolic pool, which turns over more slowly than vacuolar pools in the fungi which have been most studied, for example *Candida utilis* (Wiemken & Nurse, 1973) and *Neurospora crassa* (Weiss, 1973).

There was some evidence that the extra protein formed under conditions of excess nitrogen supply might also have a distinctive composition. Hydrolysis of the mycelial residue after extraction of soluble amino acids, and analysis of its amino acid composition, showed a greater proportion

Table 6.1. *Comparison of amino acid composition of mycelial protein and the free amino acid pool in* Serpula lacrymans. *Data from Venables & Watkinson (1989). Figures in brackets are proportions of each amino acid as a percentage of the total*

Amino acid	Amino acid composition in μg per g of dried mycelial hydrolyante												Amino acids present in free soluble form in μg per g dried mycelium							
	High N medium[a]						Low N medium[a]						High N medium[a]				Low N medium[a]			
	1 wk		2 wk		3 wk		1 wk		2 wk		3 wk		1 wk		3 wk		1 wk		3 wk	
Alanine	372	(9.6)	283	(9.4)	353	(9.6)	198	(9.6)	72	(10.5)	109	(11.7)	10.6	(3.0)	12.8	(3.2)	3.1	(4.4)	2.9	(10.4)
Glycine	403	(10.4)	311	(10.4)	370	(10.0)	231	(11.2)	78	(11.4)	115	(12.3)	8.8	(2.5)	11.1	(2.8)	3.8	(5.4)	1.4	(5.0)
Valine	202	(5.2)	202	(6.8)	252	(6.8)	145	(7.1)	51	(7.4)	76	(8.1)	1.2	(0.3)	1.4	(0.3)	0.6	(1.0)	0.3	(1.1)
Threonine	230	(5.9)	198	(6.7)	252	(6.8)	144	(7.0)	49	(7.1)	73	(7.8)	2.9	(1.0)	3.6	(1.0)	1.0	(1.4)	0.4	(1.4)
Serine	305	(7.9)	233	(7.8)	277	(7.5)	171	(8.3)	58	(8.5)	89	(9.5)	7.3	(2.1)	9.6	(2.4)	4.0	(5.6)	2.0	(7.1)
Leucine	367	(9.5)	281	(9.4)	337	(9.1)	226	(11.0)	73	(10.6)	71	(7.6)	0.7	(0.2)	1.0	(0.3)	0.6	(1.0)	0.1	(0.4)
Isoleucine	142	(3.7)	151	(5.0)	186	(5.0)	106	(5.2)	37	(5.4)	35	(3.7)	0.3	(0.1)	0.6	(0.2)	0.3	(0.5)	0.2	(0.7)
γ-aminobutyric acid	n.d.	n.d.	n.d.	n.d.	n.d.	n.d.	n.d.	n.d.	n.d.	n.d.	n.d.	n.d.	1.2	(0.3)	2.2	(0.6)	0.4	(1.0)	n.d.	(0)
Proline	226	(5.8)	166	(5.5)	199	(5.4)	131	(6.4)	44	(6.4)	56	(6.0)	1.3	(0.3)	2.9	(0.7)	0.7	(1.0)	0.2	(0.7)
Pyroglutamic acid	4	(0.0)	n.d.	n.d.	n.d.	n.d.	n.d.	n.d.	n.d.	n.d.	n.d.	n.d.	17.2	(4.9)	25.3	(6.4)	1.5	(2.1)	0.2	(0.7)
Methionine	62	(0.02)	40	(1.3)	50	(1.4)	23	(1.1)	6	(1.0)	10	(1.1)	n.d.	n.d.	n.d.	n.d.	<0.1	(0)	0.4	(1.4)
Aspartic acid[b]	514	(13.3)	401	(13.4)	483	(13.1)	312	(15.2)	98	(14.3)	139	(14.9)	28.4	(8.1)	25.3	(6.4)	11.3	(16.0)	5.1	(18.2)
Phenylalanine	164	(4.2)	125	(4.2)	148	(4.0)	91	(4.4)	28	(4.1)	41	(4.4)	0.5	(0.1)	1.0	(0.3)	0.7	(1.0)	0.3	(1.1)
Ornithine	20	(0.1)	15	(0.5)	22	(0.6)	n.d.	n.d.	n.d.	n.d.	n.d.	n.d.	41.3	(11.8)	46.4	(11.8)	4.2	(6.0)	1.0	(3.6)
Glutamic acid[b]	392	(10.1)	284	(9.5)	342	(9.3)	199	(9.7)	65	(9.5)	88	(9.4)	215.6	(61.8)	224.0	(56.9)	28.7	(40.4)	11.2	(40.0)
Lysine	324	(8.4)	211	(7.1)	300	(8.1)	57	(2.8)	20	(2.9)	23	(2.5)	6.3	(1.8)	18.2	(4.6)	3.0	(4.2)	0.4	(1.4)
Tyrosine	147	(3.8)	91	(3.0)	117	(3.2)	20	(1.0)	7	(1.0)	10	(1.1)	1.5	(0.4)	2.3	(0.6)	1.0	(1.4)	0.6	(2.2)
α-aminoadipic acid	n.d.	n.d.	n.d.	n.d.	n.d.	n.d.	n.d.	n.d.	n.d.	n.d.	n.d.	n.d.	3.6	(1.0)	6.6	(1.6)	5.6	(7.9)	1.2	(4.3)
Total amino acid	3874	n.d.	2992	n.d.	3688	n.d.	2054	n.d.	686	n.d.	935	n.d.	349		394		71		28	

[a]The culture medium contained 7.5 or 0.075 mM $(NH_4)_2SO_4$ as sole nitrogen source.
[b]Asparagine and glutamine are converted respectively to aspartic acid and glutamic acid during extraction and derivatization procedures.
Analysis was by GLC. Figures are the mean of two replicates. n.d., not detected.

of lysine in nitrogen-rich mycelial protein. A storage protein which is a lectin has recently been characterized from *Arthrobotrys oligosporus* mycelium, and accumulates following successful capture and digestion of nematodes (Rosen *et al.*, 1997). If similar storage proteins occur in *S. lacrymans*, the mycelium would have two amino acid based nitrogen stores, a protein depot and a more mobile vacuolar pool.

Discontinuity of nutrient supply could regulate foraging development through local variations in nitrogen metabolites. From the evidence cited above it is clear that nitrogen metabolites can act as morphogens as well as nutrients. This makes adaptive sense because nitrogen is both essential for growth and in short and unpredictable supply in the environment. The amount of it which the organism has sequestered is a measure of its capabilities for growth in a carbon-rich environment, and also an indicator of successful exploitation of a wood food source. Because nitrogen is translocated, there is moreover the possibility that indicator metabolites as well as nutrients for growth may be moved through the mycelium. Intracellular nitrogen enrichment following contact with a fresh food source could be part of the stimulus which is observed to induce branching from bridging hyphae. Differential nitrogen enrichment in different hyphae of a mycelium will result in discontinuities of nitrogen metabolism, which in turn could induce developmental pathways locally.

Nitrogen metabolism in relation to the coordination of development in a large mycelial system

The following account of the processes of nitrogen uptake and metabolism in fungi is intended to clarify the ways in which environmental discontinuities in nitrogen supply might affect the intracellular nitrogen status of mycelium. There have recently been major advances in our understanding of fungal nitrogen uptake and metabolism and their regulation, in the nature of subcellular compartmentation, and in the cellular processes of vesicle trafficking and signal transduction. These have come mainly from those fungi for which molecular genetic technology has been developed, mainly *A. nidulans, N. crassa* and *S. cerevisiae.* Findings which apply to all these three can be cautiously extended to other fungi in the case of primary metabolic and uptake processes, but are not safely applicable to regulatory processes, in which major differences between fungal species have been found. It is therefore important that molecular technology is developed for the study of processes regulating uptake, metabolism and compartmentalization in Basidiomycete filamentous fungi.

Processes of nitrogen metabolism which could contribute to differential intracellular levels in mycelium of cord formers in natural environments are reviewed below.

Nitrogen acquisition from the environment by extracellular hydrolases

Nutrient acquisition has an extracellular phase, which may require secretion of extracellular hydrolases, followed by uptake across the wall and plasma membrane. Clearly mycelial systems such as those of basidiomycete cord formers, which exploit volumes of many cubic metres in natural habitats by means of microscopic exploratory filaments, must deploy and regulate a sophisticated array of extracellular processes and membrane transport systems as they develop within their highly heterogeneous environments. Environmental sources of nitrogen for cord formers are soil nitrate and ammonium, and macromolecules including nucleic acids, mucopolysaccharides, proteins and glycoproteins. Of the extracellular hydrolases which enable fungi to access these nitrogen sources, proteinases are the best investigated (North, 1982; Kalisz, Wood & Moore, 1987; Beynon & Bond, 1994; Burton, Hammond & Minamide, 1994; Micales, 1992). Extracellular proteolytic activity of *S. lacrymans* is repressed by ammonium and induced by bovine serum albumin and peptone medium (Wadekar, 1995). Two extracellular proteinases have been characterized (Wadekar, North & Watkinson, 1995). The one most likely to be responsible for the digestion of wood protein is a peptone-induced pepstatin-sensitive aspartic proteinase with optimal activity at pH 4.0 or below, and secreted into the medium at all stages of growth. An activity with these characteristics can be detected by azocaseinase assay of culture medium, and is probably identical to the activity found by SDS-PAGE of culture medium samples on gelatin-containing gels, as a band with molecular weight 30 000 kDa. Interestingly, although the mycelium contains similar azocaseinase activity, extracts of mycelium do not give this band on gels. Further investigation suggested that the intracellular form might be different from the secreted form of this proteinase. The other secreted proteinase is primarily intracellular but leaks out in large amounts to the extracellular medium. Termed S1, it has the characteristics of a serine proteinase, being partly inhibited by 1,10-phenanthroline and most active around neutral pH. It becomes more abundant in the medium as the colony ages. It is the predominant intracellular proteinase in mature mycelial systems, and is induced by starvation (Wadekar, 1995; Wadekar *et al.*, 1995). In liquid culture it appears to leak from the senescent mycelium.

Different species of fungi differ in the localization of proteinase secretion in the mycelial system. Transferring whole colonies on semi-permeable membranes on to skimmed milk agar produced localized zones of clearing in the agar and demonstrated that, in *S. lacrymans* and some other brown-rot fungi, extracellular proteolytic (caseinase) activity in the external medium was greatest in the central oldest part of the colony (Venables & Watkinson, 1989*b*). It is likely that S1 is responsible for this as it leaks from older, starved mycelium, whereas the other extracellular proteinase detected, S4, is secreted from actively growing mycelium and is induced by protein and peptone under nutrient-rich conditions. When whole colonies on membranes were transferred to split plates with nutrient-rich and nutrient-poor agar media on each side, total intracellular proteinase (azocaseinase) activity was repressed on the starved side (Wadekar, 1995). Other fungi tested had different extracellular proteinases and they were differently located in whole colonies. For example, culture medium of the white-rotting, non-cord-forming, wood decay basidiomycete *Coriolus versicolor* gave five separate proteinase bands on gelatin-containing gels (Wadekar *et al.*, 1995), and the site of maximum proteinase activity in solid medium was in a narrow band just behind the colony margin (Venables & Watkinson, 1989*b*). Thus not only does the activity and type of extracellular proteinases vary from species to species, but also the way in which they are deployed spatially within the mycelial system. It is not unexpected that extracellular proteinases differ between species, since diversification to exploit many nutritional niches is a feature of the fungal kingdom. Wood-decay fungi are known to acquire nitrogen by a variety of mechanisms from a wide range of different sources (Carroll & Wicklow, 1992).

The site of uptake in the mycelium is likely to be among the features which differ between species adapted to different modes of life, but this is poorly understood at the physiological level in naturally growing fungi. Uptake is generally assumed to take place mainly at hyphal tips. Enhanced nutrient supply to parts of colonies may induce local branching, increasing the number of hyphal tips available for taking up nutrients. This has been demonstrated in some saprotrophic cord-forming Basidiomycetes adapted to exploit discontinuous nutrient resources.

These fungi must be capable of opportunistic uptake at any point in the mycelial system, and it has been shown that nutrient-rich patches in the substratum can stimulate localized hyphal branching in such fungi, for example in phosphorus and nitrogen-rich patches encountered by mycelium of *Stropharia caerulea* (Donnelly & Boddy, 1997). Such observations confirm the view that new tips are needed for effective uptake.

Uptake of soluble nitrogen compounds from the hyphal environment

Soluble sources of nitrogen utilized by fungi include ammonium and nitrate ions, amino acids, purines, pyrimidines and amides. The processes involved have been well-reviewed recently (Horak, 1986; Garrill, 1994; Griffin, 1994). Ammonium uptake is very poorly understood in spite of the fact that ammonium is both abundant in ecosystems and central to nitrogen metabolism. In *S. cerevisiae* and *N. crassa* it shows dual kinetics with a K_m of 7 μM and 8.5 mM and transport of radiolabelled methylammonium ion, which is thought to be via the same system, was found to occur by an active process with a repressible high-affinity carrier. The evidence for a specific carrier for nitrate ions is equivocal, and nitrate may simply diffuse into the hypha down the gradient produced by the intracellular activities of nitrate and nitrite reductases. Amino acid uptake is via permeases, membrane spanning proteins which operate by symport of amino acids with protons across the plasma membrane. The *N. crassa* neutral and aromatic amino acid permease has been cloned and sequenced. Unlike the animal and plant permeases that have been described, fungal permeases only transport amino acids into the cell, not out (Horak, 1986). Uptake is by symport with protons, with proton/amino acid stoichiometry of 2/1. Amino acid uptake systems exist for all protein amino acids and vary in number and specificity. *S. cerevisiae* has 16 and *N. crassa* five, exemplifying the differences which exist between differently adapted fungi. In *N. crassa* three are constitutive: the neutral and aromatic amino acid transporter, the basic amino acid transporter, and one which carries all amino acids except proline. There are inducible transporters for acidic amino acids, and one for methionine, but these are only active under nitrogen starvation conditions (Garrill, 1994).

Regulation of uptake

The numerous separate uptake systems for different nitrogen compounds are highly regulated. The preferred nitrogen sources for most fungi are ammonium and glutamate, and if these are present many other systems for the uptake and metabolism of other nitrogen compounds are inactive, owing to repression, including amino acid catabolic pathways, nitrate and nitrite reductases (Marzluf, 1993), and amino acid permeases (Griffin, 1994). In most cases there is an obvious energetic advantage in the selectivity, but the order of preference cannot be easily explained in all cases. For example, *S. cerevisiae* has a long sequence of preferences amongst the amino acids (Slaughter, 1988). The mechanisms of regulation include regulation of the expression of permease genes by repression and induction, inhibition of permease activity by the permease substrates including trans-inhibition across the plasma membrane, and possibly efflux mechanisms. In regulating nitrogen uptake by hyphae which form part of spatially extensive mycelial systems, internal compartmentation and the intracellular processes which maintain it probably play an important part in regulating the uptake of nitrogen compounds across the plasma membrane. However, at present this remains a speculation, since almost all recent knowledge of metabolite storage and transport in subcellular compartments comes from yeast, with *N. crassa* the best investigated filamentous fungus. Cytoplasmic compartments which regulate their amino acid content are the vacuoles and vesicles, the mitochondria and the cytosol (Sanders, 1990; Griffin, 1994; Markham, 1994). The vacuoles are repositories of the basic amino acids arginine, ornithine and lysine under luxury nitrogen supply (see below). Metabolism of the two amino acids arginine and proline is partitioned between the cytosol and mitochondria. The significance of subcellular compartments in metabolism is discussed below.

Nitrogen metabolism and its regulation

The central intermediate for nitrogen metabolism in all fungi is ammonia. Most fungi can assimilate nitrate by reduction to ammonia, and the usual assimilatory pathway is then formation of glutamate via NADP-glutamate dehydrogenase-catalysed combination with oxoglutarate. The 'GOGAT' pathway with assimilation of ammonium forming glutamine and subsequent transfer of the amino group is also common in fungi but is not the main route for ammonium assimilation in most species (Griffin,

1994; Van Laere, 1994). Glutamate supplies nitrogen for all the nitrogen-requiring biosynthesis of the cell. All 20 amino acids are synthesized and their synthetic and catabolic pathways are catalysed by different sets of enzymes. Proline is an exception, with much of the catabolic and synthetic pathway enzymes in common, but synthesis from glutamate, using NADPH, is in the cytosol and catabolism in the mitochondrion, producing NAD. Proline and glutamate can both pass across the mitochondrial membrane in either direction, so there is the potential for transfer of electrons into mitochondria during proline catabolism, as described by Griffin (1994). This prompts the speculation that proline metabolism could have a role in the redox sensing, which is a key feature of mycelial development. Proline as sole nitrogen source produced alterations in the morphology of coremia of *P. claviforme*, resembling those seen in oxygen-starved cultures (S. C. Watkinson, unpublished observation). Proline as sole nitrogen source for *S. cerevisiae* induced the pseudofilamentous form of growth (Gimeno *et al.*, 1992). Proline is known to have regulatory roles in metabolism (Phang, 1985) and to act as an osmoticant (Samaras *et al.*, 1995). Arginine and ornithine metabolism, like that of proline, form part of a cycle which is partitioned between mitochondrion and cytosol.

The genetic regulation of nitrogen metabolism is a fast-moving research field and a sensitive and highly complex set of controls is being revealed (Marzluf, 1993). These operate to ensure efficient use of nitrogen and maintain cellular supplies of all the essential nitrogen monomers for cell growth and development within the constraints of unpredictable environmental supply.

Most recent knowledge concerning the genes which regulate nitrogen metabolism is from gene cloning and transformation experiments with *A. nidulans*, *N. crassa* and *S. cerevisiae*. A picture is developing of large families of regulatory genes encoding proteins which act as regulators of transcription and translation of the different sets of enzymes which operate specific pathways. These act hierarchically upon each other. Pathways of nitrate assimilation and amino acid catabolism are subject to repression by ammonium, glutamate and glutamine, the preferred nitrogen sources, and to derepression by substrates including nitrate. This system is the result of the combined operation of a globally acting gene which senses nitrogen starvation, and of many pathway-specific genes such as those for nitrate and nitrite reductase, and for some amino acid and purine catabolic enzymes. These pathway-specific genes are under the positive control of the globally acting gene and their transcription is activated by the special DNA-binding protein which it encodes. The globally acting genes in *N. crassa*, *NIT2*, and *A. nidulans*, *AREA*, have been cloned and found to contain a region of 50 residues with 98% homology. This region includes a 17-amino acid DNA-binding zinc-finger motif. In the context of environmental regulation of development it is interesting that AREA protein of *A. nidulans* has sequence homology with a transcriptional activator of glutamine synthetase of yeast, gene *GLN3* encoding protein GLn3p, which senses intracellular glutamine starvation and activates the gene *GLN1* encoding the enzyme glutamine synthetase. There is also strong homology with the GATA-binding proteins mediating tissue-specific gene expression during mammalian erythropoiesis (Marzluf, 1993).

Amino acid synthesis, as well as catabolism, shows the same feature of both general, and pathway-specific control. Starvation for just one of several amino acids (induced experimentally with pathway-specific mutants or chemical inhibitors of a particular pathway) results in activation of biosynthetic pathways for many different amino acids. However, if one amino acid is present in excess, transcription of enzymes of its biosynthetic pathway is specifically repressed. The result is to maintain a balance of all 20 protein amino acids (Griffin, 1994; Hinnebusch, 1988).

Cellular compartmentation of nitrogen compounds

Nitrogen is acquired and assimilated by the processes of uptake and metabolism described, which act with great efficiency to maintain availability of all the amino acids required for biosynthesis. Amino acids and other nitrogen resources are not uniformly distributed in the cell, but are localized in compartments with different degrees of accessibility to metabolism. Long-term reserves are storage proteins and insoluble nitrogen compounds such as chitin. Shorter-term reserves, which may be more spatially mobile, are held in vacuoles, and readily available amino acids accumulate in the cytosol. All these compartments are opportunistically filled up when supply allows. Under starvation conditions supplies are released from protein and vacuolar stores into the cytosol.

The nitrogen status of particular regions in a large mycelial unit must depend not only on rates of uptake and metabolism, but also on internal partitioning and mobility. Vacuoles and cytosol are the main repositories of free soluble intracellular amino acids.

Vacuoles were once thought to be spaces in the hyphae from which cytoplasm had been lost, or at best, lysosome-like bodies with the role of sequestering hydrolases within the cell until they were required for autolysis. They are now recognized to have a more complex and dynamic role. The following description is mainly drawn from reviews by Garrill (1994), Markham, (1994) and Klionsky, Herman & Emr (1990), in which most of the findings mentioned in this section are cited. Vacuoles act as stores of nitrogen in the form of basic amino acids arginine, ornithine and lysine, concentrated by a number of specific transport proteins in the tonoplast. The vacuoles of yeast (Wiemken & Nurse, 1973) and *N. crassa* (Weiss, 1973) store arginine, the most nitrogen-rich amino acid. Most of the amino acids accumulated in vacuoles reach up to 40 times their cytosolic concentrations. They vary much more in response to environmental nitrogen supply and growth conditions than does cytosolic amino acid (Ohsumi & Anraku, 1981). Arginine constitutes 25–30% of the basic amino acid pool in *S. cerevisiae* (Durr *et al.*, 1979) and *N. crassa* (Weiss & Davis, 1977) and up to 85% when arginine is the sole nitrogen source. A similar system of vacuolar accumulation of basic amino acids appears to operate in *S. lacrymans*, where there is a differential increase of ornithine and lysine as the mycelium ages on high and low nitrogen media (Venables & Watkinson, 1989*a*; Table 6.3).

The mechanism of vacuolar amino acid transport has been well investigated in a few species. Amino acid transport proteins have been identified in vesicles of vacuolar membrane from yeast (Ohsumi & Anraku, 1981; Sato, Ohsumi & Anraku, 1984; Kitamoto *et al.*, 1988), which operate proton antiport systems for arginine, arginine–lysine, histidine, phenylalanine–tryptophan, glutamine–asparagine, and isoleucine–leucine. There are similar but fewer tonoplast amino acid permeases in *N. crassa* (Subramanian, Weiss & Davis, 1973; Davis, 1986; Garrill, 1994), indicating the likelihood of major differences between different fungal species. Vacuolar amino acid accumulation systems are driven by a proton gradient generated by a tonoplast ATPase, which in yeast differs from the plasma membrane and from the mitochondrial ATPases (Klionsky *et al.*, 1990). The tonoplast amino acid transporters have a higher K_m than does the plasma membrane, measured in yeast vacuolar preparations. This is likely to reflect the fact that plasma membrane permeases may have a scavenging role, collecting amino acids released around the hypha by extracellular proteolysis of limited amounts of protein from wood, while tonoplast permeases act to concentrate amino acids to high levels in vacuoles from an already rich source in the cytoplasm. Amino acids are released from the vacuole into the cytoplasm during nitrogen starvation. The mechanism is poorly understood, but may be active, as it is decreased by respiratory inhibitors. It is highly relevant in the role of vacuoles as mobile stores of amino acids and deserves further investigation. Fungal vacuolar pH, measured by pH-sensitive fluorochromes, is around 6.0. Vacuoles also accumulate polyphosphate and sequester cations including K^+, Ca^{2+}, Cu^{2+}, Mn^{2+}, and Fe^{2+} (Gadd, 1994; Garrill, 1994).

Vacuoles, like mammalian lysosomes, contain several hydrolytic enzymes. Their molecular genetics, synthesis, processing and cellular sorting have been very thoroughly described (reviewed by Klionsky *et al.*, 1990). They are predominantly proteinases and peptidases. Their levels vary with growth stage and nutrient supplies. They are derepressed under nitrogen and glucose starvation and reach their maximum levels as the cells approach stationary phase. They are believed to have a role in the degradation and reutilization of peptides and proteins. Yeast carboxypeptidase Y is the best investigated. Synthesized on ribosomes in the cytosol as an inactive precursor it is translocated into the vacuole directed by an N-terminal leader sequence, which is later cleaved. During processing an oligosaccharide component is added. The enzyme is activated in the vacuole by other proteinases, which specifically cleave it to remove a propeptide sequence. This feature is shared by other vacuolar hydrolases. Other vacuolar peptidases include a metallopeptidase which shares some characteristics with an intracellular peptidase activity from *S. lacrymans* (see below): both are active against leucine substrates and are metalloproteinases activated by Zn^{2+}.

On the basis of regulatory mechanisms so far described, a plausible sequence of events can be envisaged by which intracellular compartmentation could be affected by the extracellular environment. The ability to store different amino acids in vacuoles is clearly a result of the insertion of appropriate transport proteins into the vacuolar membrane. The analysis by molecular genetics of the yeast secretory system has provided evidence that the sorting of proteins and correct placing in the cell is a result of their correct structure, thus genes determine the cellular location of their products (Klionsky *et al.*, 1990). The production of a variety of vacuoles and vesicles with different metabolite transport properties could thus result from the modulation of gene expression. Environmental nutrient status could influence levels of intracellular amino acids in compartments in specific ways, via effects on uptake characteristics of compartments produced by regulating expression of transport proteins. These compartmented amino acids could act locally and specifically in development, by modulating the expression of developmental genes such as *brlA* in *A. nidulans (*Adams *et al.*, 1992). The long-distance mobility of intracellular nitrogen stores in basidiomycetes could enable local activation of developmental genes by means of alterations in local amino acid level brought about by the controlled import of vacuoles charged with amino acids elsewhere in the mycelial system.

Mechanisms for long-distance translocation of nitrogen within mycelium as amino acids

Translocation processes must exist, as discussed above, to account for bidirectional translocation, and movement of substances in mycelium in the absence of, or up, a water potential gradient. Vesicles have long been known to be moved by cytoskeletal motors and to carry cell components and wall-synthesizing enzymes to hyphal tips and points of incipient branch formation (Gooday, 1994). However, it is only recently that vacuoles have been recognized as a potential translocation pathway within mycelium. Ashford and her associates have now shown, using fluorochromes as vital stains (Butt *et al.*, 1989; Stewart & Deacon, 1995) for light microscopy of translocating hyphae, that motile systems of vacuoles and connecting microtubules are present in all major phyla of fungi (Shepherd, Orlovich & Ashford, 1993*a*,*b*; Rees, Shepherd & Ashford, 1994). In *Pisolithus tinctorius*, movement is fast enough to account for observed nutrient translocation rates, and it appears to be driven by cytoskeletal motors which are microtubule-dependent (Hyde, Perasso & Ashford, 1996). These discoveries suggest the hypothesis (Ashford, 1998) that amino acid stores in mobile vacuolar compartments could be accumulated and delivered in different parts of a mycelial system.

In a search for similar, potentially translocating subcellular compartments in *S. lacrymans*, mycelia grown over glass slides in the microcosm system described above (Fig. 6.1), were treated with various fluorochromes (Mason, 1993) as well as the pH-specific dye 2′7′-*bis*-(2-carboxyethyl)-5-(-6-)-carboxyfluorescein. Compartments of various sizes became visible, including large vacuoles which were seen moving unidirectionally, along with small vesicles moving randomly in the cytoplasm. With a view to making visible the vacuoles accumulating basic amino acids, we synthesized a fluorochrome-linked amino acid to study its localization (S.C. Watkinson & T. Pullen, unpububublished observations). Ornithine was chosen as representative of the vacuolar amino acid pool in *S. lacrymans*, and a compound designed to track ornithine-containing compartments was synthesized with ornithine covalently linked to the fluorochrome 7-nitrobenz-2-oxa-1,3-diazo-4-yl (NBD). Hyphae at the distal end of a hyphal bridge in glass slide microcosms were fed with ornithine-NBD. Fluorescence appeared in the vacuoles of some of the hyphae extending over glass, although it is not yet clear that this is due to the presence of the entire ornithine-NBD complex, or if the mycelium degraded the complex after uptake. Movement of fluorescent vacuoles was seen and the rate measured for about 10 minutes. Nocodazole had a significant effect on the initial rate of vacuolar movement, measured over 10 minutes after feeding, indicating a role for microtubules in the movement observed. This evidence, although at present unconfirmed, suggests that vacuoles of the kind described by Ashford and found to be capable of translocation, are found as expected in *S. lacrymans*.

Localized hyphal death, the *S. lacrymans* reallocation response, and intracellular proteinases

The three components of the reallocation response initiated by the contact of part of a mycelium with a fresh nutrient source are the propagated stimulus to hyphal branching, the stimulus to cord formation in connecting hyphae, and the regression of non-connecting mycelium by the selective death of those hyphae. Local regression is part of the process of nitrogen recycling with adaptive value to wood-decay fungi (Watkinson, 1984*a*,*b*; Lilly, Wallweber & Higgins, 1991). How is it activated? Autolysis in response to starvation is well-known. This process however is elicited not merely by nutrient limitation, but by a signal from the mycelium. It is clearly to the advantage of an organization to reallocate resources from unproductive units, especially if other units are known to be able to use the resources more productively. Lysis of hyphae which have not made fruitful connections with new resources is induced when a new resource is being exploited by another part of the same mycelial organization. What is sensed in order to activate lysis is likely to be relative nitrogen poverty compared with the rest of the mycelium, rather than simple local nitrogen limitation. We do not know how this information is conveyed, or how it elicits selective hyphal lysis. To investigate the mechanism of the signal pathway, one approach is to try to quantify an early response, as an assay for the signal.

Intracellular proteinases are of particular interest because they are required in this process, must be activated by the signal which instructs the mycelium to return its components to the more successful part of the system for reuse, and because they are molecules which can be assayed to measure and characterize the cell death response and the signal which elicits it. In *S. lacrymans* they have been investigated by SDS-PAGE and by the use of specific substrates. Three intracellular proteinase activities are known for *S. lacrymans*, from gelatin-containing SDS gels, termed S1, S2, and S3, characterized by different pH ranges and inhibitor sensitivities (Table 6.2). S1 has a wide pH range, its activity is enhanced in the presence of Zn^{2+} and it is more active in older and starved cultures (Wadekar, 1995). There are in addition two intracellular metalloproteinase activities, both inhibited by EDTA and phosphoramidon, of molecular mass 43 000 and 47 000 kDa respectively, and which are present in young, growing mycelium. Using chromogenic specific substrates to differentiate proteinase activities according to protocols developed by Burton *et al.* (1993, 1997) for the investigation of proteinases of *Agaricus bisporus,* four proteinase/peptidase activities were revealed in *S. lacrymans* (Watkinson *et al.*, 1996) (Table 6.3). One appeared to be a serine proteinase with optimal pH 6.0, acting on substrate *Succ*-ala-ala-pro-phe-*p-nitroanilide.* It increased in activity on transfer of previously well-grown mycelium to salts-only medium (these conditions are not exactly the same as starvation conditions, because they present the mycelium with a need to reallocate reserves rather than an overall nitrogen limitation, which single-celled organisms such as yeast experience on transfer to salts-only medium). This resembled, in pH range, inhibitor sensitivity, and activation under conditions stimulating reallocation of protein resources, the intracellular serine proteinase of *A. bisporus* (Burton *et al.*, 1993, 1997). Two metalloproteinases active against leu-*p-nitroanilide* and arg-*p-nitroanilide* with pH optima of 8.0 and 8.5 and inhibited by EDTA could be the two metalloproteinase activities S2 and S3, which are also found in actively growing mycelium and are little affected initially when mycelium is transferred to salts-only medium. They may represent enzyme activators similar to the proteolytic enzyme required to activate the biosythetic enzyme chitin synthase (Hanseler, Nylen & Rast, 1983). The activity against leu-*p-nitroanilide* was almost completely abolished by removal of cations in equilibrium dialysis, and

Table 6.2. *Intracellular proteinase and peptidase activities in* S. lacrymans *found with SDS-PAGE on gelatin-containing gels.* (*Data from Wadekar, North & Watkinson, 1995*)

Band	Mol. wt.	Inhibitors	pH	Growth phase
S1	65 000	Phenanthroline	Wide, around 7	Stationary
S2	47 000	Phenanthroline, EDTA, Phosphoramidon	Wide, around 7	Active growth
S3	43 000	As for S2	As for S2	As for S2

Table 6.3. *Intracellular proteinase and peptidase activities found in* S. lacrymans *shown with specific peptide substrates (Data of Watkinson* et al., *1996)*

Substrate	P_i	pH	Inhibitors	Growth phase
Succ-ala-ala-pro-phe-*pNA*	4.3	6.5	PMSF	Transfer to salts-only medium increased activity
Arg-*pNA*	3.4	8.5	?Aminobenzamidine*	Transfer to salts-only had no effect
Leu-*pNA*	n.d.	8.0	EDTA Phenanthroline	As for Arg-*pNA* activity
Benz-phe-val-arg-*pNA*	n.d.	<2.5	None found	Sustained increase on transfer to salts-only

*Activity against Arg-*pNA* was abolished by equilibrium dialysis to remove ions, and quantitatively restored by Mn^{2+} and Co^{2+}.

quantitatively restored by the re-addition of Mn^{2+} over a concentration range between 0 and 200 μg/ml. A fourth activity was found, which resembled none of those found by other methods, and was active against *Benz*-phe-val-arg-*p-nitroanilide*. It was maximally active at very low pH with most activity recorded at the bottom limit of the buffers available, pH 2.5. This activity rose sharply on transfer to salts-only medium, and then decreased rapidly. The low pH required for activity suggested a similarity with the aspartic proteinase S4, which is exclusively extracellular, but unlike this enzyme, activity against *Benz*-phe-val-arg-*p-nitroanilide* was not pepstatin-sensitive. Its location *in vivo* is presumably in lysosome-like compartments. *S. lacrymans* is remarkable even among the other brown-rot basidiomycetes for its rapid autolysis, elicited by starvation or by growth on nitrate as sole nitrogen source, and also for the differentiation of empty 'vessel' hyphae at an early stage of mycelial growth, suggesting a capacity to produce a type of hypha which can die in a rapid and controlled manner. Burton *et al.* (1994) found no comparable activity in their investigation of *A. bisporus* intracellular proteinases, suggesting a specific adaptive role for this enzyme in *S. lacrymans*.

In summary, the serine-type proteinase activity which degrades the substrate *Succ*-ala-ala-pro-phe-*pNA* appears to be the intracellular proteinase of choice for use as a marker of the reallocation response; it possesses the desired characteristics of activation on transfer to starvation media, and a specific assay for it is now available. It may be possible in future to use the transcription of this protein as a way of quantifying and locating reception of the long-distance signal which elicits the reallocation response. This could be done by using the strategies devised by Thurston and associates for analysis of the expression of the cellulase genes in *A. bisporus* (Chow *et al.*, 1994; Armesilla, Thurston & Yague, 1994).

The evidence for long-distance signal propagation and its possible mechanism

The remote response of selective hyphal lysis, perhaps akin to apoptosis in animals, is triggered by a signal which must be transmitted over long distances and take effect at specific sites. The mechanism by which information is sent through mycelium to coordinate development is unknown. Olsson & Hansson (1995 and Chapter 2), recently reported an action potential in fungal cords and rhizomorphs which is sensitive to stimulation by application of nutrients. They found a spontaneous action potential in mycelium of both *Pleurotus ostreatus* and *Armillaria bulbosa* (the latter famous as being the 'largest and oldest living organism' (Smith *et al.*, 1992). It has also been found in *S. lacrymans* (Olsson, personal communication). With a resting frequency of 0.5–5 Hz, and amplitude of 5–50 mV, the frequency was sharply increased 30 S after a drop of 2% malt agar, or a block of fresh beech wood (but not a Perspex block) was added 1–2 cm from the electrode, i.e. the velocity of signal propagation was about 0.5 mm per second and the response was elicited by nutrients. While the significance of electrical field around hyphal tips has been intensively investigated (Gow, 1994), intracellular long-distance action potentials in large mycelial units have not been previously observed.

Conclusions

Rayner (Rayner *et al.*, 1994; Rayner, 1997) has developed the concept of fungal mycelium as a large individual organism able to undergo a self-regulating and modular type of development (see also Chapter 1). On the available evidence, coordinated development of the mycelial unit is regulated by a complex set of interdependent decisions made as single hyphae continuously encounter the outside world, without a multicellular body to maintain homeostasis as in plants and animals. Fungi can only embark successfully on committed multihyphal developmental programmes when sufficient reserves have accumulated in the mycelium for their completion. Intracellular nitrogen levels, sensed by proteins like the global transcription regulator AREA, may signal availability of reserves for development. It is more important to sense nitrogen than carbon because the latter is far more abundant in the fungal environment.

As mycelium encounters variable nutrient supply in the environment, intracellular nitrogen reserves will vary in composition, localization and amount, as a result of the processes described in detail above. These intracellular changes may travel short distances by vacuolar movement, and also stimulate long-distance electrical signals. The whole mycelium could thus behave in a coordinated manner without central control, responding in a modular manner to its unpredictable environment.

The mechanism of long-distance signalling, and the role of the action potential, remains obscure. Olsson has suggested (personal communication) that electrical signalling observed from hyphae in contact with nutrients could be required to inhibit apoptosis, so that non-bridging parts of the mycelium, receiving an insufficient electrical impulse, die back. Electrical signalling could transmit information about the size of a food resource being colonized if the signal is proportional to nutrient concentration or to the number of hyphae signalling. It is, however, possible that the electrical potential in cords is merely a side-effect of nutrient-induced processes. Electrical signalling does not preclude the possibility that hydrostatic pressure differentials could also act within mycelium to transmit information and coordinate development.

The concept of intracellular nitrogen status as a transmissible signal for development is compatible with the hypothesis that 'radical, adaptive shifts in mycelial pattern can be explained by purely contextual, rather than genetic, changes' (Davidson *et al.*, 1996). The contextual change producing the reallocation response in cord-forming basidiomycetes is localized contact with a fresh food supply. Responsiveness of the whole system to nutrient encounters at its boundary can be plausibly envisaged as mediated by the reactions of known cellular systems to the extracellular environment.

Acknowledgements

I thank colleagues at the Department of Plant Sciences, University of Oxford, and John Brookes University, Oxford, for help with the use of fluorochromes, and Horticulture Research International for generous help with facilities during a sabbatical. Dr M.J. Carlile and Professor C.F. Thurston kindly read and advised on the manuscript, and Professor Thurston wrote the Introduction.

References

Adams, T. H., Hide, W. A., Yager, L. N. & Lee, B. N. (1992). Isolation of a gene required for programming the initiation of development by *Aspergillus nidulans*. *Molecular and Cellular Biology*, **12**, 3827–33.

Andrews, J. H. (1995). Fungi and the evolution of growth form. *Canadian Journal of Botany*, **73**, S1206–12.

Armesilla, A. L., Thurston, C. F. & Yague, E. (1994). CEL 1: a novel cellulose-binding protein secreted by *Agaricus bisporus* during growth on crystalline cellulose. *FEMS Microbiology Letters*, **116**, 293–9.

Arnebrant, K., Ek, H., Finlay, R. D. & Soderstrom, B. (1993). Nitrogen translocation between *Alnus glutinosa* (L.) Gaertn seedlings infected with *Frankia* sp. and *Pinus contorta* Doug. Ex Loud seedlings connected by a common ectomycorrhizal mycelium. *New Phytologist*, **130**, 231–42.

Ashford, A. E. (1998). Dynamic pleiomorphic vacuole systems: are they endosomes and transport compartments in fungal hyphae? *Advances in Botanical Research*, **28**, 121–59.

Beynon, R. J. & Bond, J. S. (1994). *Proteolytic Enzymes*. Oxford: Oxford University Press; IRL Press. The Practical Approach Series, eds. D. Rickwood & B. D. Hames.

Blumer, K. J. & Johnson, G. L. (1994). Diversity in function and regulation of MAP kinase pathways. *Trends in Biochemistry*, **19**, 236–40.

Boddy, L. (1993). Saprotrophic cord-forming fungi: warfare strategies and other ecological aspects. *Mycological Research*, **97**, 641–55.

Boddy, L. & Watkinson, S. C. (1995). Wood decomposition, higher fungi, and their role in nutrient distribution. *Canadian Journal of Botany*, **73**, S1377–83.

Brownlee, C. & Jennings, D. H. (1982*a*). Long distance translocation in *Serpula lacrimans*: velocity estimates and the continuous monitoring of induced perturbations. *Transactions of the British Mycological Society*, **79**, 143–8.

Brownlee, C. & Jennings, D. H. (1982*b*). The pathway of translocation in *Serpula lacrimans*. *Transactions of the British Mycological Society*, **79**, 143–8.

Burton, K. S., Hammond, J. B. W. & Minamide, T. (1994). Protease activity in *Agaricus bisporus* during periodic fruiting (flushing) and sporophore development. *Current Microbiology*, **28**, 275–8.

Burton, K. S., Partis, M. D., Wood, D. A. & Thurston, C. F. (1997). Accumulation of serine proteinase in senescent sporophores of the cultivated mushroom, *Agaricus bisporus*. *Mycological Research*, **101**(2), 146–52.

Burton, K. S., Wood, D. A., Thurston, C. F. & Barker, P. J. (1993). Purification and characterisation of a serine proteinase from senescent sporophores of the commercial mushroom (*Agaricus bisporus*). *Journal of General Microbiology*, **139**, 1379–86.

Butler, G. M. (1957). The development and behaviour of mycelial strands in *Merulius lacrymans* (Wulf.) Fr. I. Strand development during growth from a foodbase through a non-nutrient medium. *Annals of Botany*, **21**, 523–37.

Butler, G. M. (1958). The development and behaviour of mycelial strands in *Merulius lacrymans* (Wulf.) Fr. II. Hyphal behaviour during strand formation. *Annals of Botany*, **22**, 219–36.

Butt, T. M., Hoch, H. C., Staples, R. C. & St. Leger, R. J. (1989). Usc of fluorochromes in the study of fungal cytology and differentiation. *Experimental Mycology*, **13**, 303–20.

Cairney, J. W. G., Jennings, D. H. & Veltkamp, C. J. (1989). A scanning electron microscope study of the internal structure of mature linear mycelial organs of four basidiomycete species. *Canadian Journal of Botany*, **67**, 2266–71.

Carlile, M. J. (1994). The success of the hypha and mycelium. In *The Growing Fungus*, ed. N. A. R. Gow & G. M. Gadd, pp. 3–19. London: Chapman & Hall.

Carroll, G. C. & Wicklow, D. T. (1992). *The Fungal Community: Spatial Organisation and Community Structure*. New York: Marcel Dekker.

Chow, C. M., Yague, E., Raguz, S., Wood, D. A. & Thurston, C. F. (1994). The CEL 3 gene of *Agaricus bisporus* codes for a modular cellulase and is transcriptionally regulated by the carbon source. *Journal of Applied and Environmental Microbiology*, **94**, 2779–85.

Davidson, F. A., Sleeman, B. D., Rayner, A. D. M., Crawford, J. W. & Ritz, K. (1996). Context-dependent macroscopic patterns in growing and interacting mycelial networks. *Proceedings of the Royal Society of London, Series B*, **263**, 873–80.

Davis, R. H. (1986). Compartmental and regulatory mechanisms in the arginine pathways of *Neurospora crassa* and *Saccharomyces cerevisiae*. *Microbiological Reviews*, **50**, 280–313.

Dobson, J., Power, J. M., Singh, J. & Watkinson, S. C. (1993). The effectiveness of 2-aminoisobutyric acid as a translocatable fungistatic agent for the remedial treatment of dry rot caused by *Serpula lacrymans* in buildings. *International Biodeterioration*, **31**, 129–41.

Donnelly, D. P. & Boddy, L. (1998). Developmental and morphological responses of mycelial systems of *Stropharia caerulea* and *Phanerochaete velutina* to soil nutrient enrichment. *New Phytologist*, **138**, 519–31.

Donnelly, D. P., Wilkins, M. F. & Boddy, L. (1995). An integrated image analysis approach for determining biomass, radial extent and box-count fractal dimension of macroscopic mycelial systems. *Binary*, **7**, 19–28.

Dowson, C. G., Rayner, A. D. M. & Boddy, L. (1986). Outgrowth patterns of mycelial cord forming basidiomycetes from and between woody resource units in soil. *Journal of General Microbiology*, **132**, 203–11.

Dowson, C. G., Rayner, A. D. M. & Boddy, L (1988*a*). Inoculation of mycelial cord-forming basidiomycetes into woodland soil and litter. I. Initial establishment. *New Phytologist*, **109**, 335–41.

Dowson, C. G., Rayner, A. D. M. & Boddy, L (1988*b*). Inoculation of mycelial cord-forming basidiomycetes into woodland soil and litter. II. Resource capture and persistence. *New Phytologist*, **109**, 343–9.

Dowson, C. G., Springer, P. & Rayner, A. D. M. (1989). Resource relationships of foraging mycelial systems of *Phanerochaete velutina* and *Hypholoma fasciculare*. *New Phytologist*, **111**, 501–9.

Durr, M. K. U., Boller, T., Wiemken, A., Schwenke, J. & Nagy, M. (1979). Sequestration of arginine by polyphosphate in vacuoles of yeast (*Saccharomyces cerevisiae*). *Archives of Microbiology*, **121**, 169–75.

Elliot, M. L. & Watkinson, S. C. (1989). The effect of 2-aminoisobutyric acid on wood decay and wood spoilage fungi. *International Biodeterioration*, **25**, 335–71.

Gadd, G. M. (1994). Signal transduction. In *The Growing Fungus*, ed. N. A. R. Gow & G. M. Gadd, pp. 183–201. London: Chapman & Hall.

Garrill, A. (1994). Transport. In *The Growing Fungus*, ed. N. A. R. Gow & G. M. Gadd, pp. 163–83. London: Chapman & Hall.

Gimeno, C. J., Ljungdahl, P. O., Styles, C. A. & Fink, G. R. (1992). Unipolar cell division in the yeast *Saccharomyces cerevisiae* leading to filamentous growth: regulation by starvation and RAS. *Cell*, **68**, 1077–90.

Gooday, G. W. (1994). Cell membrane. In *The Growing Fungus*, ed. N. A. R. Gow & G. M. Gadd, pp. 163–83. London: Chapman & Hall.

Gow, N. A. R. (1994). In *The Growing Fungus*, ed. N. A. R. Gow & G. M. Gadd, pp. 163–83. London: Chapman & Hall.

Griffin, D. H. (1994). *Fungal Physiology*, 2nd edn. New York: Wiley-Liss.

Hanseler, E., Nylen, L. E. & Rast, D. M. (1983). Isolation and properties of chitin synthase from *Agaricus bisporus* mycelium. *Experimental Mycology*, **7**, 17–30.

Helsby, L. (1976). Structural development of mycelium of *Serpula lacrymans*. D.Phil thesis, University of Oxford.

Hennebert, G. L., Boulenger, P. & Balon, F. (1990). *Le Mérule: Science, Technique et Droit*. Brussels: Editions Ciaco.

Hinnebusch, A. G. (1988). Mechanisms of gene regulation in the general control of amino acid biosynthesis in *Sacharomyces cerevisiae*. *Microbiological Reviews*, **52**, 248–73.

Horak, J. (1986). Amino acid transport in eukaryotic organisms. *Biochimica et Biophysica Acta*, **864**, 223–56.

Hyde, G. J., Perasso, L. & Ashford, A. E. (1996). Motile membranous tubules in fungal hyphae – motility is microtubule-dependent and is not an artefact of dye loading. *Plant Physiology*, **111** (2SS), 809.

Ishikawa, T. (1989). Control of growth and differentiation by cyclic AMP in fungi. *Botanical Magazine of Tokyo*, **102**, 471–90.

Jennings, D. H. (1984). Water flow through mycelia. In *The Ecology and Physiology of the Fungal Mycelium*, ed. D. H. Jennings, & A. D. M. Rayner, pp. 55–79. Cambridge: Cambridge University Press.

Jennings, D. H. (1987). The translocation of solutes in fungi. *Biological Reviews*, **62**, 215–43.

Jennings, D. H. & Bravery, A. F. (1991). *Serpula lacrymans: fundamental biology and control strategies*. Chichester, England: John Wiley.

Jennings, L. & Watkinson, S. C. (1982). The structure and development of mycelial strands in *Serpula lacrimans*. *Transactions of the British Mycological Society*, **78**, 465–74.

Kalisz, H. M., Wood, D. A. & Moore, D. (1987). Production, regulation and release of extracellular proteinase activities in basidiomycete fungi. *Transactions of the British Mycological Society*, **88**, 221–7.

Kitamoto, K., Yoshizawa, K., Ohsumi, Y. & Anraku, Y. (1988). Dynamic aspects of vacuolar and cytosolic amino acid pools of *Saccharomyces cerevisiae*. *Journal of Bacteriology*, **170**, 2683–6.

Klionsky, D. J., Herman, P. K. & Emr, S. D. (1990). The fungal vacuole: composition, function and biogenesis. *Microbiological Reviews*, **54**, 226–92.

Lilly, W. W., Wallweber, G. J. & Higgins, S. M. (1991). Proteolysis and amino acid recycling during nitrogen deprivation in *Schizophyllum commune*. *Current Microbiology*, **23**, 27–32.

Markham, P. (1994). Organelles of filamentous fungi. In *The Growing Fungus*, ed. N. A. R. Gow & G. M. Gadd, pp. 163–83. London: Chapman & Hall.

Marzluf, G. A. (1993). Regulation of sulphur and nitrogen metabolism in fungi. *Annual Review of Microbiology*, **47**, 31–55.

Mason, W. T. (1993). *Fluorescent and Luminescent Probes for Biological Activity*. London: Academic Press.

Micales, J. A. (1992). Proteinases of the brown rot fungus *Postia placenta*. *Mycologia*, **84**, 815–22.

Moore, D. (1994). Tissue formation. In *The Growing Fungus*, ed. N. A. R. Gow & G. M. Gadd, pp. 424–65. London: Chapman & Hall.

North, M. J. (1982). Comparative biochemistry of the proteinases of eukaryotic microorganisms. *Microbiological Reviews*, **46**(3), 308–40.

Ogilvie-Villa, S., DeBusk, R. M. & DeBusk, A. G. (1981). Characterisation of 2-aminoisobutyric acid transport in *Neurospora crassa*. *Journal of Bacteriology*, **147**, 944–8.

Ohsumi, Y. & Anraku, Y. (1981). Active transport of basic amino acids is driven by a proton motive force in vacuolar membrane vesicles of *Saccharomyces cerevisiae*. *Journal of Biological Chemistry*, **256**, 2079–82.

Olsson, S. (1995). Mycelial density profiles. *Mycological Research*, **98**, 143–53.

Olsson, S. & Hansson, B. S. (1995). The action potential-like activity found in fungal mycelium is sensitive to stimulation. *Naturwissenschaften*, **82**, 30–1.

Phang, J. M. (1985). The regulatory functions of proline and pyrroline-5-carboxylic acid. *Current Topics in Cellular Regulation*, **25**, 91–132.

Rayner, A. D. M. (1997). *Degrees of Freedom*. London: Imperial College Press.

Rayner, A. D. M., Griffith, G. S. & Ainsworth, A. M. (1994). Mycelial interconnectedness. In *The Growing Fungus*, ed. N. A. R. Gow & G. M. Gadd, pp. 21–40. London: Chapman & Hall.

Rees, B., Shepherd, V. A. & Ashford, A. E. (1994). Presence of a motile tubular vacuolar system in different phyla of fungi. *Mycological Research*, **98**(9), 985–92.

Rosen, S., Sjollema, K., Veenhuis, M. & Tunlid, A. (1997). A cytoplasmic lectin produced by the fungus *Arthrobotrys oligospora* functions as a storage protein during saprophytic and parasitic growth. *Microbiology*, **143**, 2593–604.

Samaras, Y., Bressan, R. A., Csonka, L. N., Garcia-Rios, M. G., Paino-D'Urzo, M. & Rhodes, D. (1995). Proline accumulation during drought and salinity. In *Environment and Plant Metabolism*, ed. N. Smirnoff. Oxford: BIOS Scientific Publishers.

Sanders, D. (1990). Kinetic modelling of plant and fungal membrane transport systems. *Annual Review of Plant Physiology and Plant Molecular Biology*, **41**, 77–107.

Sato, T., Ohsumi, Y. & Anraku, Y. (1984). Substrate specificities of active transport systems for amino acids in vacuolar membrane vesicles of *Saccharomyces cerevisiae*. *Journal of Biological Chemistry*, **259**, 11505–8.

Shapira, R., Altman, A., Henis, Y. & Chet, I. (1989). Polyamines and ornithine decarboxylase activity during growth and differentiation in *Sclerotium rolfsii*. *Journal of General Microbiology*, **135**, 1361–7.

Shepherd, V. A., Orlovich, D. A. & Ashford, A. E. (1993*a*). A dynamic continuum of pleiomorphic tubules and vacuoles in growing hyphae of a fungus. *Journal of Cell Science*, **104**, 495–507.

Shepherd, V. A., Orlovich, D. A. & Ashford, A. E. (1993*b*). Cell-to-cell transport via motile tubules in growing hyphae of a fungus. *Journal of Cell Science*, **105,** 1173–8.

Slaughter, J. C. (1988). Nitrogen metabolism. In *Physiology of Industrial Fungi*, ed. D. R. Berry. Oxford: Blackwell.

Smith, M. L., Bruhn, J. N. & Anderson, J. B. (1992). The fungus *Armillaria bulbosa* is among the largest and oldest living organisms. *Nature*, **356**, 428–31.

Stewart, A. & Deacon, J. W. (1995). Vital fluorochromes as tracers for fungal growth studies. *Biotechnic and Histochemistry*, **70**, 57–66.

Subramanian, K. N., Weiss, R. L & Davis, R. H. (1973). Use of external, biosynthetic, and organellar arginine by *Neurospora*. *Journal of Bacteriology*, **115**, 284–90.

Trinci, A. P. J. (1995). Pure and applied mycology. *Canadian Journal of Botany*, **73**, S1–S14.

Van Laere, A. (1994). Intermediary metabolism. In *The Growing Fungus*, ed. N. A. R. Gow & G. M. Gadd, pp. 211–31. London: Chapman & Hall.

Venables, C (1988). The nitrogen economy of basidiomycete wood decay fungi. D.Phil thesis, University of Oxford.

Venables, C. E & Watkinson, S. C. (1989*a*). Medium-induced changes in patterns of free and combined amino acids in mycelium of *Serpula lacrymans*. *Mycological Research*, **92**, 273–7.

Venables, C. E & Watkinson, S. C. (1989*b*). Production and localisation of proteinases in colonies of timber decaying basidiomycete fungi. *Journal of General Microbiology*, **135**, 1369–74.

Wadekar, R. V. (1995). Regulation of proteinase activities in basidiomycete wood decay fungi. D.Phil.Thesis, University of Oxford, England.

Wadekar, R. V., North, M. J. & Watkinson, S. C. (1995). Proteolytic enzymes in two wood-decaying basidiomycete fungi, *Serpula lacrymans* and *Coriolus versicolor*. *Microbiology*, **141**, 1575–83.

Watkinson, S. C. (1971). The mechanism of mycelial strand induction in *Merulius lacrymans:* a possible effect of nutrient distribution. *New Phytologist*, **70**, 1079–88.

Watkinson, S. C. (1975*a*). Regulation of coremium morphogenesis in *Penicillium claviforme*. *Journal of General Microbiology*, **87**, 292–300.

Watkinson, S. C. (1975*b*). The relation between nitrogen nutrition and the formation of mycelial strands in *Merulius lacrymans*. *Transactions of the British Mycological Society*, **64**, 195–200.

Watkinson, S. C. (1977). The effect of amino acids on coremium development in *Penicillium claviforme*. *Journal of General Microbiology*, **101**, 269–75.

Watkinson, S. C. (1979). Growth of rhizomorphs, mycelial strands, coremia and sclerotia. In *Fungal Walls and Hyphal Growth*, Second Symposium of the British Mycological Society, ed. J. H. Burnett & A. P. J. Trinci, pp. 93–113. Cambridge: Cambridge University Press.

Watkinson, S. C. (1981). Accumulation of amino acids during the development of coremia in *Penicillium claviforme*. *Transactions of the British Mycological Society*, **76**(2), 231–6.

Watkinson, S. C. (1984*a*). Inhibition of growth and development of *Serpula lacrymans* by the non-metabolised amino acid analogue 2-aminoisobutyric acid. *FEMS Microbiology Letters*, **24,** 247–50.

Watkinson, S. C. (1984*b*). Morphogenesis of the *Serpula lacrymans* colony in relation to its function in nature. In *The Ecology and Physiology of the Fungal Mycelium*. Eighth Symposium of the British Mycological Society, ed. D. H. Jennings & A. D. M. Rayner. Cambridge: Cambridge University Press.

Watkinson, S. C. (1994). The biology and treatment of *Serpula lacrymans*. *Biodeterioration Abstracts* **8**(3).

Watkinson, S. C., Davison, E. M. & Bramah, J. (1981). The effect of nitrogen availability on growth and cellulolysis by *Serpula lacrimans*. *New Phytologist*, **89**, 295–305.

Watkinson, S. C., Wood, D. A. & Burton, K. (1996). Characterisation of developmentally regulated intracellular proteinases of *Serpula lacrymans*. *Abstracts of the 44th Harden Conference of the Biochemical Society*.

Weiss, R. L. (1973). Intracellular localisation of ornithine and arginine pools in *Neurospora*. *Journal of Biological Chemistry*, **248**, 5409–13.

Weiss, R. L. & Davis, R. C. (1977). Control of arginine utilisation in *Neurospora*. *Journal of Bacteriology*, **129**, 866–73.

Wells, J. M. & Boddy, L. (1995). Translocation of soil-derived phosphorus in mycelial cord systems in relation to inoculum resource size. *FEMS Microbiology Ecology*, **17**, 67–75.

Wells, J. M., Hughes, C. & Boddy, L. (1990). The fate of soil-derived phosphorus in mycelial cord systems of *Phanerochaete velutina* and *Phallus impudicus*. *New Phytologist*, **114**, 595–606.

Wiemken, A. & Nurse, P. (1973). Isolation and characterisation of the amino acid pools located within the cytoplasm and vacuoles of *Candida utilis*. *Planta*, **109**, 293–306.

7

Role of phosphoinositides and inositol phosphates in the regulation of mycelial branching

G. D. ROBSON

Introduction

All cells have evolved regulatory mechanisms which allow them to respond to external signals and are essential for cell multiplication and survival. In mammalian cells, an array of signal transduction cascades have been described that respond to growth factors and hormones (Mooibroek & Wang, 1988; Su & Karin, 1996). More recently, highly homologous signalling cascades have been reported in the yeasts *Saccharomyces cerevisiae* and *Schizosaccharomyces pombe*, which are involved in a variety of cellular processes including mating, hyper- and hypo-osmotic sensing, invasive filamentous growth and cell wall integrity (Nishida & Gotoh, 1993; Roberts & Fink, 1994; Waskiewicz & Cooper, 1995; Su & Karin, 1996; Cahil, Janknecht & Nordheim, 1996). The presence of such highly conserved signal transduction pathways suggests that these signal cascades may first have evolved in eukaryotic microbes and have been conserved and adapted during eukaryotic evolution (Janssens, 1987; Kincaid, 1991; Csaba, 1994; Gadd, 1995; Rasmussen *et al.*, 1996).

One of the most highly studied signal transduction pathways in mammalian cells is the phosphoinositide cycle which has been the centre of intense research since the first report that inositol 1,4,5-trisphosphate (Ins(1,4,5)P_3), acts as a second messenger, mobilizing Ca^{2+} from intracellular stores in response to a variety of growth factors, hormones and other ligands (for reviews see Berridge & Irvine, 1984; Nishizuka, 1984; Divecha & Irvine, 1995). In this chapter, current understanding of the structure and function of phosphoinositide signalling in filamentous fungi is reviewed.

Phosphoinositides

The phosphoinositides are a group of minor acidic phospholipids of which phosphatidylinositol (PtdIns) is the principal member, constituting 3–10% of the total phospholipid present in the membrane (Fig. 7.1). PtdIns is synthesized by the membrane-associated phosphatidylinositol synthase from cytidine 5′-phosphate diacylglycerol (CDP-diacylglycerol) and inositol. PtdIns synthase is encoded by a single gene PIS, which when

Phospholipid	% range	Headgroup (X)
Phosphatidylcholine	38 - 65	$-CH_2CH_2N(CH_3)_3$
Phosphatidylethanolamine	19 - 33	$-CH_2CH_2NH_2$
Phosphatidylserine	5 - 15	$-CH_2CH(COOH)NH_2$
Phosphatidylglycerol	0.4 - 2	$CH_2CH(OH)CH_2OH$
Phosphatidylinositol	3 - 10	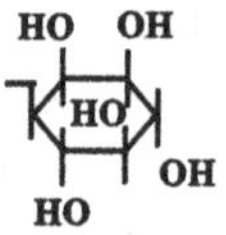 HO OH HO OH HO (inositol ring)

Phospholipid structure

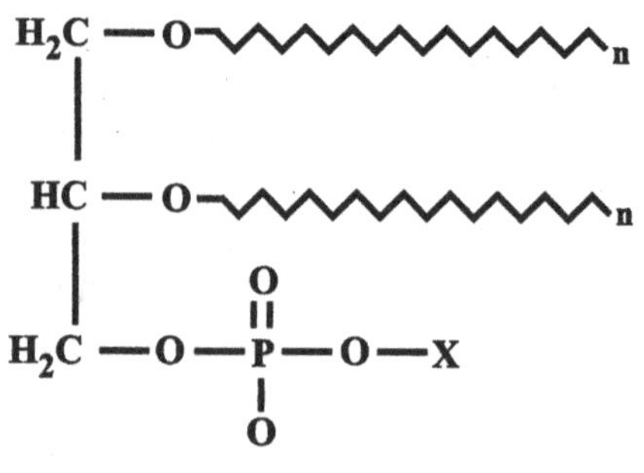

H_2C —O—(chain)$_n$
|
HC —O—(chain)$_n$
|
H_2C —O—P(=O)(O)—O—X

Fig. 7.1. Structure and composition of major phospholipids in eukaryotic membranes.

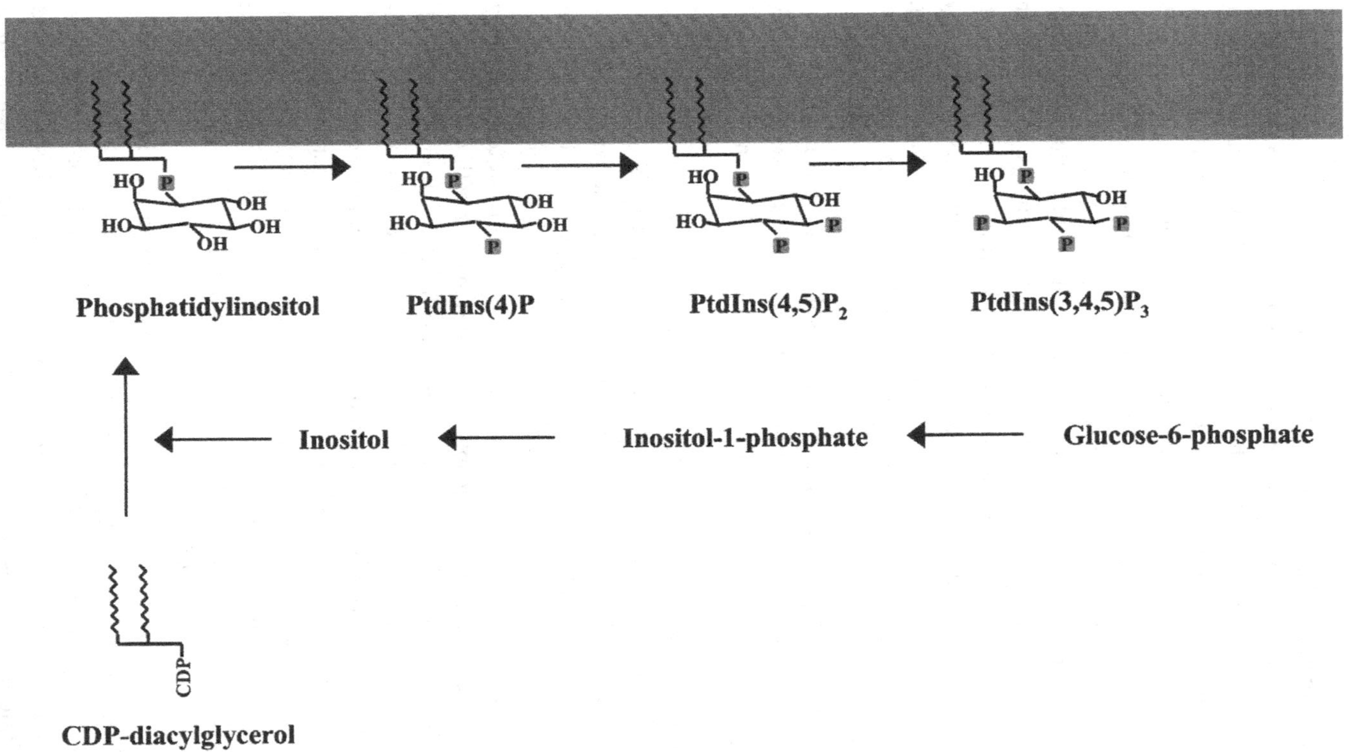

Fig. 7.2. Schematic representation of phosphoinositide biosynthesis in mammalian cells.

disrupted confers a lethal phenotye in *S. cerevisiae* (Nikawa, Kodaki & Yamashita, 1987) and represents the first committed step in the synthesis of these phospholipds. Although there appears to be no direct regulation of this gene, exogenous inositol has been shown to increase phosphatidylinositol synthesis rapidly, apparently by diverting more of the CDP-diacylglycerol precursor from phosphatidylserine biosynthesis (Kelley *et al.*, 1988). Inositol is synthesized by a two-step process from glucose-6-phosphate. The first step, catalysed by inositol-1-phosphate synthase, converts glucose-6-phosphate to inositol-1-phosphate, which is then dephosphorylated by inositol-1-phosphatase to form free inositol (Fig. 7.2). Synthesis of inositol from glucose-6-phosphate appears to be highly conserved in most eukaryotes and has been reported in a number of fungi including *S. cerevisiae* and *Neurospora crassa* (Donahue & Henry, 1981; Zsindely *et al.*, 1983).

PtdIns is itself phosphorylated by PtdIns 4′-kinase forming phosphatidylinositol-4-phosphate (PtdIns(4)P). The PtdIns 4′-kinases are a family of kinases functionally conserved in both mammalian and yeast cells (Wong, Meyer & Cantly, 1997). In *S. cerevisiae*, two genes have been cloned with 4′-kinase activity. The first, *PIK1*, is essential and appears to be predominantly associated with the nucleus whilst disruption of the second gene, *STT4*, leads to slow growth and a phenotype similar to that found in protein kinase C mutants (Flanagan *et al.*, 1993; Garcia-Bustos *et al.*, 1994; Yoshida *et al.*, 1994*a,b*). The discovery of two 4′-kinases with different functions and subcellular locations suggests a further level of control in phosphoinositide signalling with discrete compartmentalization of kinases with different cellular functions. PtdIns(4)P is further phosphorylated at the 5′ position yielding phosphatidylinositol-4,5-bisphosphate (PtdIns(4,5)P_2) in both mammalian and yeast cells and is the substrate for phosphoinositide-specific phospholipase C (Boronenkov & Anderson, 1995). In *S. cerevisiae*, disruption of Fab1p, which encodes the 5′-kinase, leads to vacuolar and growth defects, suggesting a role for PtdIns(4,5)P_2 in the regulation of vacuolar homeostasis (Yamamoto *et al.*, 1995).

In addition to its role in second messenger generation, PtdIns(4,5)P_2 has also been shown to regulate the assembly of the actin cytoskeleton by binding to a number of actin-binding proteins such as profilin in both mammalian and yeast cells. Profilin possess both PtdIns(4,5)P_2- and actin-binding domains, which are located close together so that binding of one excludes binding of the other (Goldschmidt-Clermont *et al.*, 1990;

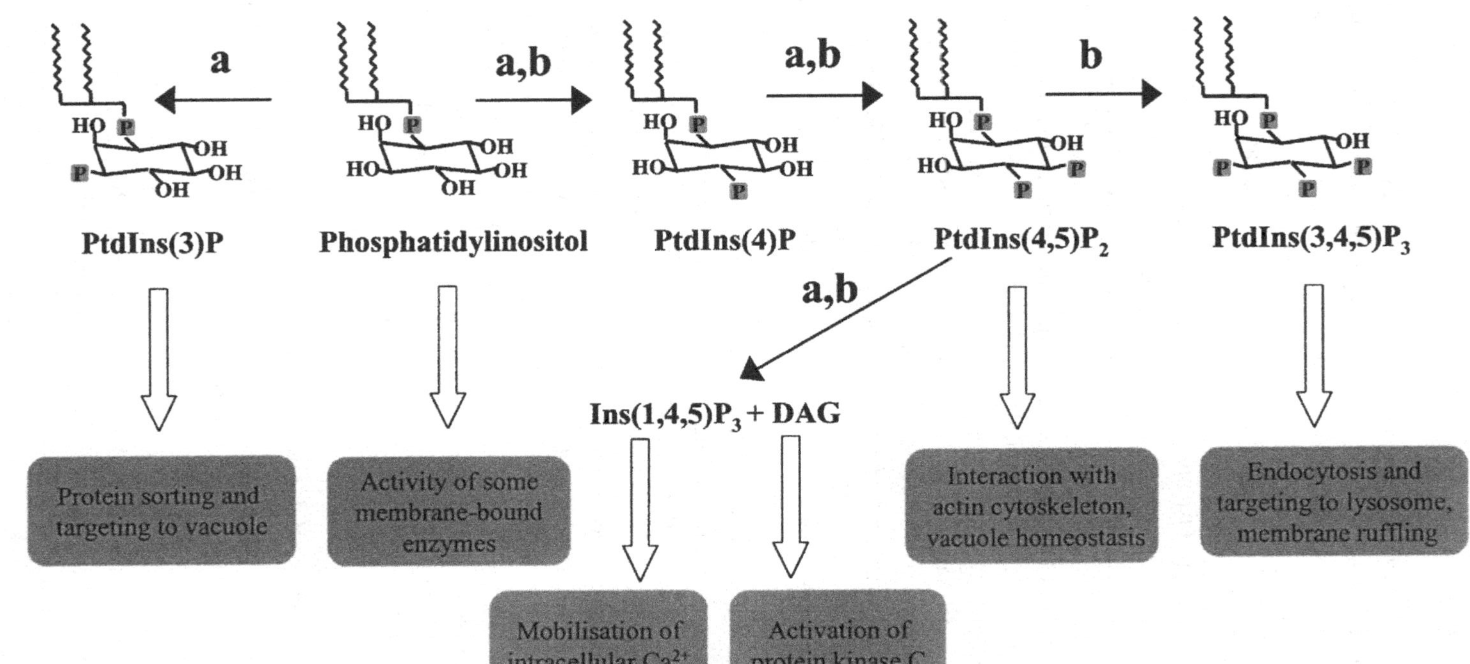

Fig. 7.3. Principal functions of the phosphoinositide cascade in eukaryotic cells; a, occurs in fungi; b, occurs in mammalian cells.

Metzler *et al.*, 1994; Sohn & Goldschmidt-Clermont, 1994). In *S. cerevisiae*, profilin has been found both in the cytosol and associated with the plasma membrane. Moreover, depletion of PtdIns(4,5)P_2 by inositol deprivation of an inositol auxotroph and by glucose starvation, leads to the mobilization of part of the membrane-associated profilin. This mobilization response could be reversed when PtdIns(4,5)P_2 levels were restored to normal (Ostrander, Gorman & Carman, 1995). More recently PtdIns 3′-kinases have been described in both mammalian and yeast cells and represent a further class of kinases with distinctive regulatory roles. In *S. cerevisiae* and *S. pombe*, PtdIns 3′-kinase is encoded by *VSPS34* and disruption of the gene leads to mis-targeting of hydrolases to the vacuole and abnormal vacuole development. PtdIns 3′-kinase differs from the mammalian homologue *P110/P85* in being insensitive to the inhibitor wortmannin and in having a strict substrate specificity for PtdIns leading to the formation of phosphatidylinositol-3-phosphate (PtdIns(3)P) (Auger *et al.*, 1989; Herman & Emr, 1990; Otsu *et al.*, 1991; Schu *et al.*, 1993; Stack & Emr, 1994; Kimura *et al.*, 1995; Takegawa, DeWald & Emr, 1995). In mammalian cells, PtdIns 3′ kinase is wortmannin sensitive and is capable of phosphorylating PtdIns, PtdIns(4)P and PtdIns(4,5)P_2, although *in vivo* PtdIns(4,5)P_2 appears to be the major substrate leading to the formation of phosphatidylinositol-3,4,5-trisphosphate (PtdIns(3,4,5)P_3) (Stephens *et al.*, 1994; Woscholski *et al.*, 1994; Brown *et al.*, 1995). Despite these differences, mammalian PtdIns 3′-kinase appears functionally similar to yeast VSPS34, where it is involved in protein targeting to the lysosome, which plays a similar function in mammalian cells to the yeast vacuole. To date, there are no reports of any PtdIns(4,5)P_2-specific 3′-kinases in yeast. Thus, the yeast and mammalian PtdIns 3′-kinases appear to play similar functions in protein targeting, with PtdIns(3,4,5)P_3 fulfilling a similar role in mammalian cells to PtdIns(3)P in yeast cells (Fig. 7.3).

Phosphoinositide-specific phospholipase C

Hydrolysis of PtdIns(4,5)P_2 by phosphoinositide-specific phospholipase C leads to the generation of Ins(1,4,5)P_3 and diacylglycerol, which both possess second messenger functions. A rise in intracellular Ins(1,4,5)P_3 leads to the specific release of intracellular Ca^{2+} from internal pools whilst diacylglycerol acts within the membrane to stimulate protein kinase C (Berridge & Irvine, 1994; Divicha & Irvine, 1995). In mammalian cells, phosphoinositide-specific phospholipase C represents a family of enzymes which are coupled to either G-protein or tyrosine kinase receptors and are stimulated by a range of agonists, leading to rapid hydrolysis of PtdIns(4,5)P_2 and rise in intracellular Ins(1,4,5)P_3 (Boyer, Hepler & Harden, 1989) (Fig. 7.4). In mammalian cells, Ins(1,4,5)P_3 triggers the release of Ca^{2+} from the endoplasmic reticulum, leading to a transient rise in intracellular Ca^{2+}. The increase in Ca^{2+} above the tightly regulated physiological levels in the cytoplasm activates an array of Ca^{2+}-dependent proteins, includ-

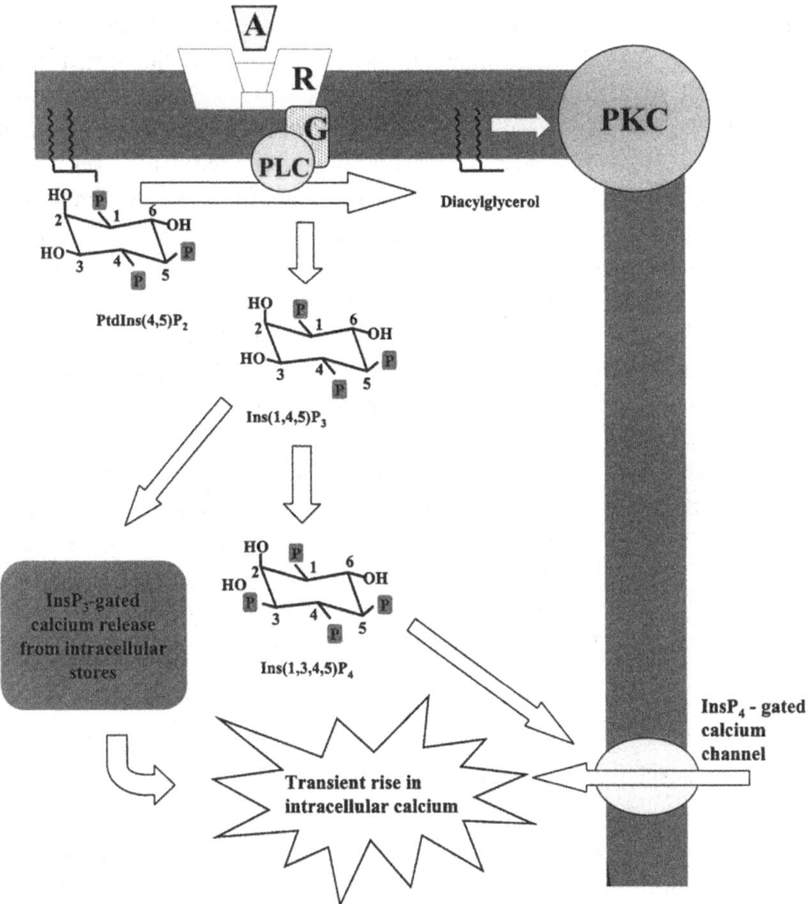

Fig. 7.4. Schematic representation of agonist stimulated inositol trisphosphate generation in mammalian cells. A, agonist; R, receptor; G, G-protein; PLC, phosphatidylinositol-specific phospholipase C; PKC, protein kinase C.

ing the ubiquitous Ca^{2+}-binding protein calmodulin (CAM), which in turn can activate a number of calmodulin-dependent enzymes including CAM-dependent protein kinases and phosphatases (Heinzmann & Hunziker, 1991; Lu & Means, 1993). Ins(1,4,5)P_3 has been shown to release Ca^{2+} specifically from the vacuoles of *S. cerevisiae*, *Candida albicans* and *N. crassa* (Cornelius, Gebauer & Techel, 1989; Schultz *et al.*, 1990; Belde *et al.*, 1993; Calvert & Sanders, 1995), which appear to act as the major intracellular Ca^{2+} store in fungi (Cornelius & Nakashima, 1987; Miller, Vogg & Sanders, 1990) and a phosphatidylinositol-specific phospholipase homologue (PLC-1) has been cloned from *S. cerevisiae*, *S. pombe* (Flick & Thorner, 1993; Yoko-o *et al.*, 1993, 1995; Andoh *et al.*, 1995) and more recently identified in a number of filamentous fungi (Jung *et al.*, 1997). Thus, a functional Ins(1,4,5)P_3 signal transduction pathway appears to exist in fungi although *in vivo* evidence of Ins(1,4,5)P_3-stimulated Ca^{2+} release is lacking and it remains unclear what factors are involved in stimulating phosphatidylinositol-specific phospholipase C activity. In *S. cerevisiae* and *S. pombe*, disruption of PLC-1 leads to a number of phenotypes depending on the strain and the medium the organism is grown in but generally include heat sensitivity and an inability to adapt to osmotic stress, suggesting that PLC-1 may be involved in relaying general stress signals across the membrane (Flick & Thorner, 1993; Payne & Fitzgerald-Hayes, 1993; Yoko-o *et al.*, 1993, 1995; Andoh *et al.*, 1995; Fankhauser *et al.*, 1995). In addition, there have been contradictory reports for a stimulation of phosphoinositide turnover and Ins(1,4,5)P_3 formation on exposure of starved cells to glucose and nitrogen, which would imply a role in nutrient sensing (Schomerus & Kuntzel, 1992; Brandao *et al.*, 1994; Prior, Robson & Trinci, 1994; Giugliano, Dennery & Rana, 1995) although a consensus view has yet to be reached.

Phosphoinositides and inositol phosphates in fungi

In filamentous fungi, all the major components of the phosphoinositide signalling pathway have been identified, including PtdIns, PtdInsP and PtdInsP_2 (Hanson, 1991; Robson *et al.*, 1991*a*; Prior *et al.*, 1993). Using HPLC to separate deacylated lipids from *N. crassa*, PtdInsP_2 was identified as PtdIns(4,5)P_2 whilst PtdInsP has been resolved and found to be composed of both PtdIns(4)P and PtdIns(3)P with levels of PTdIns(4)P being similar to PtdIns(3)P (Lakin-Thomas, 1993; Prior *et al.*, 1993).

Thus *N. crassa* more closely resembles *S. cerevisiae* and *S. pombe* in containing significant levels of both PtdInsP isomers (Hawkins, Stephens & Piggott, 1993; Stuart *et al.*, 1995) unlike mammalian cells, which contain predominantly only PtdIns(4)P and suggests that both a PtdIns 3′ and a PtdIns 4′ kinases are present. The levels of PtdIns(4,5)P_2 present in fungi as a percentage of the total phosphoinositides present is much lower than reported for mammalian cells, typically between 0.1 to 1% compared with 3.0 to 9.3% respectively. Moreover, the profile of inositol phosphates present in filamentous fungi appears complex. In *N. crassa*, evidence from HPLC separations suggests the presence of multiple isomers of InsP_2, InsP_3 and InsP_4 at similar levels, making their detection, identification and function difficult to establish (Lakin-Thomas, 1993; Prior *et al.*, 1993).

Metabolism of Ins(1,4,5)P_3 is an important control point in the Ins(1,4,5)P_3-mediated Ca^{2+}-release mechanism, serving to reduce Ins(1,4,5)P_3 levels and hence terminate the signal. Metabolism occurs by two mechanisms, phosphorylation by a specific kinase to InsP_4 and by dephosphorylation to InsP_2. InsP_2 is itself further dephosphorylated to InsP and ultimately inositol, which can then return to the phosphoinositide cycle by incorporation back to PtdIns. In mammalian cells, Ins(1,4,5)P_3 is dephosphorylated principally by a 5′-phosphomonoesterase and phosphorylated by a specific 3′-kinase leading to the formation of Ins(1,3,4,5)P_4 (Shears, 1992). The formation of Ins(1,3,4,5)P_4 itself acts as a further signal molecule, stimulating the opening of Ca^{2+} channels in the plasma membrane, prolonging the increase in intracellular Ca^{2+} (Irvine, 1992). In *S. cerevisiae*, no phosphomonoesterase activity was detected in either soluble or microsomal fractions and the phosphorylation by a soluble kinase found to produce Ins(1,4,5,6)P_4, indicating a 6′-kinase rather than a 3′-kinase described in mammalian cells (Estevez *et al.*, 1994). In *N. crassa*, dephosphorylation of Ins(1,4,5)P_3 was associated principally with the microsomal fraction and appeared to occur through the concerted action of 1′, 4′ and 5′-phosphomonoesterase activities. Ins(1,4,5)P_3 metabolism in the cytoplasm appeared to occur largely by phosphorylation by a 3′-kinase leading to the formation of Ins(1,3,4,5)P_4 identical with that found in mammalian cells, although a minor unidentified InsP_4 was also present suggesting *N. crassa* may contain two InsP 3′-kinases (Hosking, Trinci & Robson, 1997). In *C. albicans*, both an Ins(1,4,5,)P_3 kinase and Ins(1,4,5)$_3$ phosphatase activities were demonstrated *in vitro* although isomers formed were not identified (Gadd & Foster, 1997).

Studies with inositol-requiring mutants

The first indications that phosphoinositides may play an important role in hyphal morphology and branching were made by Beadle (1944) who noted that inositol-requiring strains of *N. crassa*, when grown on sub-optimal concentrations of inositol, produced small, compact mycelial pellets in liquid cultures rather than the normal spreading mycelial mat formed by the wild-type strain. This alteration in the mycelial branching pattern was later correlated to a reduction in the content of inositol-containing phospholipids (Fuller & Tatum, 1956; Shatkin & Tatum, 1961). Inositol starvation was found to lead to a rapid drop in viability and a deterioration of cellular membranes in *N. crassa* and was attributed to the uncoupling of membrane synthesis and other cellular components leading to unbalanced growth and so-called 'inositol-less death' (Shatkin & Tatum, 1961; Munkres, 1976). Similar observations were also made for inositol auxotrophs of *S. cerevisiae*, where inositol starvation was also associated with a rapid decrease in cell viability and degeneration of membranes (Atkinson, Kolat & Henry, 1977; Henry *et al.*, 1977). In addition, inositol deprivation was also associated with changes in the cell wall composition of both *S. cerevisiae* and *N. crassa*. In *N. crassa*, the glucosamine content of the wall was reduced, indicating that chitin synthase activity was lower whilst in *S. cerevisiae*, inositol starvation was associated with a decrease in both the level of mannan and glucan in the wall. The decrease in the levels of mannan was correlated to a reduction in the activity of UDP-*N*-acetylglucosamine:dolichol phosphate *N*-acetylglucosamine-1-phosphate transferase (GlcNAc-1-P), a membrane-associated enzyme which catalyses the first step in N-linked mannan synthesis (Hanson & Lester, 1982). GlcNAc-1-P transferase activity could be restored *in vitro* by the addition of PtdIns. Hence, at least for mannan synthesis, a decrease is due to a dependency of GlcNAc-1-P transferase on PtdIns for maximal activity. One of the problems associated with using inositol-requiring strains of *N. crassa* to investigate the relationship between phosphoinositides and morphology is that inositol deprivation leads to a reduction in all classes of phosphoinositides in the membrane, due to the reduction in the synthesis of PtdIns (Table 7.1). As well as acting as the precursor for PtdInsP and $PtdInsP_2$, PtdIns, which comprises 80–95% of the total phosphoinositides in the membrane, is also required for the maximal activity of a number of membrane-bound enzymes. In addition, PtdIns is also the precursor for a further class of inositol-containing lipids, the sphingolipids, which themselves play an

Table 7.1. *Effect of inositol deprivation on the phosphoinositide composition of an inositol-requiring strain of* Neurospora crassa. *Data from Lakin-Thomas (1993)*

	Phosphoinositide content (pmol/g dry weight)	
Phosphoinositide	Inositol-sufficient	Inositol-deficient
PtdIns	2500	240
PtdIns(3)P	29.5	7.3
PtdIns(4)P	29.5	5.0
PtdIns(4,5)P_2	24.3	6.9

essential role in cell viability and are also depleted during inositol deprivation (Hanson & Brody, 1978; Dickson *et al.*, 1990; Patton *et al.*, 1992; Nagiec *et al.*, 1997). Thus, morphological changes associated with inositol deprivation may be due to factors other than interference with phosphoinositide turnover.

Studies with phosphoinositide turnover inhibitors

Studies with inhibitors of phosphoinositide synthesis have also been correlated to changes in colony morphology and hyphal branching. Validamycin A, an aminoglycosidic antibiotic used in the control of *Rhizoctonia solani*, which causes sheath blight in rice plants, was found to induce profuse branching and a reduction in hyphal extension in *R. solani* and *Rhizoctonia cerealis* (Robson, Kuhn & Trinci, 1988). This alteration in the spatial pattern of the colony was correlated to a reduction in the total phosphoinositide content of the membrane, which was only slightly reversed by the addition of exogenous inositol (Robson, Kuhn & Trinci, 1989). Validamycin was reported to lead to an alteration in the wall composition of *R. solani*, suggesting that the observed changes in hyphal extension and branching may be due to an alteration in wall assembly at the hyphal tip. Hanson (1991) examined the effects of lithium chloride on the morphology and phosphoinositide composition of *N. crassa*. Studies on a range of mammalian cells have shown that hydrolysis of inositol phosphates by inositol monophosphatases is highly sensitive to lithium and that exposure leads to a reduction in phosphoinositide synthesis (Drummond, 1987; Nahorski & Potter, 1989). When grown in the presence of LiCl up to a concentration of 4 mM, the morphology of *N. crassa* changed from a spreading mat to compact pellets, similar to that found during inositol deprivation of inositol-requiring strains of *N. crassa*. This dose-dependent change in hyphal morphology was associated with a decrease in the levels of PtdIns, PtdInsP and $PtdInsP_2$. However, as with the inositol-requiring strains, the primary effects are due to a reduction in PtdIns synthesis and therefore not necessarily due to inhibition of the phosphoinositide signalling pathway.

More recently, we investigated the effect of specific phosphoinositide turnover inhibitors on the growth and morphology of *N. crassa*. Both lysocellin and piericidin B1 *N*-oxide were found to reduce hyphal extension and induce hyphal branching without affecting the overall specific growth rate (Hosking, Robson & Trinci, 1995). An inositol analogue (2*S*, 3*R*, 5*R*-3-azido-2-benzoyloxy-5-hydroxycyclohexanone), which would reduce the levels of all the phosphoinositides in the membrane, also had similar morphological effects (Table 7.2). On further investigation, lysocellin was found to reduce the levels of PtdInsP and $PtdInsP_2$ by about 70% and 90% respectively and to cause a significant increase in the levels of PtdIns *in vivo* (Table 7.3) suggesting that lysocellin specifically blocks conversion of PtdIns to PtdInsP. This was confirmed *in vitro* on isolated membrane fractions from *N. crassa*, where potent inhibition of PtdIns 4′-kinase activity was observed. Thus, in *N. crassa*, lysocellin potently blocks at least PtdIns (4)P synthesis and consequently $PtdIns(4,5)P_2$ formation. When young mycelia grown on Cellophane overlaid plates in the absence of lysocellin were transferred to plates containing lysocellin, hyphal extension was reduced and profuse branching was initiated throughout the whole mycelium, first appearing about 1 h after transfer (Fig. 7.5). Thus, interference in phosphoinositide signalling by lysocellin leads to a profound effect on hyphal growth and morphology. When the effect of exposure to lysocellin of growing hyphae was investigated, it was found that lysocellin caused a rapid and complete cessation of hyphal extension within 30 s. This was followed by a recovery after 4–7 min to a rate which was only about 30% of the rate prior to treatment (Fig. 7.6). The rapid reduction in hyphal extension rate indicates that phosphoinositide turnover must be very rapid in growing hyphae and that the rate of hyphal extension is in some way dependent on maintaining levels of PtdInsP and/or $PtdInsP_2$. If $PtdIns(4,5)P_2$ plays an important role in regulating the actin cytoskeleton in filamentous fungi, the reduction in hyphal extension and the induction of profuse branching observed throughout the mycelium on lysocellin treatment may result from a reorganization of the actin cytoskeleton following

Table 7.2. *Effect of phosphoinositide turnover inhibitors on the growth and morphology of* Neurospora crassa. *Data from Hosking, Robson & Trinci (1995)*

Concentration of inhibitor	Hyphal growth unit length (μm)	Specific growth rate (h^{-1})
Control	1443 ± 141	0.34 ± 0.07
8 μM lysocellin	993 ± 141	0.30 ± 0.05
112 μM piericidin	513 ± 48	0.33 ± 0.05
7.6 μM inositol analogue	578 ± 42	0.33 ± 0.08

Table 7.3. *Effect of lysocellin on the percentage phosphoinositide composition of* Neurospora crassa. N. crassa *was grown in the presence or absence of 1.6 μM lysocellin in a medium containing [^{3}H]-inositol and extracted phospholipids deacylated and separated by HPLC against authentic standards. Data from Hosking, Robson & Trinci (1995)*

Phosphoinositide	Absence of lysocellin	Presence of lysocellin
PtdIns	87.6 ± 0.4	96.9 ± 0.1
PtdInsP	5.2 ± 1.6	1.5 ± 0.1
$PtdInsP_2$	0.48 ± 0.1	0.05 ± 0.02

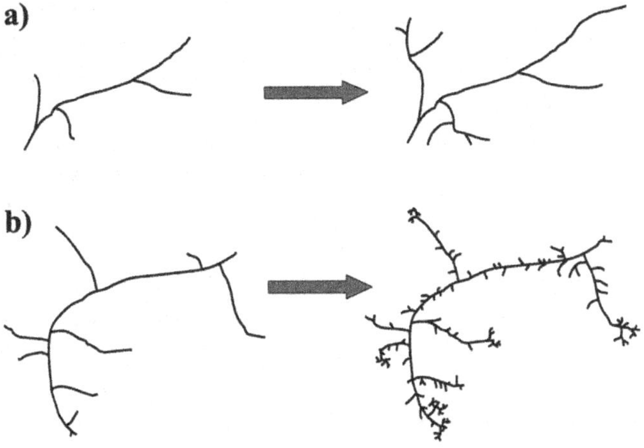

Fig. 7.5. Effect of transferring young germlings of *N. crassa* from agar medium to agar medium lacking (a) or containing (b) 4 μM lysocellin. The effect of the transfer on hyphal morphology is shown after 3 h. (Based on the data of Hosking, Robson & Trinci, 1995.)

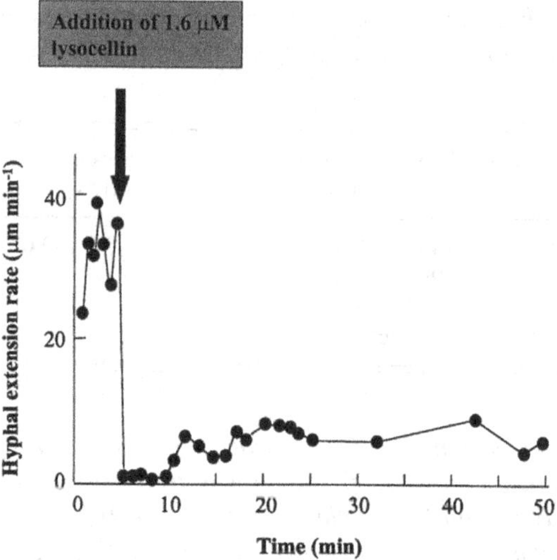

Fig. 7.6. Effect on the extension rate of a single growing hypha of *N. crassa* to exposure to 1.6 μM lysocellin. Hyphae were grown in a cellophane sandwich overlaying agar medium and lysocellin added at the time indicated. (Based on the data of Hosking, Robson and Trinci, 1995.)

the reduction in PtdIns(4,5)P_2 in the membrane. Actin is associated with sites of polarized growth and is important in both polarized hyphal extension and branch initiation (Heath, 1990). Alternatively, the reduction of PtdIns(4,5)P_2 in the membrane by lysocellin treatment would also lead to a reduction in Ins(1,4,5)P_3 and DAG levels, which may result in lower levels of intracellular Ca^{2+} and protein kinase C activity. A protein kinase C homologue has been cloned from a number of fungi including *S. cerevisiae*, *S. pombe*, *C. albicans*, *Aspergillus niger*, *Trichoderma reesei* and *N. crassa* (Antonsson *et al.*, 1994; Morawetz *et al.*, 1996; Paravicini *et al*, 1996). Disruption of the PKC locus in *S. cerevisiae* and *S. pombe* leads to an osmotically sensitive phenotype due to an increase in β-1,3-glucanase activity in the cell wal (Levin *et al.*, 1994; Paravicini *et al.*, 1996). In filamentous fungi, disruption of PKC has yet to be reported. A reduction in the levels of Ins(1,4,5)P_3 may result in a reduction in intracellular Ca^{2+}, which could also affect mycelial morphology. Growth of *F. graminearum* and *N. crassa* in a Ca^{2+}-depleted medium resulted in a more highly branched phenotype and abnormal hyphal development (Schmid & Harold, 1988; Robson, Wiebe & Trinci, 1991*b*) whilst the highly branched phenotypes of two mutants of *N. crassa*, *frost* and *spray*, could be reversed by high levels of extracellular Ca^{2+} (Dicker & Turian, 1990). However, if changes in mycelial morphology associated with a reduction in PtdIns(4,5)P_2 levels were due to a reduction, in either Ins(1,4,5)P_3 or DAG, it would imply that phosphoinositide-specific phospholipase C actively hydrolyses PtdIns(4,5)P_2 during normal hyphal growth, rather than in response to an outside stimulus.

To date, it is clear that phosphoinositide turnover is intimately linked with hyphal extension and branch induction although the exact nature of the relationship remains unclear. As more homologues of this signal cascade are identified in filamentous fungi, the nature and function of phosphoinositide signalling and its evolution from primitive to higher eukaryotes will be revealed.

References

Andoh, T., Yoko-o, T., Matsui, Y. & Tohe, A. (1995). Molecular-cloning of the PLC1(+) gene of *Schizosaccharomyces pombe*, which encodes a putative phosphoinositide-specific phospholipase C. *Yeast*, **11**, 179–85.

Antonsson, B., Montessuit, S., Friedli, L., Payton, M. A. & Paravicini, G. (1994). Protein kinase C in yeast. Characteristics of the *Saccharomyces cerevisiae* PKC1 gene product. *Journal of Biological Chemistry*, **269**, 16821–8.

Atkinson, K. D., Kolat, A. I. & Henry, S. A. (1977). Osmotic imbalance in inositol-starved spheroplasts of *Saccharomyces cerevisiae. Journal of Bacteriology*, **132**, 806–17.

Auger, K. R., Carpenter, C. L., Cantely, L. C. & Varticovski, L. (1989). Phosphatidylinositol 3-kinase and its novel product phosphatidylinositol 3-phosphate, are present in *Saccharomyces cerevisiae. Journal of Biological Chemistry*, **264**, 20181–4.

Beadle, G. W. (1944). An inositol-less mutant strain of *Neurospora crassa* and its use in bioassays. *Journal of Biological Chemistry*, **156**, 683–9.

Belde, P. J., Vossen, J. H., Borst-Pauwels, G. W. & Theuvenet, A. P. (1993). Inositol 1,4,5-trisphosphate releases Ca^{2+} from vacuolar membrane vesicles of *Saccharomyces cerevisiae. FEBS Letters*, **323**, 113–18.

Berridge, M. J. & Irvine, R. F. (1984). Inositol trisphophate, a novel second messenger in cellular signal transduction. *Nature*, **312**, 315–21.

Boronenkov, I. V. & Anderson, R. A. (1995). The sequence of phosphatidylinositol phosphate 5-kinase defines a novel family of lipid kinases. *Journal of Biological Chemistry*, **270**, 2881–4.

Boyer, J. L., Hepler, J. R. & Harden, T. K. (1989). Hormone and growth factor receptor-mediated regulation of phospholipase C activity. *Trends in Pharmacological Sciences*, **10**, 360–4.

Brandao, R. L., de Magalhaes-Rocha, N. M., Alijo, R., Ramos, J. & Thevelein, J. M. (1994). Possible involvement of a phosphatidylinositol-type signalling pathway in glucose-induced activation of plasma-membrane

H^+-ATPase and proton extrusion in the yeast *Saccharomyces cerevisiae. Biochimica et Biophysica Acta*, **1223**, 117–24.

Brown, W. J., DeWald, D. B., Emr, S. D., Plutner, H. & Balch, W. E. (1995). Role for phosphatidylinositol 3-kinase in the sorting and transport of newly synthesised lysosomal enzymes in mammalian cells. *Journal of Cell Biology*, **130**, 781–96.

Cahil, M. A., Janknecht, R. & Nordheim, A. (1996). Signalling pathways: Jack of all cascades. *Current Biology*, **6**, 16–19.

Calvert, C. M. & Sanders, D. (1995). Inositol trisphosphate-dependent and -independent Ca^{2+} mobilisation pathways at the vacuolar membrane of *Candida albicans. Journal of Biological Chemistry*, **270**, 7272–80.

Cornelius, G., Gebauer, G. & Techel, D. (1989). Inositol trisphosphate induces calcium release from *Neurospora crassa* vacuoles. *Biochemical and Biophysical Research Communications*, **162**, 852–6.

Cornelius, G. & Nakashima, H. (1987). Vacuoles play a decisive role in calcium homeostasis in *Neurospora crassa. Journal of General Microbiology*, **133**, 2341–7.

Csaba, G. (1994). Phylogeny and ontogeny of chemical signalling: origin and development. *International Review of Cytology*, **155**, 1–48.

Dicker, J. W. & Turian, G. (1990). Calcium deficiencies and apical hyperbranching in wild-type and the '*frost*' and '*spray*' mutants of *Neurospora crassa. Journal of General Microbiology*, **136**, 1413–20.

Dickson, R. C., Wells, G. B., Schmidt, A. & Lester, R. L. (1990). Isolation of mutant *Saccharomyces cerevisiae* strains that survive without sphingolipids. *Molecular and Cellular Biology*, **10**, 2176–81.

Divecha, N. & Irvine, R. F. (1995). Phospholipid signalling. *Cell*, **80**, 269–78.

Donahue, T. F. & Henry, S. A. (1981). Myo-inositol-1-phosphate synthase. Characterisation of the enzyme and identification of its structural gene in yeast. *Journal of Biological Chemistry*, **256**, 7077–85.

Drummond, A. H. (1987). Lithium and inositol lipid-linked signalling mechanisms. *Trends in Pharmacological Sciences*, **8**, 129–33.

Estevez, F., Pulford, D., Stark, M. J. R., Carter, A. N. & Downes, C. P. (1994). Inositol trisphosphate metabolism in *Saccharomyces cerevisiae*-identification, purification and properties of inositol 1,4,5-trisphosphate 6-kinase. *Biochemical Journal*, **302**, 709–16.

Fankhauser, H., Schweingruber, A. M., Edenharter, E. & Schweingruber, M. E. (1995). Growth of a mutant defective in a putative phosphoinositide-specific phospholipase C of *Schizosaccharomyces pombe* is restored by low concentrations of phosphate and inositol. *Current Genetics*, **28**, 199–203.

Flanagan, C. A., Schneiders, E. A., Emerick, A. W., Kunisawa, R., Admon, A. & Thorner, J. (1993). Phosphatidylinositol 4-kinase: gene structure and requirement for yeast cell viability. *Science*, **262**, 1444–8.

Flick, J. S. & Thorner, J. (1993). Genetic and biochemical characterisation of a phosphatidylinositol-specific phospholipase C in *Saccharomyces cerevisiae. Molecular and Cellular Biology*, **13**, 5861–76.

Fuller, R. C. & Tatum, E. L. (1956). Inositol-phospholipid in *Neurospora* and its relationship to morphology. *American Journal of Botany*, **43**, 361–5.

Gadd, G. M. (1995). Signal transduction in fungi. In *The Growing Fungus*, ed. N. A. R. Gow & G. M. Gadd, pp. 183–210. London: Chapman and Hall.

Gadd, G. M. & Foster, S. A. (1997). Metabolism of inositol 1,4,5-trisphosphate in *Candida albicans*: significance as a precursor of inositol polyphosphates and in signal transduction during the dimorphic transition from yeast cells to germ tubes. *Microbiology*, **143**, 437–48.

Garcia-Bustos, J. F., Marini, F., Stevenson, I., Frei, C. & Hall, M. N. (1994). PIK1, an essential phosphatidylinositol-4-kinase associated with the yeast nucleus. *EMBO Journal*, **13**, 2352–61.

Giugliano, E. R., Dennery, N. & Rana, R. S. (1995). Glucose metabolism stimulates phosphoinositide biosynthesis in *Saccharomyces cerevisiae. FASEB Journal*, **9**, A1308.

Goldschmidt-Clermont, P. J., Machesky, L. M., Baldassare, J. J. & Pollard, T. D. (1990). The actin-binding protein profilin binds to PIP_2 and inhibits its hydrolysis by phospholipase C. *Science*, **247**, 1575–8.

Hanson, B. A. (1991). The effects of lithium on the phosphoinositides and inositol phosphates of *Neurospora crassa. Experimental Mycology*, **15**, 76–90.

Hanson, B. A. & Brody, S. (1978). Lipid and cell wall changes in an inositol-requiring mutant of *Neurospora crassa. Journal of Bacteriology*, **138**, 461–6.

Hanson, B. A. & Lester, R. L. (1982). Effect of inositol starvation on the *in vitro* synthesis of mannan and *N*-acetylglucosaminylpyrophosphoryl-dolichol in *Saccharomyces cerevisiae. Journal of Biological Chemistry*, **151**, 334–42.

Hawkins, P. T., Stephens, L. R. & Piggott, J. R. (1993). Analysis of inositol metabolites produced by *Saccharomyces cerevisiae* in response to glucose stimulation. *Journal of Biological Chemistry*, **268**, 3374–83.

Heath, I. B. (1990). The roles of actin in tip growth of fungi. *International Review of Cytology*, **123**, 95–127.

Heinzmann, C. W. & Hunziker, W. (1991). Intracellular calcium-binding proteins: more sites than insights. *Trends in Biochemical Sciences*, **16**, 98–103.

Henry, S. A., Atkinson, K. D., Kolat, A. I. & Culbertson, M. R. (1977). Growth and metabolism of inositol-starved *Saccharomyces cerevisiae. Journal of Bacteriology*, **130**, 472–84.

Herman, P. K. & Emr, S. D. (1990). Characterisation of VPS34, a gene required for vacuolar protein sorting and vacuole segregation in *Saccharomyces cerevisiae. Molecular and Cellular Biology*, **10**, 6742–54.

Hosking, S. L., Robson, G. D. & Trinci, A. P. J. (1995). Phosphoinositides play a role in hyphal extension and branching in *Neurospora crassa. Experimental Mycology*, **19**, 71–80.

Hoskings, S. L., Trinci, A. P. J. & Robson, G. D. (1997). *In vitro* metabolism of inositol 1,4,5-trisphosphate by *Neurospora crassa. FEMS Microbiology Letters*, **154**, 223–9.

Irvine, R. F. (1992). Is inositol tetrakisphosphate the second messenger that controls Ca^{2+} entry into cells? *Advances in Second Messenger and Phosphoprotein Research*, **26**, 161–85.

Janssens, P. M. W. (1987). Did vertebrate signal transduction mechanisms originate in eukaryotic microorganisms? *Trends in Biochemical Sciences*, **12**, 456–9.

Jung, O. J., Lee, E. J., Kim, J. W., Chung, Y. R. & Lee, C. W. (1997). Identification of putative phosphoinositide-specific phospholipase C genes in filamentous fungi. *Molecules and Cells*, **7**, 192–9.

Kelley, M. J., Baillis, A. M., Henry, S. A. & Carman, G. M. (1988). Regulation of phospholipid biosynthesis in *Saccharomyces cerevisiae* by inositol. Inositol is an inhibitor of phosphatidylserine synthase activity. *Journal of Biological Chemistry*, **263**, 18078–85.

Kimura, K., Miyake, S., Makuuchi, M., Morita, R., Usui, T., Yoshida, M., Horinouchi, S. & Yasuhisa, F. (1995). Phosphatidylinositol-3 kinase in fission yeast: a possible role in stress responses. *Bioscience Biotechnology and Biochemistry*, **59**, 678–82.

Kincaid, R. L. (1991). Signalling mechanisms in microorganisms: common themes in the evolution of signal transduction pathways. *Advances in Second Messenger and Phosphoprotein Research*, **23**, 165–81.

Lakin-Thomas, P. L. (1993). Effects of inositol starvation on the levels of inositol phosphates and inositol lipids in *Neurospora crassa*. *Biochemical Journal*, **292**, 805–11.

Levin, D. E., Bowers, B., Chen, C. Y., Kamada, Y. & Watanabe, M. (1994). Dissecting the protein kinase C/MAP kinase signalling pathway of *Saccharomyces cerevisiae*. *Cellular and Molecular Biology*, **40**, 229–39.

Lu, K. P. & Means, A. R. (1993). Regulation of the cell cycle by calcium and calmodulin. *Endocrine Reviews*, **14**, 40–58.

Metzler, W. J., Bell, A. J., Ernst, E., Lavoie, T. B. & Mueller, L. (1994). Identification of the poly-L-proline-binding site on human profilin. *Journal of Biological Chemistry*, **269**, 4620–5.

Miller, A. J., Vogg, G. & Sanders, D. (1990). Cytosolic calcium homeostasis in fungi: roles of plasma membrane transport and intracellular sequestration of calcium. *Proceedings of the National Academy of Sciences, USA*, **87**, 9348–52.

Mooibroek, M. J. & Wang, S. H. (1988). Integration of signal transduction processes. *Biochemistry and Cell Biology*, **66**, 557–66.

Morawetz, R., Lendenfeld, T., Mischak, H., Muhlbauer, M., Gruber, F., Goodnight, J., de Graaf, L. H., Visser, J., Mushinski, J. F. & Kubicek, C. P. (1996). Cloning and characterisation of genes (*pkc1* and *pkcA*) encoding protein kinase C homologues from *Trichoderma reesei* and *Aspergillus niger*. *Molecular and General Genetics*, **250**, 17–28.

Munkres, K. D. (1976). Ageing of *Neurospora crassa* III. Induction of cellular death and clonal senescence of an inositol-less mutant by inositol starvation and the protective effect of dietary antioxidants. *Mechanisms of Ageing and Development*, **5**, 163–9.

Nagiec, M. M., Nagiec, E. E., Baltisberger, J. A., Wells, G. B., Lester, R. L. & Dickson, R. C. (1997). Sphingolipid synthesis as a target for antifungal drugs. Complementation of the inositol phosphorylceramide synthase defect in a mutant strain of *Saccharomyces cerevisiae* by the AURI gene. *Journal of Biological Chemistry*, **272**, 9809–17.

Nahorski, S. R. & Potter, B. V. L. (1989). Molecular recognition of inositol polyphosphates by intracellular receptors and metabolic enzymes. *Trends in Pharmacological Sciences*, **10**, 139–44.

Nikawa, J., Kodaki, T. & Yamashita, S. (1987). Primary structure and disruption of the phosphatidylinositol synthase gene of *Saccharomyces cerevisiae*. *Journal of Biological Chemistry*, **262**, 4876–81.

Nishida, E. & Gotoh, Y. (1993). The MAP kinase cascade is essential for diverse signal transduction pathways. *Trends in Biochemical Sciences*, **18**, 128–31.

Nishuzka, Y. (1984). Turnover of inositol phospholipid and signal transduction. *Science*, **225**, 1365–70.

Ostrander, D. B., Gorman, J. A. & Carman, G. M. (1995). Regulation of profilin localisation in *Saccharomyces cerevisiae* by phosphoinositide metabolism. *Journal of Biological Chemistry*, **270**, 27045–50.

Otsu, M., Hiles, I., Gout, I., Fry, M. J., Ruiz-Larrea, F., Panaotou, G., Thompson, A., Dhand, R., Hsuan, J., Totty, N., Smith, A. D., Morgan, S. J., Courtneidge, S. A., Parker, P. J. & Waterfield, M. D. (1991). Characterisation of two 85 kDa proteins that associate with receptor tyrosine kinases, middle-T/pp60c-src complexes and PI3-kinase. *Cell*, **65**, 91–104.

Paravicini, G., Mendoza, A., Antonsson, B., Cooper, M., Losberger, C. & Payton, M. A. (1996). The *Candida albicans PKC1*-gene encodes a protein kinase C homologue necessary for cellular integrity but not dimorphism. *Yeast*, **12**, 741–56.

Patton, J. L., Srinivasan, B., Dickson, R. C. & Lester, R. L. (1992). Phenotypes of sphingolipid-dependent strains of *Saccharomyces cerevisiae. Journal of Bacteriology*, **174**, 7180–4.

Payne, W. E. & Fitzgerald-Hayes, M. (1993). A mutation in *PLC1*, a candidate phosphoinositide-specific phospholipase C gene from *Saccharomyces cerevisiae* causes aberrant mitotic chromosomal segregation. *Molecular and Cellular Biology*, **13**, 4351–64.

Prior, S. L., Cunliffe, B. W., Robson, G. D. & Trinci, A. P. J. (1993). Multiple isomers of phosphatidylinositol monophosphate and inositol bis- and trisphosphates from filamentous fungi. *FEMS Microbiology Letters*, **110**, 147–52.

Prior, S. L., Robson, G. D. & Trinci, A. P. J. (1994). Phosphoinositide turnover does not mediate the effects of light or choline, or the relief of derepression of glucose metabolism in filamentous fungi. *Mycological Research*, **98**, 291–4.

Rasmussen, L., Christensen, S. T., Schousboe, P. & Wheatley, D. V. (1996). Cell survival and multiplication. The overriding need for signals: from unicellular to multicellular systems. *FEMS Microbiology Letters*, **137**, 123–8.

Roberts, R. L. & Fink, G. R. (1994). Elements of a single MAP kinase cascade in *Saccharomyces cerevisiae* mediate two developmental programs in the same cell type: mating and invasive growth. *Genes and Development*, **8**, 2974–85.

Robson, G. D., Kuhn, P. J. & Trinci, A. P. J. (1988). Effect of validamycin A on the morphology, growth and sporulation of *Rhizoctonia cerealis*, *Fusarium culmorum* and other fungi. *Journal of General Microbiology*, **134**, 3187–94.

Robson, G. D., Kuhn, P. J. & Trinci, A. P. J. (1989). Effect of validamycin A on the inositol content and branching of *Rhizoctonia cerealis* and other fungi. *Journal of General Microbiology*, **135**, 739–50.

Robson, G. D., Trinci, A. P. J., Wiebe, M. G. & Best, L. C. (1991*a*). Phosphatidylinositol 4,5-bisphosphate (PIP_2) is present in *Fusarium graminearum. Mycological Research*, **95**, 1082–4.

Robson, G. D., Wiebe, M. G. & Trinci, A. P. J. (1991*b*). Low calcium concentrations induce increased branching in *Fusarium graminearum. Mycological Research*, **95**, 561–5.

Schmid, J. & Harold, F. M. (1991). Dual roles for calcium ions in apical growth of *Neurospora crassa. Journal of General Microbiology*, **134**, 2623–31.

Schomerus, C. & Kuntzel, H. (1992). CDC25-dependent induction of inositol 1,4,5-trisphosphate and diacylglycerol in *Saccharomyces cerevisiae* by nitrogen. *FEBS Letters*, **307**, 249–52.

Schu, P. V., Takegawa, K., Fry, M. J., Stock, J. H., Waterfield, M. B. & Emr, S. D. (1993). Phosphatidylinositol 3-kinase encoded by yeast VSP34 gene is essential for protein sorting. *Science*, **260**, 88–91.

Schultz, C., Gebauer, G., Metschies, T., Rensing, L. & Jastorff, B. (1990). *Cis,cis*-cyclochexane 1,3,5-triol polyphosphates release calcium from *Neurospora crassa* via unspecific Ins 1,4,5-P_3 receptor. *Biochemical and Biophysical Research Communications*, **166**, 1319–27.

Shatkin, A. J. & Tatum, E. L. (1961). The relationship of *m*-inositol to morphology in *Neurospora crassa. American Journal of Botany*, **48**, 760–71.

Shears, S. B. (1992). Metabolism of inositol phosphates. *Advances in Second Messenger and Phosphoprotein Research*, **26**, 63–92.

Sohn, R. H. & Goldschmidt-Clermont, P. J. (1994). Profilin: at the crossroads of signal transduction and the actin cytoskeleton. *Bioessays*, **16**, 465–72.

Stack, J. H. & Emr, S. D. (1994). Vps34p required for vacuolar protein sorting is a multiple specificity kinase that exhibits both protein kinase and phosphatidylinositol-specific PI3-kinase activities. *Journal of Biological Chemistry*, **269**, 31552–62.

Stephens, L., Cooke, F. T., Walters, R., Jackson, T., Volina, S., Gout, I., Waterfield, M. D. & Hawkins, P. T. (1994). Characterisation of a phosphatidylinositol-specific phosphoinositide 3-kinase from mammalian cells. *Current Biology*, **4**, 203–14.

Stuart, J. A., Hughes, P. J., Kirk, C. J., Davey, J. & Michel, R. H. (1995). The involvement of inositol lipids and phosphates in signalling in the fission yeast *Schizosaccharomyces pombe. Biochemical Society Transactions*, **23**, 223S.

Su, B. & Karin, M. (1996). Mitogen-activated protein cascades and regulation of gene expression. *Current Opinion in Immunology*, **8**, 402–11.

Takegawa, K., DeWald, D. B. & Eme, S. D. (1995). *Schizosaccharomyces pombe* Vps34p, a phosphatidylinositol-specific PI 3-kinase essential for normal cell growth and vacuole morphology. *Journal of Cell Science*, **108**, 3745–56.

Waskiewicz, A. J. & Cooper, J. A. (1995). Mitogen and stress response pathways: MAP kinase cascades and phosphatase regulation in mammals and yeast. *Current Opinion in Cell Biology*, **7**, 798–805.

Wong, K., Meyers, D. D. R. & Cantly, L. C. (1997). Subcellular locations of phosphatidylinositol 4-kinase isoforms. *Journal of Biological Chemistry*, **272**, 13236–41.

Woscholski, R., Kodaki, T., McKinnan, M., Waterfield, M. D. & Parker, A. J. (1994). A comparison of demethoxyviridin and wortmannin as inhibitors of phosphatidylinositol 3-kinase. *FEBS Letters*, **342**, 109–14.

Yamamoto, A., DeWald, D. B., Boronenkov, I. V., Emr, S. D. & Koshland, D. (1995). Novel PI(4)P 5-kinase homologue, Fab1p, essential for normal vacuole function and morphology in yeast. *Molecular Biology of the Cell*, **6**, 525–39.

Yoko-o, T., Koto, H., Matsui, Y., Takenawa, T. & Toh-e, A. (1995). Isolation and characterisation of temperature-sensitive plcl mutants of the yeast *Saccharomyces cerevisiae*. *Molecular and General Genetics*, **247**, 148–56.

Yoko-o, T., Matsui, Y., Yagisawa, H., Nolima, H., Uno, I. & Toh-e, A. (1993). The putative phosphoinositide-specific phospholipase C gene, PLC1, of the yeast *Saccharomyces cerevisiae* is important for cell growth. *Proceedings of the National Academy of Sciences, USA*, **90**, 1804–8.

Yoshida, S., Ohya, Y., Goebi, M., Nakano, A. & Anraku, Y. (1994*a*). A novel gene *STT4*, encodes a phosphatidylinositol-4-kinase in the PCK1 protein kinase pathway of *Saccharomyces cerevisiae*. *Journal of Biological Chemistry*, **269**, 1166–72.

Yoshida, S., Ohya, Y., Nakano, A. & Anraku, Y. (1994*b*). Genetic interactions among genes involved in the STT4-PKC1 pathway of *Saccharomyces cerevisiae*. *Molecular and General Genetics*, **242**, 631–40.

Zsindely, A., Kiss, A., Schablik, M., Szabolcs, M. & Szabo, G. (1983). Possible role of a regulatory gene product upon the myo-inositol-1-phosphate synthase production in *Neurospora crassa*. *Biochimica et Biophysica Acta*, **741**, 273–8.

8

Stress responses of fungal colonies towards toxic metals

L. M. RAMSAY, J. A. SAYER AND G. M. GADD

Introduction

Fungi comprise a significant proportion of the soil microbial community as decomposer organisms and plant symbionts (mycorrhizas), playing fundamental roles in carbon mineralization and other biogeochemical cycles (Wainwright, 1988), and are often dominant in acidic soils where toxic metals may be speciated into mobile forms (Morley *et al.*, 1996). Anthropogenic activities, including fossil fuel combustion, mineral mining and processing, and production of industrial effluents and sludges, biocides and preservatives (Gadd & Griffiths, 1978; Gadd, 1992), release a variety of toxic metal species into aquatic and terrestrial ecosystems and this can have significant effects on the biota as well as resulting in metal transfer to higher organisms, plants and animals (Wainwright & Gadd, 1997). Metals and their compounds can interact with fungi in various ways depending on the metal species, organism and environment, while metabolic activity can also influence speciation and mobility. Certain mechanisms may mobilize metals into forms available for cellular uptake or leaching from the system, e.g. complexation with citric acid, other metabolites and siderophores (Francis, 1994). Metals may also be immobilized by, for example, sorption onto cell components, exopolymers, transport and intracellular sequestration or precipitation, both intra- and extracellular (Morley & Gadd, 1995; Sayer & Gadd, 1997). The apparently opposing phenomena of metal solubilization and immobilization are key components of biogeochemical cycles for toxic metals, whether indigenous or introduced into a given location, since both mobility and toxicity can be affected. Furthermore, the mobilization of essential metal ions has implications for the regulation of fungal growth, physiology and morphogenesis (Morley *et al.*, 1996; White, Sayer & Gadd, 1997).

Many metals are essential for fungal growth and metabolism, e.g. Na, K, Cu, Zn, Co, Ca, Mg, Mn, Fe, etc., but all can exert toxicity when present above certain threshold concentrations in available forms (Gadd, 1993). Other metals, e.g. Cd, Hg, Pb, have no known biological function but all can be accumulated by fungi (Gadd, 1993). Metals exert toxic effects in many ways, for example they can block the functional groups of important biological molecules such as enzymes, displace or substitute for essential metal ions, and interact with systems which normally protect against harmful effects of free radicals generated during normal metabolism (Melhorn, 1986; Gadd, 1992, 1993). However, fungi possess many properties which influence metal toxicity, including the production of metal-binding proteins, organic and inorganic precipitation, active transport and intracellular compartmentalization, while major constituents of fungal cell walls, e.g. chitin, have significant metal binding abilities (Gadd & Griffiths, 1978; Gadd, 1993). All these mechanisms are highly dependent on the metabolic and nutritional status of the organism since this will affect expression of energy-dependent resistance mechanisms as well as synthesis of wall structural components, pigments and metabolites, which gratuitously affect metal availability and organism response (Gadd, 1992, 1993). Despite this, there has been little or no attention paid to this in studies of fungal responses towards toxic metals, nor, apart from some minor exceptions, the effects metals may have at a cellular level on morphogenesis, an aspect of fungal biology of key importance in exploitation of their substrate. The purpose of this chapter is to highlight this problem by discussing the influence toxic metals and nutritional status may have on growth and morphogenesis of the fungal colony, including some key roles of fungi in transforming toxic metals between mobile and immobile phases, and the environmental and physiological significance of such phenomena.

Effects of toxic metals on growth, branching and morphogenesis

Nutritionally replete conditions

Most fungi have a potentially infinite vegetative growth capacity with limitations to growth mainly imposed by environmental conditions (Cooke & Whipps, 1993). During colony growth, hyphae of the peripheral zone deplete nutrients as they advance with any residual nutrients taken up by those hyphae immediately behind the growth zone. Hyphal density at the colony edges is constant and generation of branches occurs at a rate proportional to the number of apices present. Hence the number of apices produced by branching of leader hyphae increases exponentially as the colony expands (Trinci, 1978; Prosser, 1995) with mathematical models confirming these concepts (Ritz & Crawford, 1990; Prosser, 1995). However, the properties of these individual hyphae do not account for many of the macroscopic properties exhibited by fungal colonies, including alterations in hyphal density and extension rates within colonies dependent on the local availability of nutrients (Rayner, 1991; Rayner, Griffiths & Ainsworth, 1995), when fungi demonstrate explorative and/or exploitative strategies relative to the underlying nutritional status (Crawford, Ritz & Young, 1993; Ritz, 1995). In their natural environment, fungi rarely encounter conditions which allow exponential growth, and are usually prevented by nutritional and other abiotic environmental constraints. Major environmental factors include nutrient and water availability, temperature, pH, O_2 requirements, CO_2 tolerance, as well as salinity and toxic metals, in more extreme environments. For example, high metal concentrations can lead to reductions in fungal populations, biomass and respiration and this has been noted in soils polluted by Cu, Cd, Pb, As and Zn (Ruhling *et al.*, 1984; Babich & Stotzky, 1985; Nordgren, Baath & Soderstrom, 1983).

Although there is some information on overall changes in growth under conditions of increased toxic metal concentrations, particularly in the context of toxicity assessment (Gadd, 1986), little is known about the effects of metals on the growth of mycelia at the microscopic level. Lilly, Wallweber & Lukefahr (1992) studied some effects of Cd on the basidiomycete *Schizophyllum commune* and found significant changes in the morphology of colonies as well as a reduction in colony growth. At higher concentrations of Cd there was twisting of individual hyphae, formation of intertwined hyphal strands and looping of the individual hyphae. In addition, fewer branches were produced nearer the apex, with the distance from the hyphal apices at the colony margin to the first branch being significantly shorter than during control growth. Overall, there appeared to be a disruption of normal polarized growth of the hyphae, which may be explained by Cd affecting the mechanisms responsible for maintenance of electrochemical gradients across the apex, such as the H^+-ATPase, ion and nutrient channels, etc., which are involved in polarized growth (Wessels, 1986; Gow, 1995). In another study, Cd reduced the specific growth rate of the ectomycorrhizal fungus *Paxillus involutus* (Darlington & Rauser, 1986) while Ni increased the hyphal density of *Rhizopus stolonifer* and induced a brown discolouration of the colony (Babich & Stotzky, 1982). Despite a decrease in colony growth rates being a commonly observed manifestation of toxicity, there has been little attention paid to any concomitant changes in biomass distribution, hyphal morphogenesis or the influence of nutritional status on such a phenomenon. Most experiments are generally carried out in conditions of nutrient excess because of the assumption that fungi require considerable concentrations of energy-yielding materials in order to maintain growth (Wainwright, 1993). In our own studies, we have found that while the colony extension rates of several fungal species decreased in the presence of the toxic metals Cu, Cd and Zn (Table 8.1), such a response was highly dependent on the concentration of glucose in the media and as the amount of available carbohydrate increased, the toxicity of the metal decreased. While a contribution of metal binding by supplied substrate(s) may contribute to such an observation (Gadd & Griffiths, 1978; Gadd, 1992, 1993), it would also suggest that important processes by which fungi detoxify or can grow in the presence of toxic metals may not function under low nutrient conditions (see below).

Oligotrophic conditions

There is increasing evidence that in nature, fungi commonly exist in conditions of nutrient depletion, with soil having been regarded as an environment lacking sufficient available carbon to sustain continued microbial growth (Gray & Williams, 1971; Lynch, 1982). Despite this, many fungi can maintain rapid but sparse growth in natural soil and other nutrient-limited habitats (Wainwright *et al.*, 1991; Wainwright, 1993). Although there is a wide range of nutritional heterogeneity within the soil, e.g. from the nutrient-rich rhizosphere to habitats containing more unavailable organic material, e.g. cutin and keratin (Cooke & Rayner, 1984), it seems likely that the majority of soil fungi exist in an environment which is relatively oligotrophic (Wainwright, 1993). Mineral soil, in particular, is regarded as a poor source of available nutrients, with reports that water extracts from such soil contain only about 4 $\mu g\ ml^{-1}$ carbohydrate (Ko & Lockwood, 1967). Williams (1985) suggested that the small amount of carbon available for microbial growth in soil means that either the soil microorganisms grow slowly and continuously, or with intermittent periods of rapid growth interspersed with intervals of dormancy. Using low-carbon media, large numbers of fungi can be isolated from soil which show high growth rates in culture when grown in the presence of low levels of growth-promoting substances (Wainwright, 1988). It has been suggested by Wainwright (1993) that these organisms

Table 8.1. *Radial expansion rates of* Trichoderma viride *and* Rhizopus arrhizus *grown on low-nutrient media amended with 0.1 mM Cu, 0.1 mM Cd or 0.1 mM Zn. Glucose was added as the sole carbon source at concentrations ranging from 0.5 to 10.0% (w/v). Figures shown are mean radial expansion rates (μm h^{-1}) with the standard error of the mean for five colonies (Ramsay & Gadd, unpublished data).*

[Glucose]	*Trichoderma viride*				*Rhizopus arrhizus*			
% (w/v)	No metal	0.1 mM Zn	0.1 mM Cu	0.1 mM Cd	No metal	0.1 mM Zn	0.1 mM Cu	0.1 mM Cd
0	516.7 ± 39.7	267.0 ± 26.3	151.5 ± 19.9	92.8 ± 10.2	770.8 ± 62.7	501.1 ± 72.1	21.2 ± 6.2	281.8 ± 31.1
0.5	708.3 ± 93.6	392.6 ± 31.1	188.6 ± 24.9	121.6 ± 16.3	864.6 ± 77.1	504.8 ± 63.4	213.9 ± 25.4	384.6 ± 45.5
1.0	1135.4 ± 174.5	447.3 ± 64.5	307.1 ± 32.9	229.2 ± 19.0	1791.7 ± 204.7	699.3 ± 54.2	502.2 ± 31.6	503.6 ± 35.2
5.0	987.5 ± 82.8	486.2 ± 54.8	430.5 ± 32.6	395.6 ± 41.1	1041.7 ± 151.3	724.5 ± 52.7	668.3 ± 43.8	867.3 ± 62.8
10.0	659.7 ± 39.8	410.2 ± 67.2	405.8 ± 35.2	429.9 ± 36.8	750.1 ± 87.8	714.3 ± 52.3	794.1 ± 49.9	719.4 ± 58.6

possess characteristics that enable them to utilize low nutrient supplies efficiently, including an increased capacity to take up nutrients by possessing a high surface area and high-affinity nutrient uptake sites. This would allow an oligotrophically growing fungus to utilize new and mixed nutrient sources efficiently at low nutrient concentrations, which confers an obvious ecological advantage. Such organisms may be adapted to allow themselves to be maintained indefinitely in a hyphal form (Hirsch *et al.*, 1979). Nutrients may also be recycled through cryptic growth, where the tips of the hyphae grow at the expense of pre-formed fungal material (Schnurer & Paustian, 1986). However, *Trichoderma harizianum* growing on silica gel showed no reduction in growth after long-term oligotrophic maintenance despite there being no evidence of cryptic growth or translocation of nutrients from a food base (Wainwright, Ali & Killham, 1993). In this case, it is possible that carbon dioxide and other gases, and volatiles including methane, ethylene, benzene, as well as aldehydes, ketones and phenols, may be scavenged from the environment and act as a source of fungal nutrition (Mirocha & DeVay, 1971; Tribe & Mabadeje, 1972; Fries, 1973; Wainwright, 1993). Spores from two *Penicillium* species, which were re-introduced from carbon-rich conditions into soil, germinated to form germ tubes and hyphae which were reduced in diameter and length when compared to similar structures in carbon-rich conditions (Sheehan & Gochenaur, 1984). Cytoplasm-free hyphae have also been reported in *Alternaria* and *Cladosporium* sp. as well as many other fungi (Dickinson & Bottomley, 1980). This could be a means by which fungi conserve energy since Schnurer & Paustian (1986) have calculated that soil may lack sufficient carbon to support the measured fungal biomass when both walls and cytoplasm were formed, but contains sufficient carbon for wall production itself.

It is predictable, therefore, that the stress responses of fungi towards toxic metals will be affected by the nutritional status of the habitat or growth medium. In a low-nutrient environment, there may be a limitation to expression of both direct and indirect mechanisms of tolerance/resistance with toxicity being increased in these environments. We have also found that metals can have a significant impact on the overall length of the fungal mycelium and branching patterns (Fig. 8.1). Colonies of *Trichoderma viride* growing in low-nutrient conditions in the presence of Cu showed a decrease in the overall mycelial length and the number of branches, resulting in an extremely sparsely branched colony (Fig. 8.1a). Conversely in the presence of Cd, although the overall mycelial length was around one-third of the control length, the number of branches decreased only slightly, resulting in a colony which was very highly branched and with many aberrant branching features, including branches oriented towards the interior of the colony. In *Rhizopus arrhizus* the presence of the metals greatly reduced the overall mycelial length and the number of branches formed, with exposure to Cu resulting in a more sparsely branched colony with very few branches; in the presence of Cd short mycelial lengths with a large number of branches again resulted in a more densely branched colony (Fig. 8.1b). Both *T. viride* and *R. arrhizus* appeared to exhibit 'foraging' modes of growth on low-substrate media with sparse colonies formed (Ritz, 1995), and it would seem that both Cu and Cd are capable of disrupting this explorative growth under laboratory conditions. A large-scale mycelial-mapping technique showed that in the presence of Cu (Fig. 8.2b) and Cd (Fig. 8.2c) such disruption of normal growth resulted in alterations to the distribution of the fungal biomass. When grown on low-substrate media *T. viride* colonies showed an even distribution of biomass throughout the colony with some allocation to the periphery, suggesting an explorative growth phase (Fig. 8.2a). However, in the presence of Cu the biomass distribution was markedly altered with most being allocated to the periphery of the colony, while in the presence of Cd most biomass was located at the interior of the colony (Fig. 8.2). These results imply that toxic metals like Cu and Cd are capable of altering normal 'foraging' growth which, if manifest in natural environments, may influence success in locating nutrients as well as survival capability.

Mycorrhizal fungi have been well documented in relation to toxic metal pollution and can play an important role in establishing and maintaining plant growth in metal-polluted soils. Clover plants infected with *Glomus mosseae* vesicular-arbuscular mycorrhiza have been recovered from soil heavily contaminated with Zn and Cd (Gildon & Tinker, 1981). The presence of *Hymenoscyphus ericae* in roots of *Calluna vulgaris* on metal-contaminated soil permitted host plants to grow in sand cultures containing up to 75 mg l^{-1} Cu and 150 mg l^{-1} Zn, whereas non-mycorrhizal plants failed to grow (Bradley, Burt & Read, 1981). These symbioses rely on the ability of mycorrhizal fungi to survive toxic environments, which may depend on the plentiful carbon supply the fungus receives from the host. This may be in contrast to free-living saprotrophic fungi, which may not have the available nutritional resources to employ their detoxification mechanisms in soils of low carbon availability.

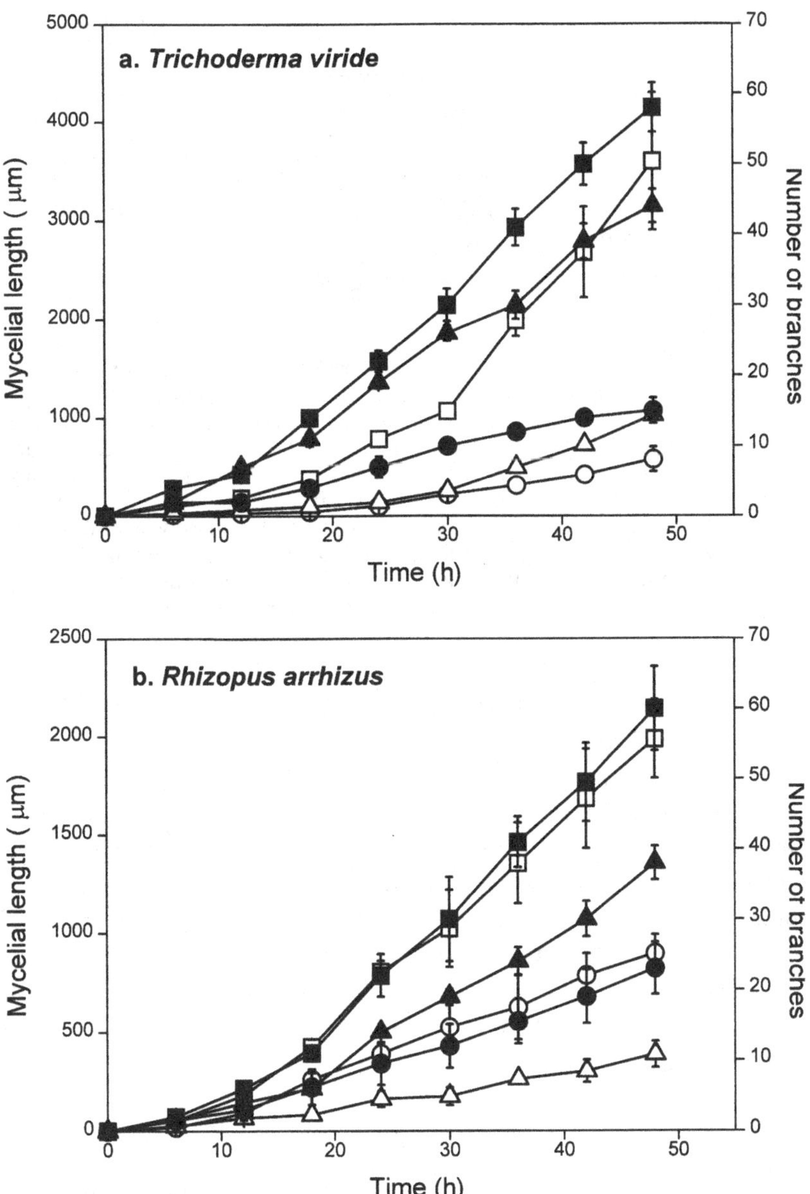

Fig. 8.1. Mean mycelial lengths and branch numbers of (a) *Trichoderma viride* and (b) *Rhizopus arrhizus* growing on low-nutrient media (control) and on low-nutrient media amended with either 0.1 mM Cu or 0.1 mM Cd. Symbols are as follows: (□) mean mycelial length of colonies growing on low-nutrient media; (■) mean number of branches of colonies growing on low-nutrient media; (○) mean mycelial length of colonies

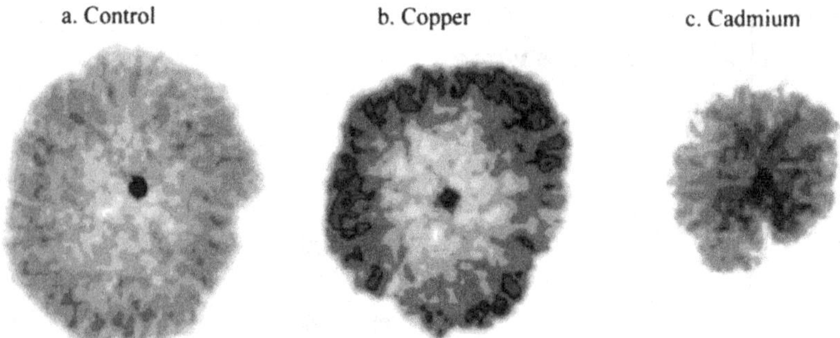

Fig. 8.2. Colonies of *Trichoderma viride* (5 days old) grown on low-substrate media. Colonies shown are (a) *Trichoderma viride* growing on low-substrate media, and on low-substrate media in the presence of (b) 0.1 mM Cu or (c) 0.1 mM Cd. Darker greys represent higher biomass density while lighter greys indicate lower biomass density. Colonies for mapping were grown from single spores and digitized colony images converted to matrices of greyscale values, which were averaged to produce matrix grids of 60 × 60 pixels. These were then mapped using the Unimap 2000 mapping program with the grids normalized and the background subtracted to produce comparable maps.

Melanization of fungal colonies in response to toxic metals

The wall is the first cellular site of interaction with metal species and possesses important biosorptive properties, dependent on physico-chemical conditions and macromolecular composition (Gadd, 1993). The wall has important protective properties, therefore, and can act as a barrier, influencing uptake of solutes including potentially toxic metal species. In some cases, wall structure and composition is affected by the presence of toxic metals and this may in turn influence colony development and morphology. The formation of melanin pigments provides the most visual example of this phenomenon. These pigments enhance the survival of many fungi in response to environmental stresses including UV, solar,

---

Fig. 8.1 (*continued*)
growing on low-nutrient media amended with 0.1 mM Cu; (●) mean number of branches of colonies growing on low-nutrient media amended with 0.1 mM Cu; (△) mean mycelial length of colonies growing on low-nutrient media amended with 0.1 mM Cd; and (▲) mean number of branches of colonies growing on low-nutrient media amended with 0.1 mM Cd. Error bars shown are the standard error of the mean of approximately 200 colonies measured and are only shown when greater than symbol dimensions.

X-ray and γ-radiation, microbial lysis and toxic metals (Bell & Wheeler, 1986; Gadd, 1993). Melanins are located in and/or exterior to cell walls and can appear as electron-dense deposits and granules on transmission electron micrographs (Gadd & Griffiths, 1980). A number of different melanin types occur in fungi (Fogarty & Tobin, 1996), with additional undefined dark pigments which may be termed 'heterogeneous' melanins (Bell & Wheeler, 1986). A variety of toxic metals can induce or accelerate melanin production in fungi, leading to blackening of colonies and chlamydospore development (Gadd & Griffiths, 1980). Chalmydospores and other melanized forms have high capacities for metal biosorption, with the majority of metal remaining within the wall structure (Gadd, 1984; Gadd & Mowll, 1985; Gadd, White & Mowll, 1987; Gadd & De Rome, 1988). In rhizomorphs of *Armillaria* sp. the highest concentrations of metals were located on the melanized outer surface (Rizzo, Blancchette & Palmer, 1992). In addition to inorganic metal forms, organometallic compounds, e.g. tributyltin chloride, are also strongly bound by fungal melanins (Gadd, Gray & Newby, 1990). Since it has been stated that a significant proportion of fungal biomass in soils is melanic (Bell & Wheeler, 1986), such interactions may be of ecological significance in polluted habitats. It is known that polymorphic *Aureobasidium pullulans*, which exhibits extensive melanization and chlamydospore development in the presence of toxic metals, can become the dominant organism on metal-polluted phylloplanes (Mowll & Gadd, 1985).

Synnema production

Synnema are defined as aerial, multihyphal structures where the apices of the component hyphae advance together and ultimately form spores (Watkinson, 1979). They are therefore concerned with the spread and survival of a given species and their formation can be triggered by a variety of external factors, e.g. light–dark transitions, low temperature, alcohols, detergents, carbon dioxide and amino acids (Watkinson, 1979). Synnema of the penicillia have received most attention and metal cations are believed to be involved in their formation. In *Penicillium claviforme* Bain. and *Penicillium clavigerum* Demelius, manganese is a requirement for synnema development (Tinnell, Jefferson & Benoit, 1977). In *Sphaerostilbe repens* Berkeley & Broome, both synnematal and rhizomorph development only occurred in the presence of calcium, with strontium (which often behaves as a calcium ion analogue) also capable of the induction of some morphological changes (Botton, 1978). Such cations may repulse electrostatic forces between hyphae, act as 'bridges', and also interact with metabolic processes (Watkinson, 1979). It has been found that tributyltin compounds can induce synnematal development in *Penicillium funiculosum* Thom, a species generally regarded as non-synnematal (Onions, Allsopp & Eggins, 1981). The structures observed were fertile along their length (2–5 mm), and conformed to standard definitions of synnemata. As in other fungi, the production of synnema results in a wider separation between the conidia and the substrate than in non-synnematal colonies, and this may aid dispersal as well as ensuring conidial formation away from potential toxicants in the substrate (Newby & Gadd, 1987). It is interesting to speculate whether such a morphological response is related to avoidance of toxicity, and whether this phenomenon is of wider significance in fungi exposed to potential toxic agents.

Role of organic acids in effecting changes in metal mobility

Solubilization

Solubilization of insoluble metal compounds is an important, but unappreciated, aspect of fungal physiology for the release of anions, such as phosphate, and essential metal cations into forms available for intracellular uptake. For insoluble toxic metal compounds, fungal solubilization may have additional implications for the organism since mobilization may result in accumulation and toxicity, while in the environment, soluble metal species may be transferred between terrestrial and aquatic ecosystems. In agriculture, most phosphate fertilizers are supplied in solid form (calcium phosphates, rock phosphate), and these have to be solubilized before becoming available to plants (Lapeyrie, Ranger & Vairelles, 1991). The soil fungus *A. niger* solubilizes rock phosphate by the production of low molecular weight organic acids, and, in one study, 3.0 g l^{-1} rock phosphate was solubilized yielding 292 μg ml^{-1} P (Vassilev *et al.*, 1995). In Canada, a formulation containing spores of *Penicillium bilaii* has been registered for application to wheat crops, since it is believed that the phosphate-solubilizing ability of the fungus will improve crop productivity (Cunningham & Kuiack, 1992).

Fungal solubilization of insoluble metal compounds, including certain oxides, phosphates, sulphides and mineral ores, can occur by several mechanisms. The proton-translocating ATPase of the plasma membrane pumps protons into the external medium, and generates a transmembrane electrochemical gradient used for the acquisition of nutrients (Serrano,

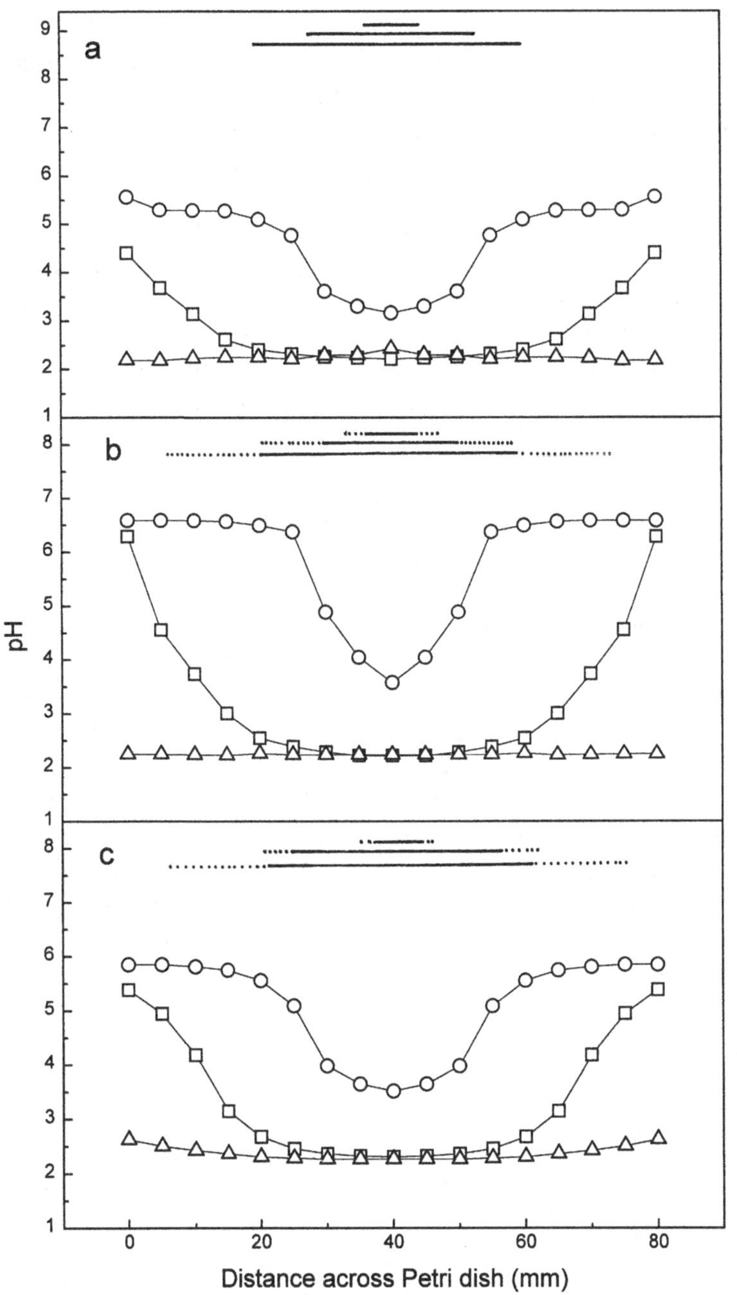
a
b
c
pH
Distance across Petri dish (mm)

1985; Gadd, 1993; Karamushka, Sayer & Gadd, 1996). Solubilization can occur by protonation of the anion of the metal compound, decreasing its availability to the cation (Hughes & Poole, 1991). The production of organic acids is a further source of protons as well as the organic acid anion, which is frequently capable of forming a complex with a metal cation and thereby altering mobility and toxicity. Many fungi produce organic acids, and this property may be employed for large-scale industrial production, for example *A. niger* is used to produce approximately 400 000 tons of citric acid per year for use in food, beverages and pharmaceuticals (Mattey, 1992). Solubilization can also occur by the production of siderophores, which are Fe(III)-specific bidentate ligands used for iron acquisition and accumulation (Watteau & Berthelin, 1994).

Colony screening assay for solubilization activity

There is current interest in 'heterotrophic leaching' of insoluble metal compounds for recycling and/or bioremediation (Burgstaller & Schinner, 1993) and an effective method of screening fungi for the solubilization of insoluble metal compounds, which also provides information on the relative toxicities of the metal compounds and tolerance of the organisms, is based on inoculating fungi onto solid medium amended with the insoluble metal compounds. After a few days growth, a halo or clear zone of solubilization around the colony indicates an active strain (Burgstaller & Schinner, 1993; Sayer *et al.*, 1995; Dixon-Hardy *et al.*, 1998). A range of metal phosphates (Al, Ca, Co, Mn, Zn) and ZnO were screened for solubilization by *A. niger* and *Penicillium simplicissimum*, two species well documented for their ability to produce organic acids (Sayer *et al.*, 1995), and of these, $Co_3(PO_4)_2$, $Zn_3(PO_4)_2$ and ZnO were found to be the most suitable compounds for screening natural fungal isolates (Sayer *et al.*, 1995). Fig. 8.3 shows a pH profile of the

---

Fig. 8.3. pH profiles of the agar under growing colonies of *Aspergillus niger* at (○) 1, (□) 4 and (△) 7 days after inoculation. *A. niger* was grown at 25°C on (a) unamended malt extract agar (MEA), (b) MEA amended with 15 mM ZnO and (c) MEA amended with 5 mM $Zn_3(PO_4)_3$. Data shown are representative of three replicate determinations, all of which gave similar results. Sizes of colonies (solid lines) and zones of solubilization (dashed lines) are indicated: the top bar shows growth/solubilization after 1 day, the middle bar after 4 days and the bottom bar after 7 days growth. Adapted from Sayer & Gadd (1997).

agar under colonies of *A. niger* unamended and amended with ZnO and $Zn_3(PO_4)_2$. As the size of the colonies increased with time, an area of lower pH in the centre of the agar increased in size, the profiles showing little variation between metal treatments. This area of low pH corresponded with the solubilization of the metal compounds (Sayer & Gadd, 1997). *A. niger* can also solubilize a wide range of insoluble metal compounds when grown on metal-amended agar, including cadmium sulphide, copper phosphate, nickel phosphate, manganese sulphide (J.A. Sayer & G.M. Gadd, unpublished work) and metal-bearing mineral ores including cuprite (CuS), rhodochrosite [$Mn(CO_3)_x$] (Sayer, Kierans & Gadd, 1997) and gypsum ($CaSO_4$) (Gharieb, Sayer & Gadd, 1998). The incidence of metal-solubilizing ability among natural soil fungal communities appears to be relatively high; in one study approximately one-third of the isolates tested were able to solubilize at least one of the test metal compounds, and approximately one-tenth were able to solubilize all three (see Table 8.2; Sayer *et al.*, 1995).

Metal immobilization by oxalate production

As mentioned earlier, liberation of mobile metal species may have implications for cellular accumulation and toxicity. The production of oxalic acid by fungi provides a means of immobilizing soluble metal ions, or complexes, as insoluble oxalates decreasing bioavailability and conferring tolerance (Sayer & Gadd, 1997). The production of insoluble calcium oxalate is a widely occurring phenomenon in many fungi, often noted in laboratory and field studies (see later). In the laboratory, colonies of *A. niger* have been shown to form oxalate crystals after 1–2 days when grown on agar amended with a wide range of metal compounds, including insoluble metal phosphates (Sayer & Gadd, 1997) and powdered metal-bearing minerals (Sayer *et al.*, 1997; Gharieb *et al.*, 1998). Fig. 8.4 shows scanning electron micrographs of copper oxalate crystals precipitated under colonies of *A. niger* on agar amended with copper phosphate (Fig. 8.4a) and the copper-bearing mineral cuprite, CuS (Fig. 8.4b). *A. niger* has been shown to produce metal oxalates with many different metals, e.g. Ca, Cd, Co, Cu, Mn, Sr and Zn (Sayer & Gadd, 1997). Morphological examination of fungal-produced oxalate crystals and comparison, where possible, with chemically synthesized oxalates, has shown clear differences in form (Vivier, Marcant & Pons, 1994; Sayer & Gadd, 1997). The use of fungal colonies in simple laboratory models

Table 8.2. *Incidence of solubilization ability among fungal strains isolated from soil, including isolates from lead- and nickel-contaminated soils, tested using different insoluble metal compounds (compounds tested individually). Numbers shown are the percentage of isolates exhibiting solubilization ability (*n *= 56). Adapted from Sayer* et al. *(1995).*

Test metal compound(s)	% of isolates exhibiting solubilizing ability
ZnO	30
$Zn_3(PO_4)_2$	14
$Co_3(PO_4)_2$	28
ZnO, $Zn_3(PO_4)_2$	4
ZnO, $Co_3(PO_4)_2$	14
$Zn_3(PO_4)_2$, $Co_3(PO_4)_2$	0
ZnO, $Zn_3(PO_4)_2$, $Co_3(PO_4)_2$	10

may be useful for future elucidation of the crystallographic properties of novel fungal metal oxalates.

Environmental significance

While some individual morphological responses of fungal hyphae and colonies towards toxic metals can be obviously linked to survival and/or dissemination, e.g. melanization and synnema formation, effects on branching may have further implications for acquisition of nutrients, and in turn, colony development and survival. This may be related to nutritional heterogeneity and limitation in the fungal environment. As described earlier, toxic effects on growth can be dependent on available substrate levels and it is tempting to speculate on the possibility of differential distribution of toxicity and survival in a nutritionally heterogeneous environment. The survival of mycorrhizas in polluted locations may provide supporting evidence for the importance of available carbon. Clearly, this is a complex problem and further work will necessitate application of simple laboratory models (Ritz, 1995) as well as field studies.

In anthropogenic terms, solubilization of insoluble toxic metal compounds in the terrestrial environment may have beneficial or adverse effects. For example, some rock phosphate fertilizers contain cadmium, and solubilization could lead to increased availability of Cd (Leyval, Surtiningsih & Berthelin, 1993). Solubilization activity in mining areas,

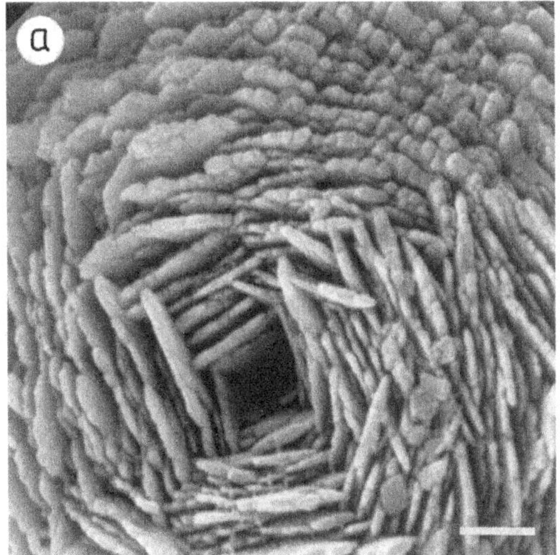

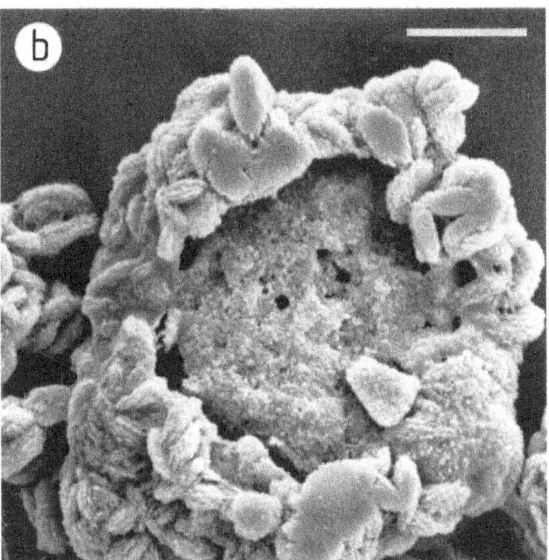

Fig. 8.4. Scanning electron micrographs of copper oxalate crystals precipitated in and purified from the agar under colonies of *Aspergillus niger* growing on malt extract agar amended with (a) 0.5 % (w/v) copper phosphate (bar marker = 1 μm) and (b) 60 mM cuprite (CuS) (bar marker = 10 μm).

or areas of heavy contamination, can also release toxic metals into the aquatic system (Francis, Dodge & Gillow, 1992; Banks, Waters & Schwabb, 1994*b*). Further, metal-contaminated mining areas are often revegetated to help stabilize soils and reduce runoff and wind erosion. The production of organic acids by rhizosphere microorganisms, the breakdown of organic matter and root exudation in such locations can influence metal mobility, especially when the soil microflora have not been completely restored (Banks *et al.*, 1994*a*). Conversely, microbial activities which increase the mobility of toxic metals may have bioremediation potential. For example, radionuclides and toxic metals may be 'heterotrophically leached' from soils, sediments or wastes with citric acid (Burgstaller & Schinner, 1993), with, if necessary, the metal–citrate complexes eventually being degraded for metal recovery (Francis, 1994).

The formation of oxalates containing potentially toxic metal cations (other than calcium) may provide a mechanism whereby oxalate-producing fungi can tolerate environments containing high concentrations of toxic metals. Most metal oxalates are insoluble, some exceptions being Na, K, Al, Li and Fe (Vivier *et al.*, 1994). Although little work has been carried out to date, some formation of toxic metal oxalates has been observed in the natural environment. Copper oxalate (moolooite) has been observed around hyphae growing on wood treated with copper as a preservative (Murphy & Levy, 1983; Sutter, Jones & Walchi, 1983, 1984). Sutter *et al.* (1983, 1984) reported that two brown rot fungi, *Poria placenta* and *Poria vaillantii*, removed copper almost completely from treated wood by the production of oxalic acid. The copper appeared on the surface of the wood and around hyphae as copper oxalate, which was reported to be non-toxic because of its insolubility. Copper oxalate has also been observed in lichens growing on copper-rich rocks, where it is thought that the precipitation of copper oxalate could be a detoxification mechanism; up to 5% (dry weight) copper was fixed in the lichen thallus as copper oxalate (Purvis & Halls, 1996).

In contrast to the formation of toxic metal oxalates, there are many reports of the formation of the calcium oxalate deposits, whewellite (calcium oxalate monohydrate) and weddellite (calcium oxalate dihydrate). Oxalate is ubiquitous in the terrestrial environment, and concentrations in the soil can reach 10^{-6}–10^{-3} M (Allison, Daniel & Cornick, 1985). The calcium oxalate crystals found around hyphae and mycorrhizal roots are thought to have a major role in detoxification (Lapeyrie, Chilvers & Bhem, 1987; Jones *et al.*, 1992). Most oxalate is secreted and intracellular crystals have not been documented in fungi (Franceschi & Loewus, 1995). Some fungi, such as brown rot fungi, often secrete oxalic acid along with cell-wall-degrading enzymes, and this is thought to assist the solubilization of pectin in membranes in the middle lamellae (Green *et al.*, 1995). Calcium oxalate is resistant to further solubilization, with only a few aerobic bacteria and fungi and anaerobic bacteria of the gastrointestinal tract known to degrade them (Morris & Allen, 1994; Allison *et al.*, 1995).

Acknowledgements

G.M. Gadd gratefully acknowledges the BBSRC (GR/J48214, 94/SPC02812), the Royal Society of Edinburgh (Scottish Office Education Department/RSE Support Research Fellowship 94/5) and NATO (Envir.Lg.950387 Linkage Grant) for some of the work described. J.A.S. and L.M.R. gratefully acknowledge receipt of a NERC postgraduate research studentship. Thanks are also due to Karl Ritz and John Crawford (SCRI) for assistance with colony mapping.

References

Allison, M. J., Daniel, S. L. & Cornick, N. A. (1995). Oxalate degrading bacteria. In *Calcium Oxalate in Biological Systems*, ed. S. R. Khan, pp. 131–68. Florida: CRC Press Inc.

Babich, H. & Stotzky, G. (1982). Nickel toxicity to microbes: effect of pH and implication for acid rain. *Environmental Research*, **29**, 334–50.

Babich, H. & Stotzky, G. (1985). Heavy metal toxicity to microbe-mediated ecologic processes: a review and potential application to regulatory policies. *Environmental Research*, **36**, 111–37.

Banks, M. K., Schwabb, A. P., Fleming, G. R. & Herrick, B. A. (1994*a*). Effects of plants and soil microflora on leaching of zinc from mine tailings. *Chemosphere*, **29**, 1691–9.

Banks, M. K., Waters, C. Y. & Schwabb, A. P. (1994*b*). Influence of organic acids on leaching of heavy metals from contaminated mine tailings. *Journal of Environmental Science and Health*, **A29**, 1045–56.

Bell, A. A. & Wheeler, M. H. (1986). Biosynthesis and functions of fungal melanins. *Annual Review of Phytopathology*, **24**, 411–51.

Botton, B. (1978). Influence of calcium on the differentiation and growth of aggregated organs in *Sphaerostilbe repens*. *Canadian Journal of Microbiology*, **24**, 1039–47.

Bradley, R., Burt, A. J. & Read, D. J. (1981). Mycorrhizal infection and resistance to heavy metals. *Nature*, **292**, 335–7.

Burgstaller, W. & Schinner, F. (1993). Leaching of metals with fungi. *Journal of Biotechnology*, **27**, 91–116.

Cooke, R. C. & Rayner, A. D. M. (1984). *Ecology of Saprotrophic Fungi*. London: Longman.

Cooke, R. C. & Whipps, J. M. (1993). *Ecophysiology of Fungi*. Oxford: Blackwell Scientific Publications.

Crawford, J. W., Ritz, K. & Young, I. M. (1993). Quantification of fungal morphology, gaseous transport and microbial dynamics in soil: an integrated framework utilising fractal geometry. *Geoderma*, **56**, 157–72.

Cunningham, J. E. & Kuiack, C. (1992). Production of citric and oxalic acids and solubilization of calcium phosphates by *Penicillium bilaii*. *Applied and Environmental Microbiology*, **58**, 1451–8.

Darlington, A. B. & Rauser, W. E. (1986). Cadmium alters the growth of the ectomycorrhizal fungus *Paxillus involutus*: a new growth model accounts for changes in branching. *Canadian Journal of Botany*, **66**, 225–9.

Dickinson, C. H. & Bottomley, D. (1980). Germination and growth of *Alternaria* and *Cladosporium* in relation to their activity in the phylloplane. *Transactions of the British Mycological Society*, **74**, 309–16.

Dixon-Hardy, J. E., Karamushka, V. I., Gruzina, T. G., Nikovska, G. N., Sayer, J. A. & Gadd, G. M. (1998). Influence of the carbon, nitrogen and phosphorus source on the solubilization of insoluble metal compounds by *Aspergillus niger*. *Mycological Research*, **102**, 1050–4

Fogarty, R. V. & Tobin, J. T. (1996). Fungal melanins and their interactions with metals. *Enzyme and Microbial Technology*, **19**, 311–17.

Franceschi, V. R. & Loewus, F. A. (1995). Oxalate biosynthesis and function in plants and fungi. In *Calcium Oxalate in Biological Systems*, ed. S. R. Khan, pp. 113–30. Florida: CRC Press Inc..

Francis, A. J. (1994). Microbial transformations of radioactive wastes and environmental restoration through bioremediation. *Journal of Alloys and Compounds*, **213/214**, 226–31.

Francis, A. J., Dodge, C. J. & Gillow, J. B. (1992). Biodegradation of metal citrate complexes and implications for toxic metal mobility. *Nature*, **356**, 140–2.

Fries, N. (1973). Effects of volatile organic compounds on the growth and development of fungi. *Transactions of the British Mycological Society*, **60**, 1–14.

Gadd, G. M. (1984). Effect of copper on *Aureobasidium pullulans* in solid medium: adaptation not necessary for tolerant behaviour. *Transactions of the British Mycological Society*, **82**, 546–9.

Gadd, G. M. (1986). Toxicity screening using fungi and yeasts. In *Toxicity Testing using Microorganisms*, Volume II, eds. B. J. Dutka & G. Bitton, pp. 43–77. Boca Raton, Florida: CRC Press.

Gadd, G. M. (1992). Metals and microorganisms: a problem of definition. *FEMS Microbiology Letters*, **100**, 197–204.

Gadd, G. M. (1993). Interactions of fungi with toxic metals. *New Phytologist*, **124**, 25–60.

Gadd, G. M. & De Rome, L. (1988). Biosorption of copper by fungal melanin. *Applied Microbiology and Biotechnology*, **29**, 610–17.

Gadd, G. M., Gray, D. J. & Newby, P. J. (1990). Role of melanin in fungal biosorption of tributyltin chloride. *Applied Microbiology and Biotechnology*, **34**, 116–21.

Gadd, G. M. & Griffiths, A. J. (1978). Microorganisms and heavy metal toxicity. *Microbial Ecology*, **4**, 303–17.

Gadd, G. M. & Griffiths, A. J. (1980). Effect of copper on morphology of *Aureobasidium pullulans*. *Transactions of the British Mycological Society*, **74**, 387–92.

Gadd, G. M. & Mowll, J. L. (1985). Copper uptake by yeast-like cells, hyphae and chlamydospores of *Aureobasidium pullulans*. *Experimental Mycology*, **9**, 230–40.

Gadd, G. M., White, C. & Mowll, J. L. (1987). Heavy metal uptake by intact cells and protoplasts of *Aureobasidium pullulans*. *FEMS Microbiology Ecology*, **45**, 261–7.

Gharieb, M. M., Sayer, J. A. & Gadd, G. M. (1998). Solubilization of natural gypsum ($CaSO_4.2H_2O$) and the formation of calcium oxalate by *Aspergillus niger* and *Serpula himantiodes*. *Mycological Research*, **102**, 825–30.

Gildon, A. & Tinker, P. B. (1981). A heavy metal tolerant strain of mycorrhizal fungus. *Transactions of the British Mycological Society*, **77**, 648–9.

Gow, N. A. R. (1995). Tip growth and polarity. In *The Growing Fungus*, ed. N. A. R. Gow & G. M. Gadd, pp. 277–99. London: Chapman & Hall.

Gray, T. R. G. & Williams, S. T. (1971). Microbial productivity in soil. *Symposia of the Society of General Microbiology*, **21**, 255–81.

Green, F., Clausen, C. A., Kuster, T. A. & Highley, T. L. (1995). Induction of polygalacturonase and the formation of oxalic acid by pectin in brown rot fungi. *World Journal of Microbiology and Biotechnology*, **11**, 519–24.

Hirsch, P., Bernhard, M., Cohen, S. S., Ensigh, J. C., Jannasch, H. W., Koch, A. L., Marshall, K. C., Matin, A., Poindexter, J. S., Rittenberg, J. S., Smith, D. C. & Veldkamp, H. (1979). Life under conditions of low nutrient concentrations. In *Strategies of Microbial Life in Extreme Environments*, ed. M. Shilo, pp. 357–72. New York: Weinheim.

Hughes, M. N. & Poole, R. K. (1991). Metal speciation and microbial growth – the hard (and soft) facts. *Journal of General Microbiology*, **137**, 725–34.

Jones, D., McHardy, W. J., Wilson, M. J. & Vaughan, D. (1992). Scanning electron microscopy of calcium oxalate on mantle hyphae of hybrid larch roots from a farm forestry experimental site. *Micron et Microscopica Acta*, **23**, 315–17.

Karamushka, V. I., Sayer, J. A. & Gadd, G. M. (1996). Inhibition of H^+ efflux from *Saccharomyces cerevisiae* by insoluble metal phosphates and protection by calcium and magnesium: inhibitory effects a result of soluble metal cations? *Mycological Research*, **100**, 707–13.

Ko, W. H. & Lockwood, J. L. (1967). Soil fungistasis: in relation to fungal spore nutrition. *Phytopathology*, **57**, 894–901.

Lapeyrie, F., Chilvers, G. A. & Bhem, C. A. (1987). Oxalic acid synthesis by the mycorrhizal fungus *Paxillus involutus*. *New Phytologist*, **106**, 139–46.

Lapeyrie, F., Ranger, J. & Vairelles, D. (1991). Phosphate solubilizing activity of ectomycorrhizal fungi *in vitro*. *Canadian Journal of Botany*, **69**, 342–6.

Leyval, C., Surtiningsih, T. & Berthelin, J. (1993). Mobilization of P and Cd from rock phosphates by rhizosphere microorganisms (phosphate dissolving bacteria and ectomycorrhizal fungi). *Phosphorus, Sulfur and Silicon*, **77**, 133–6.

Lilly, W. W., Wallweber, G. J. & Lukefahr, T. A. (1992). Cadmium absorption and its effects on growth and mycelial morphology of the basidiomycete fungus, *Schizophyllum commune*. *Microbios*, **72**, 227–37.

Lynch, J. M. (1982). Limits to microbial growth in soil. *Journal of General Microbiology*, **128**, 405–10.

Mattey, M. (1992). The production of organic acids. *Critical Reviews in Biotechnology*, **12**, 87–132.

Melhorn, R. J. (1986). The interaction of inorganic species with biomembranes. In *The Importance of Chemical Speciation in Environmental Processes*, eds. F. E. Brinckman & P. J. Sadler, pp. 85–97. Berlin: Springer-Verlag.

Mirocha, C. J. & DeVay, J. E. (1971). Growth of fungi on an inorganic medium. *Canadian Journal of Microbiology*, **17**, 1373–7.

Morley, G. F. & Gadd, G. M. (1995). Sorption of toxic metals by fungi and clay minerals. *Mycological Research*, **99**, 1429–38.

Morley, G. F., Sayer, J. A., Wilkinson, S. J., Gharieb, M. M. & Gadd, G. M. (1996). Fungal sequestration, solubilization and transformation of toxic metals. In *Fungi and Environmental Change*, ed. J. C. Frankland, N. Magan & G. M. Gadd, pp. 235–56. Cambridge: Cambridge University Press.

Morris, S. J. & Allen, M. F. (1994). Oxalate metabolizing microorganisms in sagebrush steppe soil. *Biology and Fertility of Soils*, **18**, 255–9.

Mowll, J. L. & Gadd, G. M. (1985). The effect of vehicular lead pollution on phylloplane mycoflora. *Transactions of the British Mycological Society*, **84**, 685–9.

Murphy, R. J. & Levy, J. F. (1983). Production of copper oxalate by some copper tolerant fungi. *Transactions of the British Mycological Society*, **81**, 165–8.

Newby, P. J. & Gadd, G .M. (1987). Synnema induction in *Penicillium funiculosum* by tributyltin compounds. *Transactions of the British Mycological Society*, **89**, 381–4.

Nordgren, A., Baath, E. & Soderstrom, B. (1983). Microfungi and microbial activity along a heavy metal gradient. *Applied and Environmental Microbiology*, **45**, 1829–37.

Onions, A. H. S., Allsopp, D. & Eggins, H. O. W. (1981). *Smith's Introduction to Industrial Mycology*. London: Edward Arnold.

Prosser, J. (1995). Mathematical modelling of fungal colonies. In *The Growing Fungus*, ed. N. A. R. Gow & G. M. Gadd, pp. 319–34. London: Chapman & Hall.

Purvis, O. W. & Halls, C. (1996). A review of lichens in metal-enriched environments. *Lichenologist*, **28**, 571–601.

Rayner, A. D. M. (1991). The challenge of the individualistic mycelium. *Mycologia*, **83**, 48–71.

Rayner, A. D. M., Griffiths, G. S. & Ainsworth, A. M. (1995). Mycelial interconnectedness. In *The Growing Fungus*, ed. N. A. R. Gow & G. M. Gadd, pp. 21–40. London: Chapman & Hall.

Ritz, K. (1995). Growth responses of some soil fungi to spatially heterogeneous nutrients. *FEMS Microbiology Ecology*, **16**, 269–80.

Ritz, K. & Crawford, J. W. (1990). Quantification of the fractal nature of fungal colonies of *Trichoderma viride*. *Mycological Research*, **94**, 1138–41.

Rizzo, D. M., Blancchette, R. A. & Palmer, M. A. (1992). Biosorption of metal ions by *Armillaria* rhizomorphs. *Canadian Journal of Botany*, **70**, 1515–20.

Ruhling, N., Baath, F., Nordgren, A. & Soderstrom, B. (1984). Fungi on metal contaminated soil near the Gusum brass mill, Sweden. *AMBIO*, **13**, 34–6.

Sayer, J. A. & Gadd, G. M. (1997). Solubilization and transformation of insoluble inorganic metal compounds to insoluble metal oxalates by *Aspergillus niger*. *Mycological Research*, **101**, 653–61.

Sayer, J. A., Kierans, M. & Gadd, G. M. (1997). Solubilization of some naturally occurring metal-bearing minerals, limescale and lead phosphate by *Aspergillus niger*. *FEMS Microbiology Letters*, **154**, 29–35.

Sayer, J. A., Raggett, S. L. & Gadd, G. M. (1995). Solubilization of insoluble metal compounds by soil fungi: development of a screening method for solubilizing ability and metal tolerance. *Mycological Research*, **99**, 987–93.

Schnurer, J. & Paustian, K. (1986). Modelling fungal growth in relation to nutrient limitation in soil. In *Perspectives in Microbial Ecology*, ed. F. Megusar & M. Gantar, pp. 123–30. Lljubljana, Yugoslavia: Slovene Society of Microbiology.

Serrano, R. (1985). *Plasma Membrane ATPase of Plants and Fungi*. Boca Raton, Florida: CRC Press.

Sheehan, P. L. & Gochenaur, S. E. (1984). Spore germination and microcycle conidiation of two penicillia in soil. *Mycologia*, **76**, 523 7.

Sutter, H-P., Jones, E. B. G. & Walchi, O. (1983). The mechanism of copper tolerance in *Poria placenta* (Fr.) Cke. and *Poria vaillantii* (pers.) Fr. *Material und Organismen*, **18**, 241–63.

Sutter, H-P., Jones, E. B. G. & Walchi, O. (1984). Occurrence of crystalline hyphal sheaths in *Poria placenta* (Fr.) Cke. *Journal of the Institute of Wood Science*, **10**, 19–23.

Tinnell, W. H., Jefferson, B. L. & Benoit, R. E. (1977). Manganese mediated morphogenesis in *Penicillium claviforme* and *Penicillium clavigerum*. *Canadian Journal of Microbiology*, **23**, 209–12.

Tribe, H. T. & Mabadeje, S. A. (1972). Growth of moulds on media prepared without organic nutrients. *Transactions of the British Mycological Society*, **58**, 127–37.

Trinci, A. P. J. (1978). The duplication cycle and vegetative development in moulds. In *The Filamentous Fungi*, Vol. 3, ed. J. E. Smith & D. R. Berry, pp. 132–63. London: Arnold.

Vassilev, N., Baca, M. T., Vassileva, M., Franco, I. & Azcon, R. (1995). Rock phosphate solubilization by *Aspergillus niger* grown on sugar-beet waste medium. *Applied Microbiology and Biotechnology*, **44**, 546–9.

Vivier, H., Marcant, B. & Pons, M-N. (1994). Morphological shape characterization: application to oxalate crystals. *Particle and Particle Systems Characterization*, **11**, 150–5.

Wainwright, M. (1988). Metabolic diversity of fungi in relation to growth and mineral cycling in soil – a review. *Transactions of the British Mycological Society*, **90**, 159–70.

Wainwright, M. (1993). Oligotrophic growth of fungi – stress or natural state. In *Stress Tolerance of Fungi*, ed. D. H. Jennings, pp. 127–44. New York: Marcel Dekker.

Wainwright, M., Ali, T. A. & Killham, K. (1993). Anaerobic growth of fungal mycelium from soil particles onto nutrient-free silica gel. *Mycological Research*, **98**, 761–2.

Wainwright, M., Barakah, F., Al-Turk, I. & Ali, T. A. (1991). Oligotrophic micro-organisms in industry, medicine and the environment. *Science Progress*, **75**, 313–22.

Wainwright, M. & Gadd, G. M. (1997). Fungi and industrial pollutants. In *The Mycota IV. Environmental and Microbial Relationships*, ed. D. T. Wicklow & B. E. Soderstrom, pp. 85–97. Berlin: Springer-Verlag.

Watkinson, S. C. (1979). Growth of rhizomorphs, mycelial strands, coremia and sclerotia. In *Fungal Walls and Hyphal Growth*, ed. J. H. Burnett & A. P. J. Trinci, pp. 91–113. London: Cambridge University Press.

Watteau, F. & Berthelin, J. (1994). Microbial dissolution of iron and aluminium from soil minerals and specificity of hydroxamate siderophores compared to aliphatic acids. *European Journal of Soil Biology*, **30**, 1–9.

Wessels, J. G. H. (1986). Cell wall synthesis in apical hyphal growth. *International Review of Cytology*, **104**, 37–79.

White, C., Sayer, J. A. & Gadd, G. M. (1997). Microbial solubilization and immobilization of toxic metals: key biogeochemical processes for treatment of contamination. *FEMS Microbiology Reviews*, **20**, 503–16.

Williams, S. T. (1985). Oligotrophy in soil: fact or fiction. In *Bacteria in Their Natural Environments*, ed. M. Fletcher & G. D. Floodgate, pp. 81–110. London: Academic Press.

9

Cellularization in *Aspergillus nidulans*

J. E. HAMER, J. A. MORRELL, L. HAMER, T. WOLKOW AND M. MOMANY

Introduction

Fungi form cells by inserting crosswalls called septa. Even the so-called cenocytic fungi (Chytrids and Zygomycetes) delimit sporangia, zoospores and other structures with septa. In the filamentous Basidiomycetes and Ascomycetes, septa are formed along the mycelial thallus defining cell compartments of a uniform length and nuclear content. In all true fungi, septa are formed by a similar process. A site is chosen within the cell for the assembly of the septum. Actin is recruited to this site and can be seen as a complex of dots (in the case of unicellular yeasts) (Adams & Pringle, 1984; Marks & Hyams, 1985) or as a microfilamentous belt in the case of higher fungi (Girbardt, 1979). The actin cytoskeleton facilitates the highly localized, circumferential deposition of cell wall material outside the plasma membrane. In unicellular yeast such as *Saccharomyces cerevisiae* or *Schizosaccharomyces pombe* a primary cell wall layer is deposited followed by the synthesis of a secondary septal wall on either side of the primary septum. The primary wall is enzymatically removed to allow cell separation. In the mycelia of higher fungi, septation is incomplete, leaving a complex structure called the septal pore. There is no obvious stage of secondary septal wall synthesis and no cell separation. Cell separation does occur during the septation processes that produce aerial spores.

This chapter focuses on the mechanisms controlling the pattern of septation in the filamentous fungus, *Aspergillus nidulans*. Several excellent reviews as well as chapters in this volume discuss the control of mitosis in *A. nidulans* (Doonan, 1992; Morris & Enos, 1992) and earlier reviews discuss the diversity in septal ultrastructure in other fungi (Gull, 1978; Cole, 1986). The chapter begins with a brief overview of what is known about septation in budding and fission yeast as these organisms provide important models for understanding filamentous fungal growth processes and illustrate the diversity in septation processes found within the Ascomycetes. This is followed by a discussion of physiological experiments that demonstrate the types of control mechanisms regulating septation in *A. nidulans* and a review of what has been learned from genetic and molecular genetic studies.

The problems of dividing by walls

Two major problems are faced by the septating fungal cell (or hypha). The first problem is coordination. Fungi, like all other organisms, require that cellular constituents have been appropriately increased and carefully segregated prior to the completion of cell division. Cell division (cytokinesis) must occur in coordination with nuclear division and tip growth. During cytokinesis, the cell division kinases (Cdks) must be recycled to prepare for re-entry into G1. To make matters more complicated, not all fungi have a single nucleus per cell. For example, in *A. nidulans*, tip cells contain up to 30 nuclei; sub-apical cells contain one to six nuclei; the vesicle of the conidiophore contains numerous (> 100?) nuclei; sterigmata and conidia are uninucleate, and ascospores are binucleate. Whatever the mechanisms coordinating growth and septation, they must be flexible enough to permit the formation of a wide diversity of cell types.

During septation the cell wall biosynthetic machinery must be redirected from its role in apical expansion to the circumferential synthesis of a precisely positioned wall that divides the fungal cell. Thus, once the problem of coordination is solved, how is the division site chosen and how is cell wall synthesis activated at the division site? Actin localizes to regions of cell wall growth in fungal cells (Heath, 1990) and also to sites of septation (see below). How does the relocalization of the actin cytoskeleton and cell wall biosynthetic machinery to the division site occur? The actin cytoskeleton not only plays a role in the delivery of cell wall biosynthetic enzymes and precursors to the division site, but may also participate in an actomyosin-like contraction at the division site. The central problems of coordination and division-site specification are not specific to members of the Mycota and thus answers to these questions in the genetically tractable Ascomycetes may shed light on division processes in all cells.

Septation in budding yeast

Molecular organization of fungal growth is best understood in the budding yeast *S. cerevisiae* (for a detailed review see Cid *et al.*, 1995). The problem of choosing a division site is largely solved by a molecular memory consisting of cortically located cues which restrict bud growth and septation to a site adjacent to (in haploid cells) or opposite from (in diploid cells) the previous site of budding (for review see Chant, 1996). In axial budding haploid cells, the cortical cues consist of a complex of proteins including those encoded by the *BUD3* and *BUD4* genes. The cortical cues deposited by the previous round of budding define the next division site (pre-bud site) early in G1 of the cell cycle. Components of a GTPase cascade together with other gene products direct the actin cytoskeleton and new cell wall growth to this site to produce the bud (Roemer, Vallier & Snyder, 1996). Assembly of the presumptive bud site also involves the formation of a ring of septin proteins encoded by the *CDC3, CDC10, CDC11*, and *CDC12* genes (reviewed by Longtine *et al.*, 1996). It is thought that these proteins may be involved in the reorganization of actin to the mother-bud neck. Recent studies show that septins play a direct role in the localization of a protein complex, which includes chitin synthase (Chs3p), to the site of septation (Demarini *et al.,* 1997). Septins appear to be conserved among fungi and animals and have been shown to play roles in cytokinesis in fruitflies.

The assembly of the division septum begins by the formation of a chitin ring at the presumptive bud site. Following nuclear division and cytokinesis, a thin disc of chitin is synthesized within the ring to form the primary septum. Chitin synthase III (Chs3p) is required to form the chitin ring at the incipient bud site whereas another chitin synthase (Chs2p) is required for formation of the primary septum. Secondary septal walls composed of glucan and mannans are then formed on either side of the primary septum. Following cytokinesis, the primary septum is digested away and the mother and daughter cells separate (Bulawa, 1993; Cid *et al.*, 1995). The distribution of actin patches and the major chitin synthase, Chs3p, appear first at the presumptive site, become polarized within the growing bud and later relocalize to the bud neck region during division (Lew & Reed, 1995*b*; Santos & Snyder, 1997). Relocalization of chitin synthase is dependent on a novel protein encoded by the *CHS5* gene and appears to involve vesicle mediated delivery along the actin cytoskeleton (Santos & Snyder, 1997). These studies highlight the importance of actin cytoskeleton in targeting the secretion of cell wall biosynthetic precursors and enzymes to the division site. There is as yet little evidence in budding yeast for an actomyosin-type contractile mechanism at the division site. Although budding may be superficially different from septum formation in other fungi, processes such as cortical signalling, relocalization of the actin cytoskeleton and targeted cell wall synthesis are common to all septation processes.

In *S. cerevisiae* the coordination of growth and division is maintained by highly synchronized budding and nuclear division cycles (Hartwell, 1971; Hartwell *et al.*, 1974; Pringle & Hartwell, 1981). Bud growth and entry into the nuclear division cycle is delayed until yeast cells reach a requisite size and only a single bud is initiated during each nuclear division cycle. It has been proposed that this size control reflects the need for a threshold level of G1 cyclins needed to to activate the cell cycle (Nasmyth, 1996). A series of 'checkpoints' monitor, reinforce and maintain this temporal coordination. For example, yeast cells delay entry into G2 if proper polarization of the bud site has not occurred (Lew & Reed, 1995*a*). Defects in chromosome metabolism and segregation are known to not only delay passage through the nuclear division cycle (Hartwell & Weinert, 1989; Murray, 1995; Wells, 1996), but also the completion of anaphase, cytokinesis and septation (Yamamoto, Guacci & Koshland, 1996; Yang *et al.*, 1997). Although a biochemical understanding of these checkpoint pathways awaits further investigation, they are likely to involve conserved components and function in all fungal cells.

Septation in fission yeast

Cells of the fission yeast, *S. pombe*, divide symmetrically following the formation of a centrally located division septum (for review see (Robinow & Hyams, 1989). The process is diagrammed in Fig. 9.1. Like budding yeast, mitosis and septation are also tightly coupled and the position of the septum is dictated by a circumferential actin ring, which appears to undergo condensation and may actually form a contractile ring. Although fission yeasts contain birth scars, they do not appear to play a role in determining the plane of division. There is no evidence for a cortical cue in determining the site of septation, rather the majority of evidence suggests that the division site is dictated by the position of the premitotic nucleus (Chang & Nurse, 1996; Chang, Wollard & Nurse, 1996).

Mitosis is initiated in *S. pombe* by the activation of the cell division kinase $p34^{cdc2}$, which directs the characteristic cytoskeletal rearrangements seen in dividing cells (for review see Simanis, 1995). At the initiation of mitosis, actin moves from the growing poles to the centre of the rod-shaped cell, where it assembles into an equatorial ring which is positioned over the dividing nucleus (Marks & Hyams, 1985). Ingrowth of the

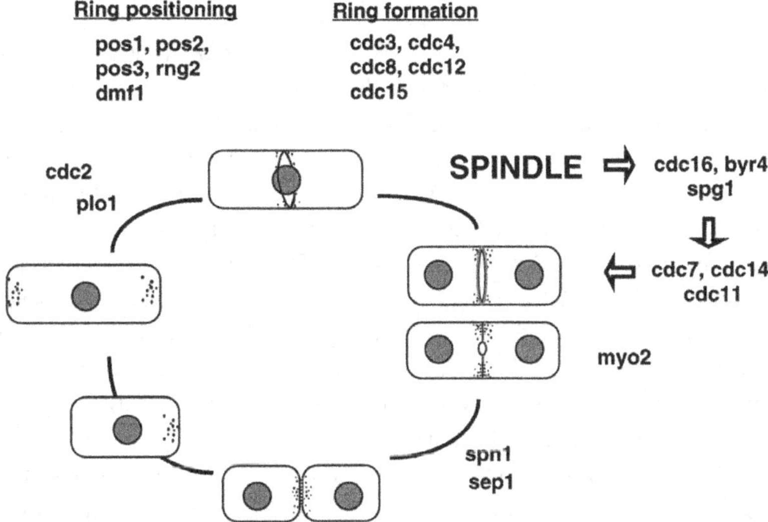

Fig. 9.1 Septation in the fission yeast *Schizosaccharomyces pombe*. Nuclei are shaded and actin localization is shown in black. Gene designations are described in the text. A premitotic cell contains actin at both cell poles. Actin relocalizes to form a ring prior to the onset of mitosis. A mitotic spindle is required for the activation of septal wall synthesis. In this diagram a constricting actin ring is shown to form coincident with the completion of primary wall synthesis. Removal of the secondary wall results in cell separation. A recently divided cell contains actin at only one pole.

primary septum from the cell cortex occurs as the cell cycle progresses. This is followed by synthesis of secondary septa on both sides of the primary septum. During these latter stages, the actin distribution changes from a ring to a cluster of centrally located dots which correlate with sites of septal wall deposition. The temporal order and relationship of primary and secondary wall deposition with actin condensation is not well-defined. Although actin ring formation is independent of the mitotic spindle, the deposition of septal wall material is dependent on the integrity of the mitotic spindle (Chang *et al.*, 1996). Treatments or mutations which disrupt the mitotic spindle prevent septal wall deposition. Following the synthesis of the secondary septum, the primary septum is dissolved away to allow cell separation.

Conditional septation mutants of *S. pombe* have begun to define genes involved in actin ring formation and actin relocalization (for review see Fankhauser & Simanis, 1994*b*; Simanis, 1995). Mutations in the *cdc3*, *cdc4*, *cdc8*, *cdc12*, *dmf1/mid1*, *rng2* or *cdc15* genes, alter or prevent the formation or position of the actin ring (Chang *et al.*, 1996). Among the genes defective in these mutants are those that encode proteins that interact with the actin cytoskeleton, localize to the actin ring, and are conserved in other eukaryotes. These include profilin (*cdc3*) (Balasubramanian *et al.*, 1994), tropomyosin (*cdc8*) (Balasubramanian, Helfman & Hemmingsen, 1992), and a protein with similarity to myosin light-chains (*cdc4*) (McCollum *et al.*, 1995). The *cdc12* gene encodes a member of the FH1–FH2 class of proteins (discussed below). Several members of this class of proteins play roles in cytokinesis and cell polarity in other organisms. Cdc12p localizes to the medial region of the cell during mitosis and interacts genetically and *in vitro* with the *cdc3* product, profilin (Chang, Drubin & Nurse, 1997).

During interphase cdc15p and dmf1/mid1p localize to the nucleus and become hyperphosphorylated during mitosis (Fankhauser *et al.*, 1993*b*; Sohrmann *et al.*, 1996). Both proteins then localize to the plane of division just prior to, or coincident with, the actin ring. Analysis of *dmf1* mutants suggests the gene product plays a role in stabilizing the position of the actin ring whereas *cdc15* is required for ring formation. Double mutant analysis suggests that the *cdc15* gene product may interact with the products of the *cdc3* and *cdc4* genes to bring about actin rearrangements during cytokinesis. The discovery of proteins that move from the nucleus to the plane of division provides support for the nuclear signalling model. Three novel genes, *pos1*, *pos2* and *pos3*, have little or no effect on septation *per se*, but cause actin rings and septa to form near cell poles (Edamatsu & Toyoshima, 1996). Nuclei do not appear to be aberrantly positioned in these mutants and thus these gene products may define positional (cortical ?) cues which direct the location of the actin ring.

Septal wall growth is directed by products of the *spg1*, *byr4*, *cdc7*, *cdc11*, *cdc14*, and *cdc16* genes, and requires an intact mitotic spindle. Mutations in the above genes permit assembly of an actin ring but either prevent or perturb the formation of the division septum. *cdc16* and *byr4* are negative regulators of septation (Fankhauser *et al.*, 1993*a*; Song *et al.*, 1996) and mutations in either gene cause cells to undergo multiple rounds of septation and also appear to arrest cells late in mitosis, suggesting these gene products are also required to complete mitosis. The product of the *cdc16* gene together with the GTPase encoded by the *spg1* gene are proposed to serve as part of a G-protein signalling cascade for septal wall formation (Schmidt *et al.*, 1997). Genetic evidence suggests that products of the *cdc7*, *cdc11*, and *cdc14* genes are also involved in this pathway (Marks, Fankhauser & Simanis, 1992; Fankhauser & Simanis, 1993; Fankhauser & Simanis, 1994*a*; Schmidt *et al.*, 1997). *spg1* interacts with a protein kinase encoded by the *cdc7* gene (Schmidt *et al.*, 1997). The location and substrates of this signalling pathway await discovery.

S. pombe can undergo multiple rounds of septation. Mutations in *byr4* or *cdc16* causes multiple rounds of septation and a failure to complete mitosis. Overexpression of *cdc7* causes a similar phenotype in which cells complete mitosis and then form multiple septa without undergoing cell separation. It is not clear if multiple actin rings form or if septal wall formation is normal during multiple rounds of septation. In contrast with these multi-septation phenotypes, over-expression of *spg1* or over-expression of a protein kinase encoded by the *plo1* gene causes septa to form without the requirement for mitosis (Okhura, Hagan & Glover, 1995; Schmidt *et al.*, 1997). Thus, elevated activity of these gene products obviates the requirement for mitosis, suggesting that *plo1* and *spg1* are important in coordinating growth and septation.

Cell separation is perhaps the least understood stage in fission yeast cell division. The septum of *S. pombe* is primarily composed of β-glucan and thus a β-glucanase must be involved in dissolving the primary septum. Interestingly cell separation does not occur in multi-septating mutants and it has been suggested that cell separation may be regulated independently or that precocious septa may be formed improperly (see discussion in Schmidt *et al.*, 1997). Cell separation is affected by mutations in the *sep1* gene (Sipiczki, Grallert & Miklos, 1993) and by depletion of the type II-like myosin encoded by the *myo2* gene (May *et al.*, 1996; Kitayama, Sugimoto & Yamamoto, 1997). Interestingly myo2p localizes to the actin ring and undergoes a progressive restriction during division suggesting that the failure to undergo cell separation in the *myo2* mutant may reflect a failure to constrict the actin ring (Kitayama *et al.*, 1997). Finally a deletion of the septin encoding gene, *spn1*, delays cell separation. Spn1p localizes to the division plane in anaphase and persists through cell separation, but is undetectable throughout the remainder of the cell cycle (Longtine *et al.*, 1996).

As in *S. cerevisiae*, cell cycle checkpoints play important roles in coordinating nuclear division and septation in *S. pombe*. Nuclear division is delayed if the actin ring is inappropriately positioned (Chang *et al.*, 1996), and septal wall synthesis will not occur if the spindle is compromised. However, there are also many differences between the fission and budding processes. Several of the *S. pombe* septation genes are either absent from the *S. cerevisiae* genome or are represented by only weakly related sequences. In the case of *spg1,* the related gene in *S. cerevisiae* (*TEM1*) could not functionally complement *spg1*$^-$ mutants (Schmidt *et al.*, 1997). Some of these genes are required for septal wall synthesis while others appear to be necessary for actin ring assembly. Taken together these findings suggest that fungi have evolved diverse mechanisms for septation. The well-coordinated septation and division cycle of *S. pombe* provides a framework for addressing the more complex septation patterns seen in filamentous ascomycetes.

Septation in *Aspergillus nidulans*

Growth and septation patterns in filamentous fungi are different from those observed in unicellular yeasts. In mycelia, growth is indeterminate and cell division is incomplete and thus 'daughters' never separate and formal cell linages never form. Septa may be dispensable at times, and as previously mentioned, patterns of division and septation can change dramatically throughout the haploid life cycle.

Ascomycetes, including *A. nidulans*, contain regularly spaced simple septa. Ultrastructural analysis has shown that these septa have several electron-dense layers perforated by a central pore (Hunsley & Gooday, 1974; Momany & Hamer, 1997). Spheroid, electron-dense vesicles, termed Woronin bodies, are often seen associated with the central pore (Richle & Alexander, 1965; Bracker, 1967; Wergin, 1973; Collinge & Markham, 1985). These Woronin bodies appear to plug the septal pore in older or injured hyphae. Much of the information about the fungal septum is from reports detailing its ultrastructure and very little is known about the biochemical nature of the septal components.

The physiological role of the septum in filamentous fungi is also poorly understood. It is thought that the construction of crosswalls may provide mechanical strength to the growing hypha. The septa may also serve to localize damage to the tip cell or a few adjacent compartments following various types of injury. Gull (1978) has argued that the main function of septa may be to allow the fungus to differentiate specialized cell types. Specific cellular environments required for the development of new cell types may require compartmentalization by septa. In *A. nidulans*, septa partition growing tip cells from quiescent subapical cells (see below). These subapical cells can re-establish growth by branching and have been proposed to serve as a 'stem cell' for tip cells and for the formation of reproductive structures (Timberlake, 1990).

Septation in *A. nidulans* can be observed during spore germination or during tip cell growth in mycelia (Fig. 9.2). The uninucleate spores of *A. nidulans* undergo the first two nuclear divisions as the spore swells and establishes a polar growth axis for germ tube emergence (for reviews see d'Enfert, 1997 or Harris, 1997). The nuclei then migrate into the elongating germ tube and following a second round of nuclear division, the four nuclei become evenly spaced along the length of the hypha. The term predivisional cell has been used to distinguish a mitotically active multinucleate cell that has not formed a septum (Harris, 1997). In this case the predivisional cell is a germling with two to four nuclei. A third round of mitosis occurs to produce an eight-nuclei germling and a septum is produced at the basal end of the germ tube. The term post-divisional cell distinguishes hyphal cells and germlings which have completed the formation of a septum. In a majority of germlings five nuclei will reside on the tip side of the germling while three nuclei will reside on the spore side.

In a growing tip cell, nuclear division proceeds as a parasynchronous wave initiating in the tip proximal nucleus and progresses basally (Robinow & Caten, 1969; Clutterbuck, 1970; Fiddy & Trinci, 1976). Tip cells may contain up to 30 nuclei. Closely following this wave of mitosis, one or more septation events occur in the basal region of the

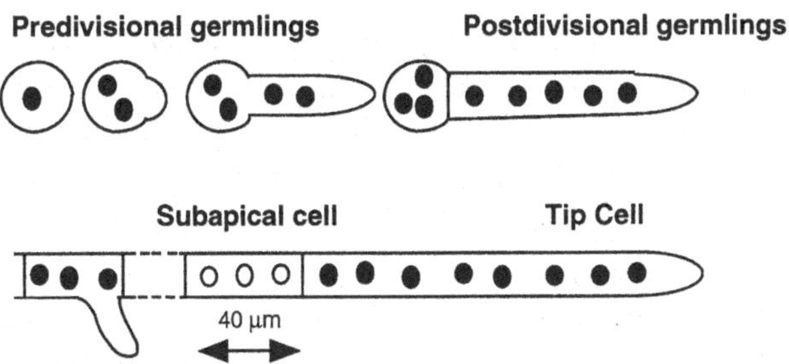

Fig. 9.2 Spore germination, nuclear division and septation in *Aspergillus nidulans*. Early nuclear divisions in a germinating spore (predivisional germling) do not activate septation. A septum is formed at the basal end of the germ tube after the cell reaches a critical size (postdivisional germling). Septation is dependent on mitosis. In a growing hyphal cell (shown below) dividing nuclei (black) are confined to the tip compartment or to branching cell establishing new tips. Tip cells can contain many nuclei. Nondividing nuclei (shown in white) are confined to nonbranching subapical cells (or intercalary compartments). These cells are approximately 40 µm in length.

tip cell. Septa are generally formed within 40 μm of the previous septum and are never formed near the tip (Trinci & Morris, 1979; Wolkow, Harris & Hamer, 1996; Harris, 1997). Following septation, nuclear division and growth are arrested in the penultimate or subapical cell. These cells contain an average of three to four nuclei, average 40 μm in length and will remain in an arrested growth state until the formation of a branch. Branch formation reinitiates the nuclear division cycle and re-establishes a growing tip.

After a growth period of approximately 15–20 hours in rich media, *A. nidulans* sub-apical cells develop developmental competence and can initiate conidiophore formation (Axelrod, Gealt & Pastushok, 1973). Conidiophores differentiate as an aerial branch that emerges at a 90° angle towards the apical side of a subapical cell compartment. This branch develops an apical vesicle and multiple nuclear divisions fill the branch and vesicle with nuclei (Timberlake, 1993). No septa are formed during this process. As nuclei fill the vesicle, conidiogenous cells, called sterigmata, bud from the surface of the vesicle and receive a single nucleus. These cells then bud in a unipolar fashion to form a second uninucleate cell layer of phaillide cells. Phaillide cells undergo multiple rounds of unipolar budding and nuclear division to produce acropetal chains of uninucleate spores.

Septation is dependent on mitosis and a critical cell size

During the unicellular sporulation process, mitosis and cell division are strictly coordinated in a manner analogous to the budding process of *S. cerevisiae*. However, mitosis is also required for septation during mycelial growth. Agents which block germlings in the nuclear division cycle before or during mitosis, such as hydroxyurea (S-phase) and the antimicrotubule drug, benomyl (M-phase), also block septation in *A. nidulans* (Harris, Morrell & Hamer, 1994; Momany & Hamer, 1997). In addition, several mitotic mutants blocked in either S-phase or at the G2/M boundary can form substantial germ tubes but never undergo septation (Wolkow *et al.*, 1996). Finally, block and release experiments with predivisional germlings demonstrate that the third round of mitosis is necessary to activate septation (Harris *et al.*, 1994; Fig. 9.3).

The delay of septation in predivisional cells could be due to a requirement that cells attain a critical number of nuclei to support mycelial growth. However, numerous filamentous fungi contain uninucleate cell compartments, and a small fraction of *A. nidulans* cell compartments also

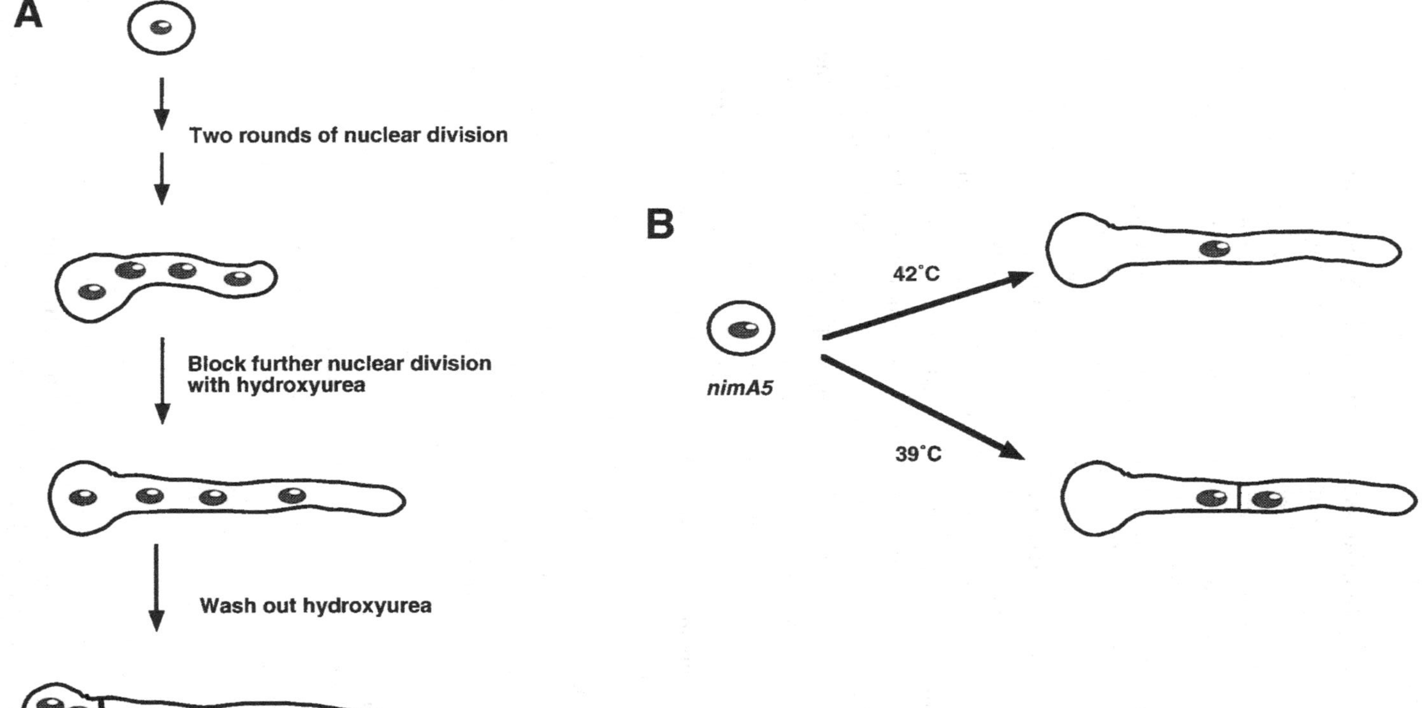

Fig. 9.3 The relationship between mitosis, cell growth and septation in *A. nidulans*. A. Block and release experiments with hydroxyurea demonstrate that mitosis is required for septation (Harris *et al.*, 1994). During the hydroxyurea block, cell growth continues but septa do not form. B. Experiments with the mitotic mutants *nimA5* demonstrate that a single nuclear division can activate septation provided it occurs in an elongated cell (Wolkow *et al.*, 1996). Septa form in binucleate *nimA5* germlings with nuclei positioned on either side.

contain single nuclei (unpublished observations). Alternatively, predivisional cells may require a minimum cell size before mitosis can activate the formation of a septum. Support for this latter hypothesis comes from the observation that a single nuclear division could activate septation if it occurred in a germinated spore with an elongated germ tube (Wolkow *et al.*, 1996; see Fig. 9.3). *nimA* encodes a protein kinase required for the G2–M transition (Osmani, May & Morris, 1987; Osmani *et al.*, 1991). At restrictive temperature *nimA5* germlings produce a germling with a single nucleus and an elongated germ tube. At a semi-restrictive temperature *nimA5* germlings still produce an elongated germ tube but are delayed in completing mitosis. Binucleate *nimA5* germlings were found to contain a single septum positioned between the two nuclei. Another nuclear division mutant, *hfaB3*, also produced septated, binucleate germlings at the semi-restrictive temperature. These findings demonstrate that a single mitosis is sufficient to activate septation in *A. nidulans* and that septation occurs between dividing nuclei. Thus, in wildtype cells of *A. nidulans* the coordination problem could be solved by the requirement for both a minimal cell size and nuclear division to activate septation. Modification of the cell size control mechanism would permit septum formation to occur in cells with different numbers of nuclei.

This requirement for a critical cell size is reminiscent of size control mechanisms regulating the cell cycle in uninucleate cells. In bacterial, yeast and cultured animal cells, the onset of division is more strictly governed by cell size than by the absolute time of growth (Tyson, 1985). This phenomena, referred to as size control, ensures that all necessary components are present in sufficient quantity before division occurs. In fission yeast, the cell size is controlled at both the G1–S and G2–M boundaries (Nurse, 1975, 1990). Mutations in the *wee1* gene abolish the size control at the G2–M boundary and cells progress into mitosis and septation with a smaller cell size. The absolute time to complete mitosis, G1 and S-phases are unchanged in a *wee1* mutant (Sveiczer, Novak & Mitchison, 1996). *wee1* encodes a highly conserved tyrosine protein kinase whose substrate is the tyrosine at position 15 on the universal cell cycle regulatory protein kinase, $p34^{cdc2}$. Phosphorylation of this tyrosine inhibits $p34^{cdc2}$ kinase activity and delays entry into mitosis. The *wee1* kinase is negatively regulated by another protein kinase encoded by the *nim1* gene, however the mechanism by which *nim1* senses the size of the *S. pombe* cell is not clear. Dephosphorylation of tyrosine-15 on $p34^{cdc2}$ is mediated by a phosphatase encoded by the *cdc25* gene.

Mutations in *cdc25* causes cells to become quite large but remain blocked at the G2–M boundary (Nurse, 1990).

Some of the components of this regulatory system have also been identified in *A. nidulans* (Doonan, 1992; Ye *et al.*, 1996; Osmani & Xe, 1997; Ye *et al.*, 1997). The role of these genes in regulating cell size at septation was studied by observing the phenotypes of mutants germinated at semi-restrictive temperatures. Interestingly, mutations in *nimX*^{cdc2} and *nimT*^{cdc25} , which reduce the level of tyrosine phosphorylated p34, caused the formation of large multinucleate cells (> 16 nuclei) without septa at semi-restrictive temperatures. In contrast, mutations that prevented tyrosine phosphorylation of p34^{cdc2}, accelerated the onset of septation and germlings septated at smaller sizes. These results provide evidence that the size control at septation in *A. nidulans* may involve a regulatory circuit analogous to the size control at the G2–M boundary in fission yeast (X. S. Ye, S. McQuire, T. Wolkow, J. E. Hamer and S. A. Osmani, unpublished results; Harris & Kraus, 1997).

Septum positioning is dependent on nuclear positioning

In a growing hypha of *A. nidulans*, the placement of septa is very regular, with an average compartment length of 38 μm (Wolkow *et al.*, 1996). These hyphal compartments usually contain three nuclei, but the number of nuclei can vary from one to eight (Trinci & Morris, 1979). The mechanisms which regulate the positions of septa are unknown. One possibility is that septation sites may be determined by the positions of nuclei. Classical micromanipulation experiments in the basidiomycete *Trametes versicolor* provide evidence for this idea (Girbardt, 1979). However, another possible mechanism is suggested from experiments with *sepA2*, a temperature-sensitive aseptate mutant of *A. nidulans* (Trinci & Morris, 1979). At 37°C, *sepA2* mutants grew and formed aseptate microcolonies. When shifted to permissive temperature (28°C) septa were formed at near regular intervals throughout the mycelium. These findings suggested that septal primordia were already positioned along the growing mycelium at 37°C, and were activated when the colony was shifted to the permissive temperature. New thermo-sensitive aseptate mutants were isolated and shown to behave in a similar manner (Harris *et al.*, 1994). Thus septal positioning could also be determined by cortically positioned cues deposited in the walls of the growing mycelium at regular intervals. Once a critical cell size is attained, mitosis could activate division at one of these predetermined sites.

To distinguish between these two hypotheses (cortical or nuclear signalling), we took advantage of the availability of thermosensitive mutations in nuclear distribution (*nud*) genes (Fig. 9.4) (Wolkow *et al.*, 1996). Hyphal compartment lengths were measured in several *A. nidulans* temperature-sensitive mutants defective in nuclear distribution (*nudA2*, *nudC3* and *nudF7*; for review see Beckwith, Roghi & Morris, 1995). Each of these mutations defined a different gene product involved in the distribution and positioning of nuclei within hyphae. When grown at a semi-restrictive temperature, the nuclei of these mutants distributed unevenly along the hyphae, producing clumps of nuclei in some areas and leaving other areas devoid of nuclei. Although the process of septum formation appeared normal, the compartment sizes produced in the *nud* mutants were extremely variable, ranging from 5 μm to more than 100 μm, suggesting that the abnormal positioning of nuclei affected the positions of septa. The formation of cell compartments much shorter than 40 μm, provided evidence against a cortical marking model for septation.

In a second experiment nuclei were displaced during germination using nuclear division block and release experiments with hydroxyurea. In germlings, septa are positioned invariantly at the basal end of the emerging germ tube. During the block, nuclei could still migrate into the

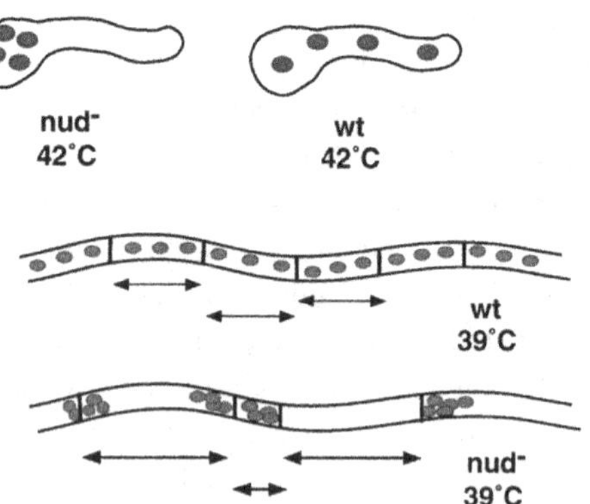

Fig. 9.4 Nuclear distribution (nud⁻) mutants demonstrate that nuclear positioning determines the position of septa. nud⁻ mutants cannot move nuclei into an extending germ tube at restrictive temperature. At semi-restrictive temperatures, the nuclei of nud⁻ mutants appear in clumps throughout the mycelium. As a result, septa become abnormally positioned and intercalary compartment sizes (arrows) become highly variable (Wolkow *et al.*, 1996).

growing germ tube; upon release of the block, nuclear division was displaced and, predictably, septa formed further up the emerging germ tube (Fig. 9.4). Together these two results provide compelling evidence that septum positioning is determined by the positions of nuclei. Given these results, how are the experiments with the *sepA* mutant to be interpreted ?

The *sepA* gene has been cloned (see below) and the temperature-shift experiments (shift-down) repeated by Harris *et al.* (1997). In these experiments nuclear division was arrested during the shift to permissive temperature. When *sepA* mutants were shifted in the presence of mitotic inhibitors, septation was completely blocked. Thus when *sepA2* mutants are returned to permissive temperature, nuclear division is require to activate septation and requires the *sepA* gene product. These findings further support a model where nuclei signal the positions of septa. This signal must arise after the benomyl-sensitive step in mitosis and suggests that dividing nuclei activate the formation of the division septum in *A. nidulans*.

Septum formation involves an invaginating actin ring dependent on microtubules

As stated in the introduction, the involvement of actin in septum formation in fungi is well-documented and actin has been localized to the septa of yeasts (Adams & Pringle, 1984; Marks & Hyams, 1985; Alfa & Hyams, 1990), filamentous ascomycetes (Robertson, 1992; Harris *et al.*, 1994; Bruno *et al.*, 1996) and basidiomycetes (Runeberg, Raudaskoski & Virtanen, 1986). In *A. nidulans*, septation can be reversibly blocked in germlings using the actin depolymerizing agent cytochalasin A, suggesting that this process requires filamentous actin (Harris *et al.*, 1994). In filamentous fungi and yeasts there is a striking difference in the pattern of actin staining during division (compare Figs. 9.1 and 9.5). In yeasts, actin relocalizes from growing poles to the plane of division during septation. However in *A. nidulans*, actin is maintained at the growing tip throughout septation (Harris *et al.*, 1994; Momany & Hamer, 1997). It is not clear if there is a localized loss of actin staining in cell areas adjacent to the division site.

The occurrence of actin arrays at fungal septation sites is probably required for localized deposition of cell wall material. However, it may also play a contractile role in pinching in the plasma membrane. Detailed studies of division in synchronized germlings have documented the dynamics of actin ring formation and chitin deposition at the septum in *A. nidulans* (summarized in Fig. 9.5, Momany & Hamer, 1997). These studies indicate the presence of faint actin rings positioned over mitotic nuclei, suggesting that actin ring assembly begins before the completion of nuclear division. After mitosis, the actin ring appears to constrict and invaginate. Invagination of the actin ring occurs coincidentally with the formation of a circumferential ring of Calcofluor-staining material (chitin). Confocal microscopy reveals that the invaginating actin ring forms an hourglass structure with the Calcofluor ring localized over the mid-point of the invagination. Later the actin hourglass appears to condense and becomes shorter in diameter. Coincidentally the chitin ring

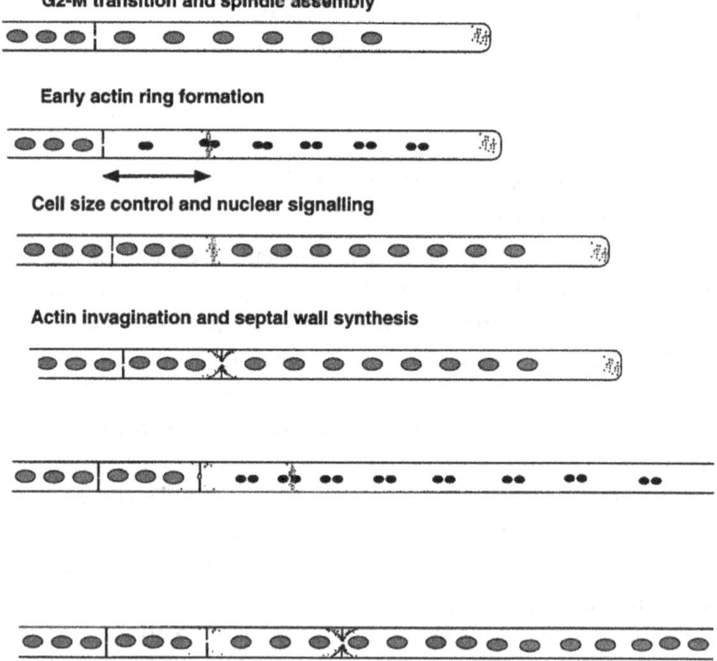

Fig. 9.5 Growth, nuclear division and septation in growing hyphae of *A. nidulans*. Actin (black dots) and interphase nuclei (shaded) and condensed mitotic nuclei (black) are shown. During septation actin accumulates as a ring either near mitotic nuclei or between interphase nuclei. Actin then forms an invaginating ring, which later condenses into a smaller ring. Septal wall synthesis occurs coincidently with actin invagination and condensation. Actin then dissipates leaving an incomplete septal wall containing a central pore. Actin remains at the tip throughout the septation process (Momany & Hamer, 1997).

becomes thicker and more densely staining. Finally the actin dissipates into a series of medial spots leaving just the chitinous ring.

These observations have led to the suggestion that septal wall synthesis occurs as the actin ring invaginates. It can be speculated that as actin invaginates it may pull the membrane inward and cell wall material, including chitin, may be deposited behind the leading edge of the invagination. Chitin synthases are membrane-bound enzymes that receive monomer on the cytoplasmic face of the membrane and extrude polymer to the outer face (Cabib, Bowers & Roberts, 1983). Without the actin invagination, activation of cell wall biosynthetic enzymes could lead to a general thickening of the septal wall rather than the highly localized cell wall synthesis observed during septation. Chitin synthases may be activated by membrane stress (Gooday & Schofield, 1995) and thus the pulling inward of the plasma membrane by actin invagination could serve to activate chitin synthesis at the septum. Localization of chitin synthases and actin ring components is needed to understand this process more fully.

The position of the cleavage furrow of animal cells is formed perpendicular to the axis of the mitotic spindle (Rappaport, 1986; Conrad & Schroeder, 1990; Satterwhite & Pollard, 1992). This positioning is dictated by a signalling mechanism requiring either midzone or astral microtubules (for discussion see Oegema & Mitchinson, 1997). By comparison, microtubules are dispensable for bud site selection in *S. cerevisiae* and for actin ring formation in *S. pombe*. However, intact microtubules are required for the completion of septum formation in *S. pombe*. The addition of benomyl to dividing *A. nidulans* germlings arrested actin ring formation and prevented actin rings from undergoing invagination (Momany & Hamer, 1997). None of these effects occurred in a strain carrying a β-tubulin mutation that conferred resistance to benomyl. *A. nidulans* may employ a microtubule-dependent signal to initiate actin ring formation and invagination. Time-course studies on the dynamics of microtubules at the time of septation are needed to determine which populations of microtubules (astral, spindle or cytoplasmic) are present during septation.

Summary of physiological controls on septation in Aspergillus

In *A. nidulans* a threshold cell size controls the ability of a mitotic division to activate actin ring formation. The cell size control could be mediated by the titration of a septation inhibitor or the accumulation of a septation activator. This activator or inhibitor may promote the ability of *nimX*^{cdc2} kinase to activate actin ring formation in response to signals arising from microtubules. Actin ring assembly is followed by the invagination of the actin ring in conjunction with septal wall deposition. In growing hyphae, we have proposed that a septation inhibitor exists in a tip at high concentration, restricting the activation of cytokinesis to nuclear divisions that occur in the basal end of the tip cell (Wolkow *et al.*, 1996). The time required to form a septum from the first appearance of an actin ring to the dissipation of the actin hourglass structure is approximately 2–3 h at 28°C (Momany & Hamer, 1997). In contrast, the nuclear division cycle is approximately 100 min at this temperature (Bergen & Morris, 1983). The differences in timing accounts for the observation that multiple rounds of septation occur during vegetative hyphal growth (Clutterbuck, 1970). In the next section is a discussion of the genetic analysis of septation in *A. nidulans* and a glimpse of the molecular identities of some of the gene products regulating this process is provided.

Genetic analysis of septation in *Aspergillus*

Ron Morris first described septation (*sep*) mutants of *A. nidulans* (Morris, 1976). A temperature sensitive (Ts) mutant screen identified five mutations in three genes: *sepA1*, *sepA2*, *sepB2*, *sepB3* and *sepD5* (a mutation designated *sepC4* was later found to be an allele of *sepB* and renamed; Harris & Hamer, 1995). The *sep* mutants were distinct from all other classes of Ts mutants identified by Morris, in that they underwent several rounds of nuclear division but failed to form septa. A second screen generated 1150 Ts mutants and identified an additional allele of *sepA*, and new genes designated *sepE*, *sepI*, *sepJ*, *sepG* and *sepH*. Only single alleles were isolated in most cases, suggesting that additional *sep* genes await genetic identification (Harris *et al.*, 1994).

Three phenotypic classes were obvious from microscopic examination of the Ts phenotypes. Mutations in *sepB*, *sepE*, *sepI* and *sepJ* arrested germlings with 8–16 nuclei and no septa. When growth-arrested cells were shifted to permissive temperature (shift-down) no septa were formed within two hours. These findings suggested that an important regulatory step occurred near the time of septation and that mutations in *sepB*, *E*, *I* and *J* prevented germlings from septating and growing past this point. Mutations in *sepA* did not arrest growth or nuclear division at restrictive temperature but caused the formation of wide hyphae with dichotomous tips (branching tip cells). Similarly mutations in *sepD*, *G* and *H* also did not arrest growth but did prevent septation. Shift-down experiments with *sepA*, *D*, *G*, and *H* showed that the aseptate phenotype could be readily reversed. These findings demonstrate that septa are not strictly required for growth in *A. nidulans*. Double mutant analysis has largely confirmed that *sepB*, *sepE*, *sepI* and *sepJ* act before *sepA*, *sepD*, *sepG* and *sepH*. No double mutant phenotypes resulted in suppression, however several double mutants could not be obtained, suggesting that these mutant combinations may produce synthetic lethal phenotypes (Morrell, 1997).

Predivisional cell death

A distinct feature of *sepB*, *I* and *J* mutants is that they accumulate long, stringy nuclei characteristic of aneuploids (Harris *et al*., 1994; Harris & Hamer, 1995; Morrell, 1997). *sepB* enodes a gene with weak similarity to an *S. cerevisiae* gene implicated in chromosome segregation, called *CTF4*. However, unlike *CTF4*, *sepB* is an essential gene and its encoded polypeptide lacks structural motifs required for *CTF4* function (Harris & Hamer, 1995). Loss of *sepB* function leads to increasingly abnormal nuclear morphology, defects in chromosome segregation, transient increases in the chromosome mitotic index, aneuploidy, a failure to septate and cell death. The cell death phenotype of *sepB* mutations is surprising because the vast majority of nuclear division mutants of *A. nidulans* can resume growth when returned to permissive temperature. Interestingly, despite a range of defects in chromosome metabolism in *sepB* mutants, mitotic spindles appeared to form normally, and *sepB* mutants can arrest and resume mitosis at restrictive temperature in response to the addition and removal of mitotic inhibitors.

The above results suggest that the effect of *sepB* on septation may be indirect. In the absence of *sepB*, predivisional cells may accumulate a type of cellular damage that activates a checkpoint response to arrest growth and division and eventually trigger cell death. Support for this hypothesis comes from a recent finding that if a *sepB* mutation is combined with mutations in genes involved in DNA-damage checkpoint controls, the aseptate phenotype is rescued. Other chemicals (for example, mutagens) and mutations which induce DNA damage also delay septation (Harris & Kraus, 1997). In the case of *sepB*, the *sepB*-induced damage would not be repaired when cells are maintained at 42°C and thus cell death would ensue. Thus the DNA-damage checkpoint in *A. nidulans* appears to block septation and may trigger cell death in the predivisional tip cell. Such a regulatory mechanism would prevent tip cells from segregating damaged nuclei to subapical cells. The specific type of damage that accumulates in *sepB* mutants is not known and its relationship to septation is currently under investigation. *sepJ* and *sepI* mutants also accumulate aberrant nuclei, fail to septate and eventually die at restrictive temperature (S. A. Harris, T. A. Wolkow and J. E. Hamer, unpublished results). Studies of these gene products may reveal how cell death is triggered in fungal cells and if predivisional cell death is an important regulatory pathway for maintaining the integrity and durability of the fungal mycelium.

SepA *and actin ring formation*

sepA mutant phenotypes (wide, aseptate hyphae; dichotomous branching) are reminiscent of cells treated with dilute concentrations of cytochalasin A (Harris *et al.*, 1994). *sepA* encodes a large protein with carboxy-terminal FH1 (formin homology 1) and FH2 (formin homology 2) domains (Harris *et al.*, 1997). The formins are a family of proteins identified in mice as playing roles in limb morphogenesis (Jackson-Grusby, Kuo & Leder, 1992). Proteins with FH1 and FH2 domains have been identified from numerous organisms and many of these genes have specific functions in cytokinesis. Among these proteins, similarities are largely confined to the FH1 and FH2 regions. The FH1 region is proline rich and studies in yeast (Evangelista *et al.*, 1997), *Drosophila* (Manseau, Calley & Phan, 1996) and human cells (Watanabe *et al.*, 1997), demonstrate that FH1/FH2 domain proteins interact with the actin-binding protein profilin. The amino-terminal regions of the *S. cerevisiae* formins, encoded by the *BNI1* and *BNR1* genes, are also capable of interacting with active forms of small GTPases (Rho and Cdc42) (Kohno *et al.*, 1996; Evangelista *et al.*, 1997; Imamura *et al.*, 1997). Localization of these FH1/FH2 proteins in budding and fission yeast show that they localize to sites of septation and also to sites of polarized growth in budding yeast. A model proposed for *S. cerevisiae* suggests that Bni1p may act as part of a multimeric complex that serves to localize the actin cytoskeleton to sites of polarized growth (Evangelista *et al.*, 1997).

The phenotypes of conditional *sepA* mutants are consistent, with a role for *sepA* in cell polarity and septation. Interestingly a *sepA* deletion (*sepA5EBm*) mutant was viable at 28°C but was thermosensitive for growth at 42°C (Harris *et al.*, 1997). At 28°C the deletion mutant was dramatically delayed in forming septa as a result of a delay in actin ring formation. Conidiation occurred at 28°C but conidiophore primary stalk and vesicles became depolarized and branched. At 42 °C the deletion mutant could germinate, undergo several rounds of nuclear division but remained aseptate with defects in cell polarity and eventually ceased growth. These phenotypes suggest *sepA* plays roles in both maintaining cell polarity and promoting septation. By analogy with other FH1/FH2 proteins, both events are most likely mediated through the interactions between *sepA* and components regulating the assembly of the actin cytoskeleton. The thermosensitive phenotype of the *sepA* deletion mutant suggests that *sepA* is dispensable for growth at 28°C but does act as part of a thermally labile complex.

Nuclear signalling and actin ring formation

As mentioned previously, shift-down experiments showed that the aseptate phenotypes in *sepA2, sepD5, sepH1* and *sepG1* were readily reversible when strains were shifted to permissive temperature after overnight growth at 42°C. However, if mitosis is blocked in these mutants during the shift-down experiments, no septa are formed (J. Marhoul, K. Bruno & J. E. Hamer, unpublished results). Consistent with this observation, *sepA2, sepG1* and *sepH1* mutants lack actin rings at restrictive temperature but are able to maintain actin arrays at apical tips (unpublished results). Together these results suggest that the *sepA* encoded formin, and the products of *sepG, sepH* (and possibly *sepD*) genes function between the completion of mitosis and the assembly of the actin ring. *sepD5, sepG1* and *sepH1* mutants are only moderately reduced in growth rate and have no observable defects in cell polarity. These results suggest that these gene products are unlikely to encode structural components of the actin cytoskeleton. The *sepH* gene has recently been cloned and encodes a protein highly related to the cdc7 kinase of *S. pombe* (Morrell, 1997). As mentioned earlier this protein kinase has been shown to be part of a signalling pathway for septation in *S. pombe*. The pathway is dispensable for actin ring formation but necessary for septal wall synthesis. In *A. nidulans sepH* mutants appear to lack actin rings, thus it can be speculated that this same pathway regulates actin ring formation in *A. nidulans*.

Summary

The transitions in the cell cycle from G1 to S phase and from G2 to M phase have been studied in considerable detail. Comparatively less is known about the transition from mitosis to cytokinesis and the return to G1. Central to this understanding is the ability to identify, isolate and study the proteins involved in regulating cytokinesis. Considerable cytological data strongly supports the idea that septating fungi assemble a contracting actin ring as an intermediate in the septation process and studies in *A. nidulans* and *S. pombe* demonstrate a requirement for microtubule-based signals in order to complete septation successfully. For cell biologists, these findings suggest that genetic studies of septation in fungi will be an important complement to ongoing biochemical studies of cytokinesis in embryos, eggs and cultured cells.

For mycologists, the study of septation offers the opportunity to understand the regulation of mycelial growth. Septation results in the cessation of growth and nuclear division in subapical cells. These subapical cells provide a reservoir of growth potential for the extending mycelium. Mycelia of many filamentous fungi grow indefinitely, and the cell cycle arrest of subapical cells may be an important adaptation for dealing with adverse growth conditions. In most fungi, the nuclei within subapical cells ultimately give rise to new branches of growth and various types of spores. Preliminary results in *A. nidulans* suggest that a novel regulatory control may prevent septation and trigger cell death when the predivisional cells become irreversibly damaged. We speculate that such a regulatory mechanism would prevent damaged nuclei from being segregated to subapical cell compartments. Further studies of this process may yield important clues to controlling fungal growth and triggering fungal cell death.

A cell size control may regulate the onset of septation in *A. nidulans* and this size control may provide the necessary flexiability to account for the multinucleate state of hyphal cell compartments. Although the wee1 kinase in *S. pombe* has been studied in some detail and homologues identified in numerous species, little is known about how wee1 activity is modulated by cell size. During sporulation, *A. nidulans* cells become uninucleate, suggesting that this size control mechanism must be modified during asexual reproduction. Studies of septation should provide insights into how fungi control simple morphogenetically triggered changes to their cell cycle as well as insight into how they control cell size.

Finally, the current genetic and cytological studies of septation in *A. nidulans* may eventually offer insights into the functioning of the septal pore complex. The phenotypes of septal pore complex mutants cannot be predicted with confidence and the identification of components of this complex will likely depend on biochemical approaches and novel genetic screens that can rapidly identify and localize gene products. Studies of the septal pore complex may eventually allow us to understand how biochemical information is relayed between fungal cells – undoubtedly an important regulatory process for the fungal colony.

Acknowledgements

The authors wish to thank Steve Harris (University of Connecticut) for continuously sharing ideas about septation and communicating unpublished results. The authors also thank Drs John Marhoul and Kelly Kinch for kindly reviewing and editing the manuscript. Research in the Hamer laboratory on septation is supported by National Institutes of Health, General Medical Sciences Program.

References

Adams, A. E. M. & Pringle, J. R. (1984). Relationship of actin and tubulin distribution to bud growth in wildtype and morphogenetic mutants of *Saccharomyces cerevisiae*. *Journal of Cell Biology*, **98**, 943–5.

Alfa, C. E. & Hyams, J. S. (1990). Distribution of tubulin and actin through the cell division cycle of the fission yeast *Schizosaccharomyces japonicus* var. *versatilis*, a comparison with *Schizosaccharomyces pombe*. *Journal of Cell Science*, **96**, 71–7.

Axelrod, D. E., Gealt, M. & Pastushok, M. (1973). Gene control of developmental competence in *Aspergillus nidulans*. *Developmental Biology*, **34**, 9–15.

Balasubramanian, M. K., Helfman, D. M. & Hemmingsen, S. M. (1992). A new tropomyosin essential for cytokinesis in the fission yeast *S. pombe*. *Nature*, **360**, 84–7.

Balasubramanian, M. K., Hirani, B. R., Burke, J. D. & Gould, K. L. (1994). The *Schizosaccharomyces pombe cdc3+* gene encodes a profilin essential for cytokinesis. *Journal of Cell Biology*, **125**, 1289–302.

Beckwith, S. M., Roghi, C. H. & Morris, N. R. (1995). The genetics of nuclear migration in fungi. In *Genetic Engineering*, Vol. 17, ed. J. K. Setlows, pp. 165–79. New York: Plenum Press.

Bergen, L. G. & Morris, N. R. (1983). Kinetics of the nuclear division cycle of *Aspergillus nidulans*. *Journal of Bacteriology* **156**, 155–60.

Bracker, C. E. (1967). Ultrastructure of fungi. *Annual Review of Phytopathology*, **5**, 343–74.

Bruno, K. S., Aramayo, R., Minke, P. F., Metzenberg, R. L. & Plamann, M. (1996). Loss of growth polarity and mislocalization of septa in a Neurospora mutant altered in the regulatory subunit of cAMP-dependent kinase. *EMBO Journal*, **15**, 5772–82.

Bulawa, C. E. (1993). Genetics and molecular biology of chitin synthesis in fungi. *Annual Review of Microbiology*, **47**, 505–34.

Cabib, E., Bowers, B. & Roberts, R. L. (1983). Vectorial synthesis of a polysaccharide by isolated plasma membranes. *Proceedings of the National Academy of Sciences, USA*, **80**, 3318–21.

Chang, F., Drubin, D. & Nurse, P. (1997). cdc12p, a protein required for cytokinesis in fission yeast, is a component of the cell division ring and interacts with profilin. *Journal of Cell Biology*, **137**, 169–82.

Chang, F. & Nurse, P. (1996). How fission yeast fissions in the middle. *Cell*, **84**, 191–4.

Chang, F., Wollard, A. & Nurse, P. (1996). Isolation and characterization of fission yeast mutants defective in assembly and placement of the contractile ring. *Journal of Cell Science*, **109**, 131–42.

Chant, J. (1996). Septin scaffolds and cleavage planes in Saccharomyces. *Cell*, **84**, 187–90.

Cid, V. J., Duran, A., Ray, F. D., Snyder, M. P., Nombela, C. & Sanchez, M. (1995). Molecular basis of cell integrity and morphogenesis in *Saccharomyces cerevisiae. Microbiological Reviews*, **59**, 345–86.

Clutterbuck, A. J. (1970). Synchronous nuclear division and septation in *Aspergillus nidulans. Journal of General Microbiology*, **60**, 133–5.

Cole, G. T. (1986). Models of cell differentiation in conidial fungi. *Microbiological Review*, **50**, 95–132.

Collinge, A. J. & Markham, P. (1985). Woronin bodies rapidly plug septal pores of severed *Penicillium chrysogenum* hyphae. *Experimental Mycology*, **9**, 80–5.

Conrad, G. W. & Schroeder, T. E. (1990). Cytokinesis: mechanisms of furrow formation during cell division. *Annals of the New York Academy of Science*, **582**, 325.

d'Enfert, C. (1997). Fungal spore germination: insights from the molecular genetics of *Aspergillus nidulans* and *Neurospora crassa. Fungal Genetics and Biology*, **21**, 163–72.

Demarini, D. J., Adams, A. E. M., Farers, H., Devirigilio, C., Valle, G., Chuang J. S., & Pringle, J. R. (1997). A septin-based heirarchy of proteins required for localized deposition of chitin in the *Saccharomyces cerevisiae* cell wall. *Journal of Cell Biology*, **139**, 75–93.

Doonan, J. R. (1992). Cell division in *Aspergillus nidulans. Journal Cell Science*, **103**, 599–611.

Edamatsu, M. & Toyoshima, Y. Y. (1996). Isolation and characterization of pos mutants defective in correct positioning of septum in *Schizosaccharomyces pombe. Zoological Science*, **13**, 235–9.

Evangelista, M., Blundell, K., Longtine, M. S., Chow, C. J., Adames, N., Pringle, J. R., Peter, M. & Boone, C. (1997). Bni1p, a yeast formin linking Cdc42p and the actin cytoskeleton during polarized morphogenesis. *Science*, **276**, 118–22.

Fankhauser, C., Marks, J., Raymond, A. & Semanis, V. (1993*a*). The *S. pombe cdc16* gene is required both for maintenance of $p34^{cdc2}$ kinase activity and regulation of septum formation: a link between mitosis and cytokinesis. *EMBO Journal*, **12**, 2697–704.

Fankhauser, C., Reymond, A., Cerutti, L., Utzig, S., Hofmann, K. & Simanis, V. (1993*b*). The S. pombe *cdc15* gene is a key element on the reorganization of F-actin at mitosis. *Cell*, **82**, 435–44.

Fankhauser, C. & Simanis, V. (1993). The *Schizosaccharomyces pombe cdc14* gene is required for septum formation and can also inhibit nuclear division. *Molecular Biology of the Cell*, **4**, 531–9.

Fankhauser, C. & Simanis, V. (1994*a*). The cdc7 protein kinase is a dosage dependent regulator of septum formation in fission yeast. *EMBO Journal*, **13**, 3011–9.

Fankhauser, C. & Simanis, V. (1994b). Cold fission: splitting the pombe cell at room temperature. *Trends Cell Biology*, **4**, 531–9.

Fiddy, C. & Trinci, A. P. J. (1976). Mitosis, septation, branching and the duplication cycle in *Aspergillus nidulans*. *Journal of General Microbiology*, **97**, 169–94.

Girbardt, M. (1979). A microfilamentous septal belt (FSB) during induction of cytokinesis in *Trametes versicolor* (L. ex Fr.). *Experimental Mycology*, **3**, 215–28.

Gooday, G. W. & Schofield, D. A. (1995). Regulation of chitin synthesis during growth of fungal hyphae: the possible participation of membrane stress. *Canadian Journal of Botany*, **73**, s114–21.

Gull, K. (1978). Form and function of septa in filamentous fungi. In *The Filamentous Fungi. Developmental Mycology*, ed. J. E. Smith & D. R. Berrys, pp. 78–93. New York: John Wiley and Sons.

Harris, S. D. (1997). The duplication cycle in *Aspergillus nidulans*. *Fungal Biology and Genetics*, **22**, 1–12.

Harris, S. & Hamer, J. E. (1995). *sepB*: an *Aspergillus nidulans* gene involved in chromosome segregation and the initiation of cytokinesis. *EMBO Journal*, **14**, 5244–57.

Harris, S. D., Hamer, L., Sharpless, K. E. & Hamer, J. E. (1997). The *Aspergillus nidulans sepA* gene encodes an FH1/2 protein involved in cytokinesis and the maintenance of cellular polarity. *EMBO Journal*, **16**, 3474–83.

Harris, S. D. & Kraus, P. R. (1998). Regulation of septum formation in *Aspergillus nidulans* by a DNA damage checkpoint pathway. *Genetics*, **148**, 1055–67.

Harris, S. D., Morrell, J. L. & Hamer, J. E. (1994). Identification and characterization of *Aspergillus nidulans* mutants defective in cytokinesis. *Genetics*, **136**, 517–32.

Hartwell, L. H. (1971). Genetic control of cell division cycle in yeasts. IV. Genes controlling bud emergence and cytokinesis. *Experimental Cell Research*, **69**, 265–76.

Hartwell, L. H., Culotti, J., Pringle, J. R. & Reid, B. J. (1974). Genetic control of the cell division cycle in yeast. *Science*, **183**, 46–51.

Hartwell, L. H. & Weinert, T. A. (1989). Checkpoints: controls that ensure the order of cell cycle events. *Science*, **246**, 629–34.

Heath, I. B. (1990). The roles of actin in tip growth of fungi. *International Review of Cytology*, **123**, 95–127.

Hunsley, D. & Gooday, G. W. (1974). The structure and development of septa in *Neurospora crassa*. *Protoplasma*, **82**, 125–46.

Imamura, H., Tanaka, K., Hihara, T., Umikawa, M., Kamaei, T., Takahashi, K., Sasaki, T. & Takai, Y. (1997). Bni1p and Bnr1p: downstream targets of the Rho family small G-proteins which interact with profilin and regulate actin cytoskeleton in *Saccharomyces cerevisiae*. *EMBO Journal*, **16**, 2745–55.

Jackson-Grusby, L., Kuo, A. & Leder, P. (1992). A variant of limb deformity transcript expressed in the embryonic mouse limb defines a novel formin. *Genes and Development*, **6**, 29–37.

Kitayama, C., Sugimoto, A. & Yamamoto, M. (1997). Type II myosin heavy chain encoded by the *myo2* gene composes the contractile ring during cytokinesis in *Schizosaccharomyces pombe. Journal of Cell Biology*, **137**, 1309–19.

Kohno, H., Tanaka, K., Mino, A., Umikawa, M., Imamura, H., Fujiwara, T., Qadota, H., Watanabe, T., Ohya, Y. & Takai, Y. (1996). Bni1p implicated in cytoskeletal control is a putative target of Rho1p small GTP binding protein in *Saccharomyces cerevisiae. EMBO Journal*, **15**, 6060–8.

Lew, D. J. & Reed, S. I. (1995*a*). A cell cycle checkpoint monitors cell morphogenesis in budding yeast. *Journal of Cell Biology*, **129**, 739–49.

Lew, D. J. & Reed, S. I. (1995*b*). Cell cycle control of morphogenesis in budding yeast. *Current Opinion in Cell Biology*, **5**, 795–804.

Longtine, M. S., DeMarini, D. J. Valencik, M. L., Al-awar, O. A., Fares, H. & Virgilio, C. D. (1996). The septins: role in cytokinesis and other processes. *Current Opinion Cell Biology*, **8**, 106–19.

Manseau, L., Calley, J. & Phan, H. (1996). Profilin is required for posterior patterning of the Drosophila oocyte. *Development*, **122**, 2109–16.

Marks, J., Fankhauser, C. & Simanis, V. (1992). Genetic interactions in the control of septation in *Schizosaccharomyces pombe. Journal of Cell Science*, **101**, 801–8.

Marks, J. & Hyams, J. S. (1985). Localization of F-actin through the cell division cycle of *S. pombe. European Journal of Cell Biology*, **39**, 27–32.

May, K. M., Watts, F. Z., Jones, N. & Hyams, J. S. (1996). Cloning and characterization of the fission yeast myo2+ gene. *Molecular Biology of the Cell*, **7**, 196a.

McCollum, D., Balasubramanian, M., Pelcher, L. E., Hemmingsen, S. M. & Gould, K. L. (1995). *Schizosaccharomyces pombe cdc4+* gene encodes a novel EF-hand protein essential for cytokinesis. *Journal of Cell Biology*, **130**, 651–60.

Momany, M. & Hamer, J. E. (1997). The relationship of actin, microtubules, and crosswall synthesis during septation in *Aspergillus nidulans. Cell Motility and the Cytoskeleton*, **38**, 373–84.

Morrell, J. (1997). A molecular genetic analysis of septation in *Aspergillus nidulans*. Ph.D. thesis, Purdue University.

Morris, N. R. (1976). Mitotic mutants of *Aspergillus nidulans. Genetics Research, Cambridge*, **26**, 237–54.

Morris, N. R. & Enos, A. P. (1992). Mitotic gold in a mold: *Aspergillus* genetics and the biology of mitosis. *Trends in Genetics*, **8**, 32–7.

Murray, A. W. (1995). The genetics of cell cycle checkpoints. *Current Opinion in Genetics and Development*, **5**, 5–11.

Nasmyth, K. (1996). At the heart of the budding yeast cell cycle. *Trends in Genetics*, **12**, 405–12.

Nurse, P. (1975). Genetic control of cell size at cell division in yeast. *Nature*, **256**, 547–51.

Nurse, P. (1990). Universal control of M-phase onset. *Nature*, **346**, 134–67.

Oegema, K. & Mitchinson, T. J. (1997). Rapport rules: cleavage furrow induction in animal cells. *Proceedings of the National Academy of Sciences, USA*, **94**, 4817–20.

Okhura, H., Hagan, I. M. & Glover, D. M. (1995). The conserved *Schizosacharomyces pombe* kinase plo1, required to form a bipolar spindle, the actin ring, and septum, can drive septum formation in G1 and G2 cells. *Genes and Development*, **9**, 1059–73.

Osmani, A. H., O'Donnell, K., Pu, R. T. & Osmani, S. A. (1991). Activation of the *nim*A protein kinase plays a unique role during mitosis that cannot be bypassed by absence of the *bim*E checkpoint. *The EMBO Journal*, **10** 2669–79.

Osmani, S. A., May, G. S. & Morris, N. R. (1987). Regulation of the mRNA levels of *nim*A, a gene required for the G2-M transition in *Aspergillus nidulans*. *The Journal of Cell Biology*, **104**, 1495–504.

Osmani, S. A. & Xe, X. S. (1997). Targets of checkpoints controlling mitosis: lessons from lower eukaryotes. *Trends in Cell Biology*, **7**, 283–8.

Pringle, J. R. & Hartwell, L. H. (1981). *The Saccharomyces Cerevisiae Cell Cycle*. Cold Spring Harbor, NY: Cold Spring Harbor Laboratory.

Rappaport, R. (1986). Establishment of the mechanism of cytokinesis in animal cells. *International Review of Cytology*, **105**, 245–81.

Richle, R. E. & Alexander, J. V. (1965). Multiperforated septations, Woronin bodies and septal plugs in *Fusarium*. *Journal of Cell Science*, **24**, 489–96.

Robertson, R. W. (1992). The actin cytoskeleton in hyphal cells of *Sclerotium rolfsii*. *Mycologia*, **84**, 41–51.

Robinow, C. F. & Caten, C. E. (1969). Mitosis in *Aspergillus nidulans*. *Journal of Cell Science*, **5**, 403–13.

Robinow, C. F. & Hyams, J. S. (1989). General cytology of fission yeasts. In *Molecular Biology of the Fission Yeast*, ed. A. Nasim, P. Young & B. F. Johnstons, pp. 273–330. San Diego: Academic Press.

Roemer, T., Vallier, L. G. & Snyder, M. (1996). Selection of polarized growth sites in yeast. *Trends in Cell Biology*, **6**, 434–41.

Runeberg, P., Raudaskoski, M. & Virtanen, I. (1986). Cytoskeletal elements in the hyphae of the homobasidiomycete *Schizophyllum commune* visualized with indirect immunofluorescence and NBD-phallacidin. *European Journal of Cell Biology*, **41**, 25–32.

Santos, B. & Snyder, M. (1997). Targeting of chitin synthase 3 to polarized growth sites in yeast requires Chs5p and myo2p. *Journal of Cell Biology*, **136**, 95–110.

Satterwhite, L. L. & Pollard, T. D. (1992). Cytokinesis. *Current Opinion in Cell Biology*, **4**, 43–52.

Schmidt, S., Sohrmann, M., Hofmann, K., Woollard, A. & Simanis, V. (1997). The Spg1 GTPase is an essential dosage-dependent inducer of septum formation in *Schizosaccharomyces pombe*. *Genes and Development*, **11**, 1519–34.

Simanis, V. (1995). The control of septum formation and cytokinesis in fission yeast. *Seminars in Cell Biology*, **6**, 79–87.

Sipiczki, M., Grallert, B. & Miklos, I. (1993). Mycelial and syncytial growth in the *Schizosaccharomyces pombe* induced by novel septation mutations. *Journal of Cell Science*, **104**, 485–93.

Sohrmann, M., Frankhauser, C., Brodbeck, C. & Simanis, V. (1998). The *dmf1/mid1* gene is essential for correct positioning of the division septum in fission yeast. *Genes and Development*, **10**, 2707–19.

Song, K., Mach, K. E., Chen, C. Y., Reynolds, T. & Albright, C. F. (1996). A novel suppressor of *ras1* in fission yeast, *byr4*, is a dosage-dependent inhibitor of cytokinesis. *Journal of Cell Biology*, **133**, 1307–19.

Specht, C. A., Liu, Y., Robbins, P. W., Bulawa, C. E., Iartchouk, N., Winter, K. R., Riggle, P. J., Rhodes, J. C., Dodge, C. L., Culp, D. W. & Borgia, P. T. (1996). The *chsD* and *chsE* genes of *Aspergillus nidulans* and their roles in chitin synthesis. *Fungal Genetics and Biology*, **20**, 153–67.

Sveiczer, A., Novak, B. & Mitchison, J. M. (1996). The size control of fission yeast revisited. *Journal of Cell Science*, **109**, 2947–57.

Timberlake, W. E. (1990). Molecular genetics of Aspergillus development. *Annual Review of Genetics*, **24**, 5–36.

Timberlake, W. E. (1993). Translational triggering and feedback fixation in the control of fungal development. *Plant Cell*, **5**, 1453–60.

Trinci, A. P. J. & Morris, N. R. (1979). Morphology and growth of a temperature-sensitive mutant of *Aspergillus nidulans* which forms aseptate mycelia at non-permissive temperatures. *Journal of General Microbiology*, **114**, 53–9.

Tyson, J. J. (1985). The coordination of cell growth and division–intentional or incidental. *BioEssays*, **2**, 72–7.

Watanabe, N., Madaule, P., Reid, T., Ishizaki, T., Watanabe, G., Kakizuka, A., Saito, Y., Nakao, K., Jockusch, B. M. & Narumiya, S. (1997). p140mDia, a mammalian homolog of drosophila diaphenous, is a target protein for Rho small GTPase and is a ligand for profilin. *EMBO Journal*, **16**, 3044–56.

Wells, W. A. (1996). The spindle assembly checkpoint: aiming for a perfect mitosis every time. *Trends in Cell Biology*, **6**, 228–34.

Wergin, W. P. (1973). Development of Woronin bodies from microbodies in *Fusarium oxysporum* f. sp. *lycopersici*. *Protoplasma*, **76**, 249–60.

Wolkow, T. D., Harris, S. D. & Hamer, J. E. (1996). Cytokinesis in *Aspergillus nidulans* is controlled by cell size, nuclear positioning and mitosis. *Journal of Cell Science*, **109**, 2179–88.

Yamamoto, A., Guacci, V. & Koshland, D. (1996). Pds1p, an inhibitor of anaphase in budding yeast, plays a critical role in the APC and checkpoint pathway(s). *Journal of Cell Biology*, **133**, 99–110.

Yang, S. S., Yeh, E., Salmon, E. D. & Bloom, K. (1997). Identification of a mid-anaphase checkpoint in budding yeast. *Journal of Cell Biology*, **136**, 345–54.

Ye, X. S., Fincher, R. R., Tang, A., O'Donnell, K. & Osmani, S. A. (1996). Two S-phase checkpoint systems, one involving the function of both BIME and Tyr15 phosphorylation of $p34^{cdc2}$, inhibit NIMA and prevent premature mitosis. *EMBO Journal*, **15**, 3599–610.

Ye, X. S., Fincher, R. R., Tang, A. & Osmani, S. A. (1997). The G_2/M DNA damage checkpoint inhibits mitosis through Tyr15 phosphorylation of $p34^{cdc2}$ in *Aspergillus nidulans*. *EMBO Journal*, **16**, 182–92.

10

Genetic control of polarized growth and branching in filamentous fungi

G. TURNER AND S. D. HARRIS

Introduction

One of the most characteristic features of the fungal mycelium is its highly polarized mode of growth. Mycologists have devoted considerable effort towards understanding the basic mechanisms underlying hyphal elongation and branching, and these subjects have been reviewed extensively (Trinci, 1979; Prosser, 1983; Trinci, Wiebe & Robson, 1994; Gow, 1994; Trinci *et al.*, Chapter 5, this volume). One approach which has been employed to investigate these mechanisms is the identification and characterization of mutants defective in hyphal morphogenesis. Such mutants are relatively easy to detect since they typically cause severe alterations in colony morphology. Indeed, a useful benefit of early experiments in the biochemical genetics of *Neurospora crassa* was the generation and description of a large collection of colonial mutants (Murray & Srb, 1962). Although these mutants have been characterized to a limited extent, the nature of the affected genes is in most cases unknown. Recently, with rapid progress being made in understanding cellular morphogenesis in the yeasts *Saccharomyces cerevisiae* and *Schizosaccharomyces pombe* (Snell & Nurse, 1994; Simanis, 1995; Roemer, Vallier & Snyder, 1996), there has been renewed interest in understanding the genetic basis of filamentous growth in fungi. Since molecular genetic analyses in the model filamentous fungi *Aspergillus nidulans* and *N. crassa* have yielded considerable insight into metabolic control, development, and mitosis (Bennett & Lasure, 1991; Martinelli & Kinghorn, 1994), it is reasonable to presume that this approach will also contribute towards achieving a thorough understanding of hyphal morphogenesis. From their studies on *Neurospora* mutants, Murray & Srb (1962) speculated that at least some processes controlling form may be widely distributed throughout the fungi and perhaps other organisms. Studies with yeast have already verified this statement, and to some degree, provide an instructive paradigm for the analysis of hyphal morphogenesis. With this in mind, this chapter focuses on genetic approaches for the analysis of hyphal morphogenesis, particularly the establishment of hyphal polarity and branch formation, with reference to related studies in yeasts.

The establishment of cellular polarity in yeast

Despite their distinct cell shapes and evolutionary distance, the unicellular yeasts *S. cerevisiae* and *S. pombe* exhibit patterns of polarized morphogenesis which share a number of features in common. First, during their respective cell cycles, both yeasts undergo a period of polarized growth during which cell wall deposition is directed to a specific site on the cell surface. In *S. cerevisiae*, this site is the tip of the new bud, whereas in *S. pombe*, growth occurs at both ends of the rod-shaped cell. Second, both yeasts subsequently undergo cell 'depolarization' prior to mitosis and septum formation. Third, upon completion of cytokinesis, each yeast employs a specific mechanism to re-establish cellular polarity and enter a new period of polarized growth. Finally, *S. cerevisiae* and *S. pombe* also exhibit polarized morphogenesis when cells of opposite mating type undergo conjugation. In this case, growth and cell wall deposition are confined to the tip of the mating projections.

Much has been learned about the molecular events underlying polarized morphogenesis in *S. cerevisiae* and *S. pombe* (for recent reviews, see Snell & Nurse, 1993; Chant, 1994; Kron & Gow, 1995; Roemer *et al.*, 1996). The establishment of cellular polarity in yeast cells is absolutely dependent upon the presence of an intact actin cytoskeleton (Novick & Botstein, 1985; Kobori *et al.*, 1989), which presumably facilitates the transport and delivery of components needed for localized cell wall deposition. Reflecting this role, the actin cytoskeleton is organized asymmetrically in cells undergoing polarized growth (Adams & Pringle, 1984; Marks & Hyams, 1985); clusters of actin patches typically localize to sites of active cell wall deposition and are subtended by filaments which run parallel to the axis of polarization. In contrast, although microtubules are aligned along the same axis (Kilmartin & Adams, 1984; Marks, Hagan & Hyams, 1986), they are not strictly required for polarized morphogenesis (Toda *et al.*, 1983; Jacobs *et al.*, 1988). Instead, their orientation along this axis serves to ensure that divided nuclei are faithfully segregated. A third cytoskeletal element, the septin-associated neck filaments, has also been characterized in budding and fission yeast (Longtine *et al.*, 1996). Although not required for the establishment of cellular polarity, the septins appear to function in polarized morphogenesis by defining and organizing sites of localized cell wall deposition. Thus, the morphogenetic events underlying polarized growth in yeast cells are primarily carried out by the different cytoskeletal elements and their associated interacting proteins. The more recent challenge has been to understand how the cytoskeleton and other cellular constituents become organized in a polarized manner in yeast cells.

Determinants of spatial organization in yeast cells

Yeast cells display clearly defined patterns of polarized morphogenesis. Following the completion of mitosis and cell division, both budding and fission yeast cells utilize spatial cues to specify the new axis of cell polarization. In addition, the site at which mating projections form during conjugation is potentially defined by external cues (i.e. gradients of mating pheromone). Considerable progress has been made towards understanding the nature of these positional cues and the means by which they are localized in both *S. cerevisiae* and *S. pombe*.

Depending on their cell type, budding yeast cells display two distinct patterns of polarized morphogenesis (for review, see Chant, 1994). Cells of mating type a or α exhibit an axial budding pattern, in which mother and daughter cells form buds adjacent to the previous division site. In contrast, a/α cells exhibit a bipolar budding pattern, in which daughter cells typically bud at the pole distal to the division site, whereas mother cells bud at either pole. A set of genes, termed the bud site selection genes, functions solely to specify the appropriate budding pattern in yeast cells (Chant & Herskowitz, 1991; Chant *et al.*, 1991; Fujita *et al.*, 1994). The axial budding pattern is controlled by the products of the *BUD3*, *BUD4*, *BUD10*, and *AXL1* gene products (Fujita *et al.*, 1994; Chant *et al.*, 1995; Sanders & Herskowitz, 1996; Halme *et al.*, 1996). Notably, Bud3p and Bud4p specify the new bud site by interacting with the septin-associated neck filaments which are located at the previous division site. In this manner, a pre-existing morphological landmark is used as a template to provide positional information. The specific function of the Bud10p and Axl1p is less clear, although Axl1p appears to be a target of the machinery which confers cell-type specificity upon the axial budding pattern. The bipolar budding pattern is controlled by a separate group of genes, including *BUD6*, *BUD7*, *BUD8*, and *BUD9* (Zahner, Harkins & Pringle, 1996; Amberg *et al.*, 1997). The phenotypes of *bud8* and *bud9* mutants suggests that they may represent the distal and proximal pole landmarks, respectively. In addition, mutations in a number of different genes implicated in the control of actin cytoskeletal organization also disrupt the bipolar budding pattern (Zahner *et al.*, 1996; Yang, Ayscough & Drubin, 1997). This suggests that proper localization of the bipolar landmarks requires a functional actin cytoskeleton. A third group of gene products, Bud1p, Bud2p, and Bud5p (Bender & Pringle, 1989; Chant *et al.*, 1991; Park, Chant & Herskowitz, 1993), constitutes a GTPase regulatory module which 'reads' the axial or bipolar positional information and tranduces it to a signal transduction module which controls polarity establishment (Fig. 10.1; Michelitch & Chant, 1996; Park *et al.*, 1997).

Fission yeast cells must re-establish cellular polarity following the completion of mitosis and cell division (for review, see Nurse, 1994). Initially, growth is unipolar in that it is confined to the end of the cell which existed prior to division (i.e. the 'old end'). Following progression into the cell cycle, cells switch to bipolar growth, whereby cell wall deposition occurs at both the 'old' and 'new' ends (Mitchison & Nurse, 1985). Failure to define the axis of cell polarization properly leads to the formation of misshapen cells, which permits ready identification of genes required for spatial patterning in fission yeast (Snell & Nurse, 1993; Verde, Mata & Nurse, 1995). One of these genes, *tea1*$^+$, encodes a product which localizes to the cell poles and appears to function as a polar landmark (Mata & Nurse, 1997). Notably, localization of tea1p to the ends of the cell requires the presence of intact microtubules. Upon the completion of mitosis, tea1p is transported by growing cytoplasmic microtubules and deposited at the cell poles. Furthermore, maintenance of tea1p localization, and thus definition of the growth zone, depends upon the continuous presence of cytoplasmic microtubules. Hence, although budding and fission yeast utilize polar landmarks to define sites of localized cell wall deposition, different cytoskeletal systems are utilized to position the markers. The localization of tea1p may also explain the asymmetric pattern of polarized growth in fission yeast cells, since greater amounts of the protein appear to be present at the 'old end' of a new cell.

Transduction of positional information to the cytoskeleton

To trigger polarized morphogenesis, the positional information which dictates the axis of cell polarization must ultimately be relayed to the cytoskeleton. In yeast cells, the key modulator of cytoskeletal organization is the highly conserved Rho/Rac-related GTPase Cdc42p (Johnson

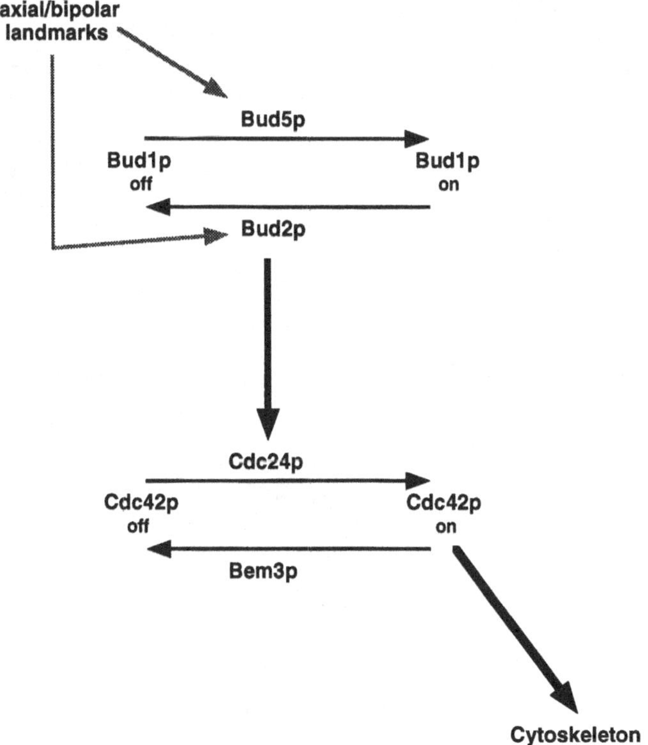

Fig. 10.1. Outline of the regulatory circuit which controls bud emergence in *S. cerevisiae*. Spatial landmarks are thought to promote cycling of Bud1p between its GTP-bound and GDP-bound states by recruiting Bud2p (a GTPase-activating protein) and/or Bud5p (a guanine nucleotide exchange factor) to the incipient bud site. Current evidence suggests that the conversion of Bud1p between these two conformations is sufficient to activate Cdc24p. Once activated, Cdc24p (a guanine nucleotide exchange factor) converts Cdc42p to its active GTP-bound state, and in turn, Cdc42p interacts with various effectors to trigger polarization of the cytoskeleton. Cdc42p activity is subsequently down-regulated by Bem3p (a GTPase-activating protein). For details, see Zheng *et al.*, 1995, Michelitch & Chant, 1996, and Park *et al.*, 1997.

& Pringle, 1990; Miller & Johnson, 1994). A remarkably similar regulatory circuit controls the localized activation of Cdc42p in budding and fission yeast (Figs. 10.1, 10.2; Zheng, Cerione & Bender, 1994; Chang *et al.*, 1995). Moreover, it appears that positional information is directly communicated to Cdc42p via this circuit in *S. cerevisiae* (Fig. 10.1; Zheng, Bender & Cerione, 1995; Park *et al.*, 1997). In particular, an effector of Bud1p is the guanine nucleotide exchange factor Cdc24p,

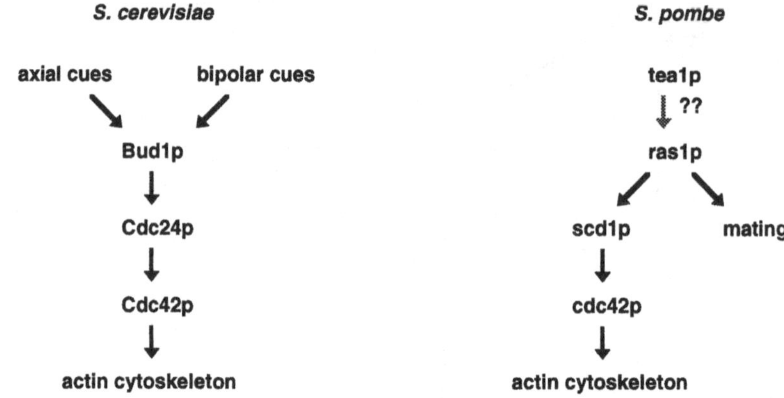

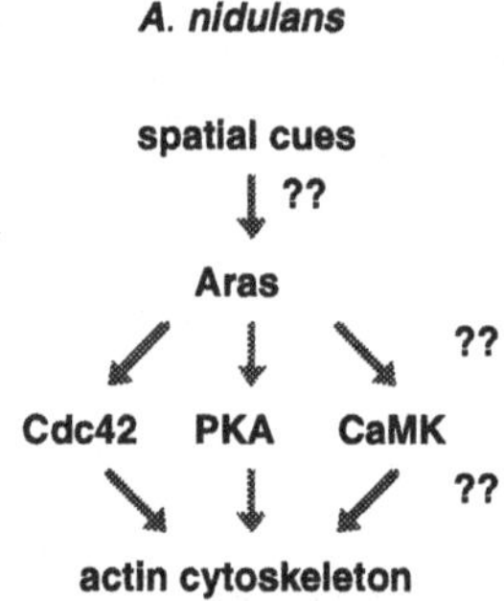

Fig. 10.2. Similar regulatory circuits control polarized morphogenesis in *S. cerevisiae* and *S. pombe*, and perhaps in *A. nidulans*. In yeast cells, localized activation of a ras-related GTPase (Bud1p/ras1p) by spatial landmarks leads to activation of Cdc42p and polarization of the cytoskeleton. The link between tea1p and ras1p in *S. pombe* is speculative. For details, see Chang *et al.*, 1995, and Park *et al.*, 1997. In *A. nidulans*, the establishment of hyphal polarity appears to be controlled by *Aras* activity (Som & Kolaparthi, 1994). All of the other links are at this time speculative.

which stimulates conversion of Cdc42p from its inactive GDP-bound state to its active GTP-bound state. Thus, positional cues presumably promote localized activation or cycling of the Bud1p GTPase, which in turn causes activation of Cdc42p. Once activated, Cdc42p recruits and interacts with a variety of effectors to organize both the actin and septin cytoskeletons. It remains to be determined whether the analogous regulatory circuit responds to positional signals such as tea1p in *S. pombe*.

How does Cdc42p function to regulate cytoskeletal organization? The best-characterized effectors of Cdc42p are the PAK-related protein kinases; Ste20p and Cla4p in *S. cerevisiae*, and pak1p in *S. pombe* (Leberer *et al.*, 1992; Cvrckova *et al.*, 1995; Ottilie *et al.*, 1995). Although pak1p is required for organization of the actin cytoskeleton, and Cla4p is needed for proper localization of the septin-associated neck filaments, the means by which these functions are accomplished still remain unclear. Futhermore, a variety of genetic observations support the notion that additional Cdc42p effectors must exist (Cvrckova *et al.*, 1995). Recently identified candidates include members of the WAS and FH families of proteins (Evangelista *et al.*, 1997; Li, 1997), each of which possess domains capable of interacting with components of the actin cytoskeleton.

Polarized morphogenesis and cellular regulatory networks

The ability of a yeast cell to reproduce efficiently is undoubtedly dependent upon the coordinate regulation of several cellular processes. For example, the conservation of cell shape and mass requires the coordination of polarized morphogenesis with both cell cycle progression and cell proliferation. Insight into the nature of the regulatory networks which link these distinct processes has recently been obtained in *S. cerevisiae*. In budding yeast, cell cycle transitions are controlled by the association of different classes of regulatory cyclins with a single cyclin-dependent kinase ([cdk]; for review, see Nasmyth, 1996). One such complex functions in late G1 to orchestrate multiple cellular events, including those which underlie the onset of S phase and the establishment of cellular polarity (Cross, 1995). Notably, this cdk complex directly promotes the polarization of the actin cytoskeleton (Lew & Reed, 1993), although the mechanism by which this is accomplished remains to be elucidated. Moreover, acting via a signal transduction module, the same complex activates a transcription factor which controls the expression of separate sets of genes required for DNA replication and localized cell wall biosynthesis (Igual, Johnson & Johnston, 1996; Madden *et al.*, 1997). Later in the cell cycle, a second cdk complex which triggers events required for mitosis also influences morphogenesis by directly promoting actin depolarization, and as a result, isotropic growth (Lew & Reed, 1993). It is becoming increasingly apparent that the patterns of polarized morphogenesis displayed by yeast cells are governed by regulatory networks such as this, which coordinate morphogenesis with other cellular processes.

The availability of sequenced genomes coupled with the use of efficient approaches for assessing global patterns of regulation (Oliver, 1997) should faciltate the deciphering of these networks and how they influence polarized morphogenesis.

The establishment of hyphal polarity in *Aspergillus nidulans*

The growth and reproduction of hyphal cells undoubtedly represents the most vivid example of polarized morphogenesis in fungi. In these cells, cell surface expansion and cell wall deposition are solely confined to the hyphal tip (Fig. 10.3). This mode of growth requires the long-range vectorial transport of vesicles laden with components needed for membrane and cell wall biosynthesis, which in turn is supported by a highly polarized cytoskeleton (for review, see Wessels, 1993; Gow, 1994). The overall pattern of polarized morphogenesis in hyphal cells differs dramatically from that displayed by yeast cells. For example, hyphal cells attain their characteristic tubular shape by retaining fixed axes of polarization and not undergoing isotropic growth. Furthermore, mitosis and cell division are not accompanied by cell depolarization as in yeasts. Instead, localized cell wall deposition occurs simultaneously at both the tip and in basal regions where division septa form. Despite these differences, recent observations hint that the molecular mechanisms underlying polarized morphogenesis in hyphal cells and yeast cells are to some extent conserved (Harris *et al.*, 1997). Thus, the unique aspects of hyphal morphogenesis presumably arise from differences in the manner by which these morphogenetic activities are regulated.

The cytoskeleton and polarized growth in A. nidulans

The genetic and molecular tractability of *A. nidulans* has recently been exploited to initiate a multi-faceted dissection of the molecular mechanisms underlying polarized morphogenesis in hyphal cells. In *A. nidulans*, polarized morphogenesis is initiated when dormant conidia undergo germination (Fig. 10.3). Following an initial period of isotropic swelling, conidia switch to a polarized mode of growth and produce an elongating germ tube. In swollen spores, both the actin cytoskeleton and cytoplasmic microtubules display a uniform organization. In particular, the spore contains a 'basket' of cytoplasmic microtubules, and possesses randomly localized cortical actin patches (S. Harris, unpublished observations). The organization of the cytoskeleton changes dramatically upon germ

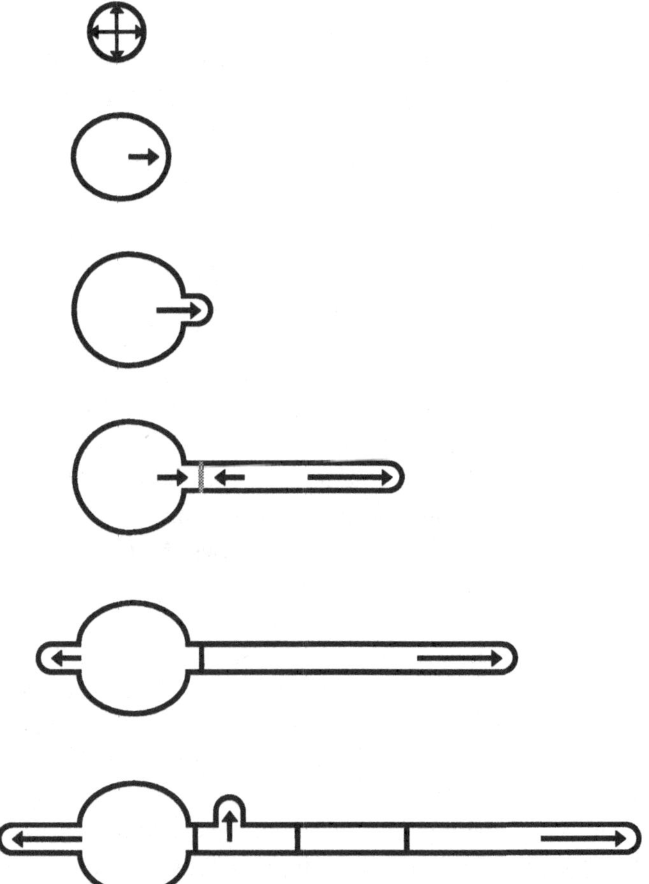

Fig. 10.3. Polarized morphogenesis in *A. nidulans* hyphae. Spores undergo an initial period of isotropic swelling, which is followed by the establishment of hyphal polarity and the emergence of a germ tube. Following growth to an appropriate cell size and the completion of at least one round of nuclear division, septum formation occurs at the base of the hypha (Wolkow *et al.*, 1996). Note that at this point, hyphal cells must coordinate two distinct sites of polarized morphogenesis. Additional sites of polarized growth are formed by subsequent rounds of germ tube emergence and by hyphal branching. Arrows depict active sites of cell surface expansion and cell wall deposition.

tube emergence. Dense clusters of actin patches localize to the immediate hyphal tip, and both actin filaments and cytoplasmic microtubules tend to run parallel to the axis of cell polarization (Osmani *et al.*, 1988*a*; Harris *et al.*, 1994). The timing of cytoskeletal reorganization relative to the switch from isotropic swelling to apical growth has not yet been determined, but it presumably precedes formation of the germ tube.

Although the functional role of actin has never been directly assessed by a genetic approach in *A. nidulans*, numerous observations indicate that the actin cytoskeleton is required for polarized morphogenesis. First, cellular polarity is lost when hyphal cells are treated with cytochalasin A (Harris, Morrell & Hamer, 1994), which is an inhibitor of actin polymerization. Second, mutations in the *sepA* gene, which prevent organization of the actin cytoskeleton at incipient septation sites, also cause severe defects in the establishment and maintenance of hyphal polarity (Harris *et al.*, 1994, 1997). Notably, SepA is a member of the FH family of proteins, which are capable of interacting with GTPases such as Cdc42p to promote localized reorganization of the actin cytoskeleton (Frazier & Field, 1997). Third, germ tube emergence and hyphal growth require an unconventional myosin (MyoA) similar to those which typically transport cargo such as secretory vesicles (McGoldrick, Gruver & May, 1995). Consistent with this role, MyoA is required for secretion in *A. nidulans.* In contrast, cytoplasmic microtubules appear to be largely dispensable for polarized morphogenesis in *A. nidulans* hyphae. Pharmacological or mutational disruption of microtubule organization causes slight alterations in cell shape, but does not prevent the establishment of hyphal polarity (Oakley & Morris, 1980; Doshi *et al.*, 1991). However, the alignment of nuclei within multinucleate hyphal cells does depend upon the integrity of cytoplasmic microtubules (Oakley & Morris, 1980). The localization of cytoplasmic dynein heavy chain to the immediate hyphal tip suggests that proper nuclear distribution requires anchoring of microtubules within this region (Xiang, Roghi & Morris, 1995).

The molecular analysis of cytoskeletal function in *A. nidulans* suggests that the bulk of vesicle transport to the hyphal tip is conducted along actin filaments. However, a role for cytoplasmic microtubules in this process cannot be discounted. In *A. nidulans*, as in other filamentous fungi, the orchestrated delivery of secretory vesicles to the hyphal tip is thought to be facilitated by a secretory granule termed the Spitzenkorper (Bartnicki-Garcia, Hergert & Gleitz, 1989), which is typically located a short distance behind the zone of cell wall deposition (Howard, 1981; Reynaga-Pena, Gierz & Bartnicki-Garcia, 1997). Long-range transport of vesicles to the vicinity of the hyphal tip is presumed to be microtubule-dependent in filamentous fungi (Steinberg & Schliwa, 1993). It is possible that microtubules perform a similar function in *A. nidulans*, and that actin filaments are utilized to transport vesicles from the Spitzenkorper to specific sites of cell wall deposition at the immediate tip. In this case, the failure of microtubule disruption to prevent polarized morphogenesis may indicate the presence of a redundant long-range transport system which utilizes actin filaments.

Signal transduction and polarized growth in A. nidulans

Employing yeast as a paradigm, it would seem reasonable to assume that the changes in cytoskeletal organization which accompany spore germination in *A. nidulans* are controlled by GTPase signalling modules. The functional characterization of an essential *A. nidulans* Ras homologue (*Aras*) has yielded results which are consistent with this notion. Notably, a constitutively active allele of *Aras* ($Aras^{G17V}$) prevents the switch from isotropic spore swelling to polarized hyphal growth (Som & Kolaparthi, 1994). How does *Aras* regulate the establishment of hyphal polarity in *A. nidulans*? It may transduce spatial cues (see below) to the cytoskeleton via a Cdc42p GTPase signalling module as in yeasts (Fig. 10.2). Alternatively, *Aras* may control polarized morphogenesis by modulating the activity of cAMP-dependent protein kinase A (PKA), which is known to control polarized growth directly in filamentous fungi such as *N. crassa* and *Ustilago maydis* (Gold *et al.*, 1994; Bruno *et al.*, 1996*b*). Finally, *Aras* may regulate polarized morphogenesis in *A. nidulans* spores by an entirely novel signalling mechanism (i.e. via Ca^{2+}-mediated signalling).

Detailed physiological experiments have clearly established that Ca^{2+} plays a key role in the regulation of polarized morphogenesis in filamentous fungi (for review, see Jackson & Heath, 1993). Influx of Ca^{2+} at the hyphal tip, coupled with release of intracellular Ca^{2+} from stored pools, is thought to promote the formation of tip-high gradients, which may regulate vesicle transport and other aspects of cell wall biosynthesis (Hyde & Heath, 1997). The molecular analysis of Ca^{2+}-mediated signalling processes in *A. nidulans* has yielded results which are consistent with this hypothesis. In particular, mutational inactivation of a Ca^{2+}/calmodulin-dependent protein kinase (CaMK) prevents the establishment of hyphal polarity in germinating spores (Dayton & Means, 1996). Moreover, a similar effect is observed when a constitutively active CaMK is overexpressed (Dayton *et al.*, 1997). These observations suggest that the switch to polarized growth requires the establishment of a precise

Ca^{2+} gradient. Although targets of Ca^{2+}-mediated signalling have not yet been identified in *A. nidulans*, attractive possibilities include proteins which control the organization and/or function of the actin cytoskeleton.

Temporal control of polarized morphogenesis in A. nidulans *conidia*

Concomitant with the complex morphogenetic events underlying spore germination, *A. nidulans* conidia undergo multiple rounds of nuclear division unaccompanied by cell division (Harris *et al.,* 1994). As a result, a multinucleate hyphal cell is generated from a uninucleate spore. The observation that polarized morphogenesis is not prevented when nuclear division is inhibited suggests that these two processes occur independent of each other in germinating conidia (Morris, 1976; Bergen & Morris, 1983). Indeed, the point at which morphogenesis is first coordinated with nuclear division in a hyphal cell could conceivably be the first cell division (Harris *et al.*, 1994). However, recent observations refute this notion by revealing the existence of a temporal link between polarized morphogenesis and nuclear division in germinating spores. Specifically, the switch from isotropic swelling to polarized growth was found to be coupled to the completion of the first mitosis in the spore (S. Harris, unpublished observations). Moreover, mutations and pharmacological inhibitors which arrest nuclear division typically cause severe delays in the switch to polarized growth (S. Harris, unpublished observations).

A possible mechanism for coupling the establishment of hyphal polarity to nuclear division was revealed by the existence of an inverse correlation between the activity of the NimA protein kinase and polarized growth. The NimA kinase is required for mitosis in *A. nidulans* (Osmani, Pu & Morris, 1988*b*), and is most active during late G2 and early M phase (Ye *et al.*, 1995). Furthermore, exit from mitosis into interphase of the next cell cycle depends upon inactivation and proteolytic destruction of NimA (Pu & Osmani, 1995). Experiments with strains producing different levels of active NimA kinase show that polarized growth does not occur in germinating conidia possessing high kinase activity (S. Harris, unpublished observations). Conversely, the timing of the switch to polarized growth is advanced in conidia in which kinase activity is low (S. Harris, unpublished observations). These observations suggest that phosphorylation of specific substrates by the NimA kinase may inhibit polarized growth. These substrates could conceivably be gene products required for polarity establishment. Alternatively, the established ability of NimA to promote chromosome condensation (Osmani *et al.*, 1988*a*; O'Connell, Norbury & Nurse, 1994) may indirectly inhibit polarized growth by preventing transcriptional activation of polarity genes.

Spatial control of polarized morphogenesis in A. nidulans *conidia*

Following the establishment of hyphal polarity and the emergence of a germ tube, *A. nidulans* conidia are able to repeat the entire process and form a second hypha (Fig. 10.3). Typically, a single spore is capable of producing up to three distinct hyphae in sequential fashion. However, since hyphae do not detach from their founding spore, a novel axis of polarized growth must be specified with each cycle of polarity establishment. Is there a defined spatial pattern in which these axes are specified, or, are they generated at random? Recent observations support the former possibility and suggest that germinating *A. nidulans* conidia resemble yeast cells in displaying specific spatial patterns of polarized morphogenesis (S. Harris, unpublished observations). Specifically, examination of a population of conidia possessing a single germ tube revealed that the second germ tube formed at an angle of 180° relative to the first with a frequency of approximately 80%. In the remaining conidia, the second germ tube formed at an angle of 90° relative to the first. Notably, the second germ tube formed at a random site in fewer than 1% of the examined conidia ($n = 500$). The bias in the selection of the site at which the second germ tube emerges suggests that an existing axis of polarized growth can influence the positioning of subsequent axes. Conceivably, the site at which the first germ tube forms may be selected at random in an otherwise apolar spore. The position of all subsequent germ tubes would then be fixed by this initial 'polarizing' event. Alternatively, swollen spores may resemble budding yeast cells in possessing localized spatial cues which determine all possible axes of polarized growth. Distinguishing between these hypothesis will require the identification of a spatially localized marker in ungerminated conidia.

To identify potential mechanisms underlying the normal pattern of germ tube emergence in *A. nidulans* conidia, mutants defective in polarized morphogenesis were examined to determine if the pattern was disrupted. Notably, mutations which inactivate the FH protein SepA lead to a completely random pattern of germ tube emergence (S. Harris, unpublished observations). This observation suggests that the ability of SepA to promote localized reorganization of the actin cytoskeleton is required for the positioning of spatial cues which mark the axis of polarized growth. Intriguingly, another FH family member, Bni1p, is required for the establishment of the bipolar budding pattern in *S. cerevisiae* (Zahner *et al.*, 1996). Thus, a conserved actin-dependent mechanism is apparently utilized to position morphological landmarks in both budding yeast cells and hyphal cells. In contrast, loss of functional microtubules does not disrupt the normal pattern of polarized morphogenesis in *S. cerevisiae* or *A. nidulans* (S. Harris, unpublished observations).

Hyphal growth and branching

Hyphal growth, septation and branching of *A. nidulans* under defined conditions have been described quantitively in some detail in studies by Clutterbuck (1970), Fiddy & Trinci (1976*a*), and Prosser & Trinci (1979). Such growth is best observed either following emergence of germ tubes from conidiospores, or at the edge of a colony growing on a solid medium. Older mycelium, which becomes densely tangled and where events are more difficult to follow microscopically, eventually develops conidiophores and conidia. Normally, a single germ tube emerges from the conidium, and after a period of unbranched growth, a second germ tube emerges from the opposite side. After the appearance of septa, branches begin to emerge from intercalary compartments, and these branches become new leading hyphae (Fig. 10.4). Generally, the leading hyphal compartment has no branch, and each intercalary compartment has a single branch. However, this pattern is not strictly maintained, so branches are sometimes seen in the apical compartment, and more than one branch is sometimes seen arising from an intercalary compartment. Careful measurements of multiple parameters, including compartment size, number and position of branches, number of nuclei per compartment, and hyphal extension rates, led to the concept of the duplication cycle (Fig. 10.4; Trinci, 1979; Trinci *et al.*, 1994). In a leading hypha at the colony margin, the apical compartment, containing approximately 50 nuclei, is divided into about half its length by formation of several septa in the rear of the compartment. The new, shorter apical compartment then extends at a linear rate, without division of the nuclei, until it doubles in length. A parasynchronous wave of mitosis, possibly triggered by the increased ratio of cytoplasmic volume per nucleus, doubles the number of nuclei, and the process begins again. Some differences are also seen between hyphae of very young colonies, recently arisen from germinating conidia (undifferentiated mycelium), and hyphae in the peripheral growth

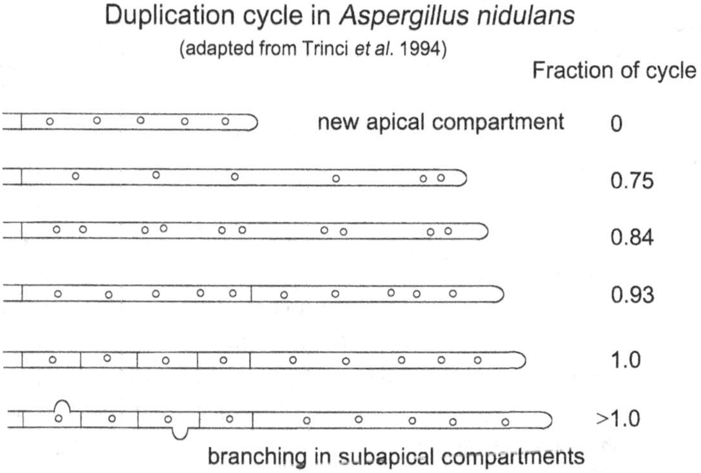

Duplication cycle in *Aspergillus nidulans*
(adapted from Trinci *et al.* 1994)

Fraction of cycle

new apical compartment 0

0.75

0.84

0.93

1.0

>1.0

branching in subapical compartments

Fig. 10.4. Duplication cycle and branch formation in *Aspergillus nidulans*. A new apical compartment extends by linear growth without nuclear division until a critical volume is reached. A parasynchronous round of mitosis within the apical compartment is followed shortly thereafter by septation, which effectively halves the length of the apical compartment. The cycle then reinitiates and branching begins in subapical compartments, leading to an average of approximately one branch per compartment. About 50 nuclei are found in the apical compartment, and about four in the intercalary compartments (modified from Trinci *et al.*, 1994).

zone of older colonies (differentiated mycelium) (Fiddy & Trinci, 1976*a*). For example, the apical compartment, which oscillates in length with each duplication cycle, gradually increases in mean length as a germling develops into an older colony. The rate of extension of the hypha increases during this transition until it reaches a maximum, linear growth rate.

The position of branch formation in hyphal cells is apparently not chosen at random, as there is some preference for the branch to develop towards the anterior of the cell. There have been attempts to explain the control of branch formation in terms of vesicle supply and demand (Trinci, 1979). Components required for hyphal extension are thought to undergo long-range vectorial transport to the hyphal tip. Any factor which perturbs tip growth may cause vesicle supply to exceed demand at the tip, and would presumably lead to localized accumulation of vesicles in distal regions of the hypha. Formation of lateral branches may then occur at quasi-random sites where the concentration of vesicles exceeds a given threshold. However, this idea must be reconciled with the observations that specific morphological landmarks are utilized to designate new growth sites in both budding and fission yeast (Chant *et al.*, 1995; Sanders & Herskowitz, 1996; Mata & Nurse, 1997). Analyses of the duplication cycle in *N. crassa* and *Geotrichum candidum* have raised the possibility that branch formation may be correlated with septation (Trinci, 1973; Fiddy & Trinci, 1976*b*; Steele & Trinci, 1977*a*). For example, *G. candidum* forms complete septa and exhibits a regular pattern of branching, whereby branch formation tends to occur close to the leading septum. In contrast, *A. nidulans* forms incomplete septa and displays a branching pattern which is less regular. Moreover, there appears to be some correlation between septation and branch frequency. In particular, when hyphal elongation is perturbed in *N. crassa cot-1* mutants, both the branching frequency and the frequency of septation increase.

It appears that a primary branch and its supporting compartment(s) initially grow exponentially, and this eventually reaches a maximum linear rate of branch extension, similar to the situation seen in germlings. The combination of linear tip growth and regular branching lead to an exponential increase in total hyphal length, and the ratio of hyphal length to tip number remains approximately constant. This ratio has been called the hyphal growth unit, G (Caldwell & Trinci, 1973), and is a useful way of monitoring the frequency of branching, since an increase in branching frequency leads to a decrease in G (average length of hypha which supports each tip). It is likely to be hyphal volume rather than hyphal length which eventually determines branch frequency, and even a small change in hyphal diameter will lead to a marked effect on volume. While there is data from some fungi to indicate that variations in temperature can alter duplication time without changing G (Trinci, 1973), it is safer to define growth conditions and growth media when quoting G values. For example, growth under conditions of nutrient limitation or starvation is likely to lead to sparser mycelium with a lower frequency of branching.

Measurements of G values have been used to characterize morphological mutants of *Fusarium graminearum* which appear spontaneously during continuous culture (Wiebe *et al.*, 1992). The appearance of stable mutants with lower G values points to genetic control of branching, and raises the question of how this is mediated and which gene(s) are involved. Although *F. graminearum* is not readily amenable to genetic analysis, complementation studies have demonstrated that mutations in a number of genes can lead to a hyperbranching phenotype (Wiebe *et al.*,

1992). Identification and characterization of genes controlling branching frequency would be easier in those fungal species which have been used extensively as model genetic systems, namely *A. nidulans* and *N. crassa.*

Morphological mutants in A. nidulans

While isolation of mutants with altered colonial morphology is extremely easy in filamentous fungi, and some of these have been used as convenient, visual markers for use in genetic analyses, few of these mutants have been characterized in detail with respect to the parameters discussed above. Mutants designated *moA*, *moB*, *moC*, and *coA* have been isolated and mapped in *A. nidulans* (Pontecorvo *et al.*, 1953; Bainbridge, 1966, 1970). *coA* and *moA* have small colonial phenotypes on solid medium, and *moC96* has a mild hyperbranching phenotype leading to compact colonies, though growth rate in liquid medium is almost the same as wild-type (Bainbridge & Trinci, 1969). This mutant therefore resembles some of the spontaneously arising morphological mutants of *F. graminearum.* Unfortunately, *moA* and *moB* strains are no longer available. The temperature-sensitive, septation deficient *sepA1* mutation causes an abnormal dichotomous branching pattern, though the G value is not different from the wild type (Trinci & Morris, 1979). Interestingly, a more recent screen for further *sep* mutations led to the isolation of a new allele, *sepA3*, which exhibits a hyperbranching phenotype (Harris *et al.*, 1994). Mutations at other *sep* loci do not appear to have this effect. This observation demonstrates that septation is not essential for branching, but suggests a link between these two processes.

Morphological mutants in N. crassa

Numerous morphological mutants have been identified and characterized in *N. crassa* (Murray & Srb, 1962; Garnjobst & Tatum, 1967; Scott, 1976; Perkins *et al.*, 1982). Many of these display abnormal hyphal growth, and a few possess abnormal branching patterns, the most studied being *cot-1*. Only a very limited number of the many morphological mutants described have been studied in detail in recent years, but these are beginning to yield interesting results, and there is clearly scope for identifying more of the genes controlling morphology. For example, *frost* and *spray* mutants show hyperbranching at the tips. Although the genes responsible have not been isolated, the abnormalities can be corrected on high calcium, and treatment of the wild-type with the calcium channel blocker verapamil leads to a similar phenotype as seen in these mutants (Dicker & Turian, 1990).

The first real step in molecular biology of polarized morphogenesis in filamentous fungi was made with the cloning of the *cot-1* gene (Yarden *et al.*, 1992). The phenotype of the *cot-1* mutant, isolated by Mitchell & Mitchell (1954), has been described in some detail by Trinci and co-workers in studies on fungal growth kinetics (Steele & Trinci, 1977*a*; Collinge, Fletcher & Trinci, 1978). While growing normally at the permissive temperature of 25°C, a shift to the non-permissive temperature (37°C) leads to a near cessation of hyphal tip elongation, and new branches appear randomly along the length of the hyphae. The new branches extend at a very slow rate, and a dramatic increase in septation is also seen. This appears to be a defect in hyphal elongation rather than loss of polarity, and is consistent with the hypothesis that branching occurs when secretory vesicle supply outpaces the ability of the tip to incorporate them. The *cot-1* gene was cloned by complementation of the mutant and restoration to wild-type using a cosmid library of wild-type DNA (Yarden *et al.*, 1992). Sequencing and database searches revealed that the gene product possesses homology with the cAMP-dependent serine/threonine protein kinase family. Completion of the *S. cerevisiae* genome sequence has since revealed that *cot-1* is closer to a gene of unknown function, *KNQ1*, than to other protein kinases. It seems likely that *cot-1* will be present in many other fungi, and its presence in another Pyrenomycete, the phytopathogen *Colletotrichum trifolii*, was recently demonstrated (Buhr *et al.*, 1996). Notably, the *C. trifoli* gene was able to partly complement the *N. crassa cot-1* mutation.

Deletion of *cot-1* is not lethal in *N. crassa*, but does cause a severe phenotype similar to that of the Ts mutant. Interestingly, ectopic expression of the first third of the gene in the wild-type background promotes a partial *cot-1* phenotype (i.e. a dominant negative phenotype). One explanation for this would be that the truncated gene product is competing with normal gene product, perhaps for binding to another protein. The *cot-1* protein kinase might be a component of a signalling pathway involved in hyphal elongation, and the challenge is to find the other components. One approach to this is to obtain and identify suppressors of *cot-1*, but this approach has to have led into a different direction, from hyphal branching to nuclear migration.

Plamann *et al.* (1994) isolated partial suppressors of *cot-1*, known as *ropy* mutants (*ro*) because of their curled hyphae and asymmetric nuclear disribution. Similar mutants had been isolated many years earlier (Garnjbost & Tatum, 1967). Three genes identified in the initial study have been shown to be components of cytoplasmic dynein or dynactin. Dynein is a microtubule-based motor protein which transports nuclei and other membranous organelles. Dynactin, another multisubunit complex, is needed for dynein-mediated vesicle transport *in vitro*. *ro-1* encodes the largest subunit of the dynein complex [equivalent to *nudA* of *A. nidulans*, (Xiang, Beckwith & Morris, 1994)], *ro-3* encodes p150^{Glued}, the largest subunit of the dynactin complex, and *ro-4* is equivalent to arp1 (actin-related protein) of the dynactin complex. The relatively simple approach of selecting suppressors of *cot-1* has been extended to isolate more than 1000 *ro* mutants, which fall into 23 complementation groups (Bruno *et al.*, 1996*a*). More than a third of the mutants are further alleles of *ro-1*, but this identification of new *ro* loci should help dissect elements of the dynein/dynactin complexes and other interacting proteins. This line of study also raises questions about the roles of microfilaments (F-actin) and microtubules in the movement of nuclei and secretory vesicles. The reason why mutations in dynein/dynactin components lead to partial suppression of the *cot-1* phenotype is not known.

Both tip extension and branch formation are forms of polar growth, and an apolar mutant, *mcb*, has been isolated in *N. crassa* (Bruno *et al.*, 1996*b*). This temperature-sensitive mutant loses polarity at the non-permissive temperature, such that conidia swell without forming germ tubes, and hyphae grown at 25°C and shifted to 37°C cease polar growth and begin to swell. In addition, the *mcb* mutant shows aberrant placement and alignment of septa. Cloning revealed that *mcb* encodes the catalytic subunit of cAMP-dependent protein kinase A, implicating the PKA pathway in polarized morphogenesis and septal positioning. Selection for suppressors of *mcb* led to isolation of a mutation in the gene encoding adenylate cyclase, *cr-1*. The lack of detectable levels of cAMP in this mutant suggests that the *mcb* phenotype results from uncontrolled expression of PKA. Equivalent mutations in the yeast PKA do not result in loss of polarity.

Hyperbranching mutants in A. nidulans

Compact morphogical mutants can be obtained following mutagenesis and plating at the normal laboratory growth temperature (37°C), the method used to previously isolate the *mo* mutants (Bainbridge, 1966), or by use of temperature-sensitive mutants. The temperature-sensitive mutant collection of Harris *et al.* (1994) has provided a wealth of material for searching for different mutations affecting essential processes. These 1200 strains were screened visually under the light microscope at 26°C and 42°C (the non-permissive temperature) for those which showed hyperbranching, leading to compact colonial growth (S. Pollerman and G. Turner, unpublished). Twenty-five strains were selected and checked by sexual crosses for 1:1 segregation of the observed phenotype, and a subset of these have been further investigated. While hyperbranching is readily observable, G values can be measured to give a quantifiable phenotype. Some of the mutants currently being studied are illustrated in Fig. 10.5. A range of phenotypes is observed, with different degrees of hyperbranching. At 42°C, G is 68 for the wild-type strain, and ranges from 14 to 39 for the mutants. All of the mutants are recessive, and of ten mutants tested so far, most are in different complementation groups. *hbr-2* failed to complement *hbr-10*, despite marked difference in phenotypes. *hbr-2* is severely hyperbranching, and *hbr-10* weakly hyperbranching. The diploid made between these two strains is temperature sensitive, with its own phenotype, closer to that of *hbr-2*. Nevertheless, when crossed, the mutations carried by strains recombine freely, showing no linkage, i.e. they are non-complementing unlinked mutations. This phenomenon has been previously reported in *S. cerevisiae* (Stearns & Botstein; 1988, Vinh *et al.*, 1993) and more recently amongst certain *ropy* mutants of *N. crassa* (Bruno *et al.*, 1996*a*). Various models have been proposed to explain this phenomenon, including physical interaction of the gene products, or their contribution to independent pathways that are both required for successful completion of a process. *hbr-4* and *hbr-6* also fail to complement, but in this case recombination frequency was low in a sexual cross (approximately 1%), suggesting that they might be allelic.

Careful examination of the *hbr* mutants under light microscopy gives some indication of why these mutants have a hyperbranching phenotype. Strikingly, some of the most severely branching mutants appear to have closely spaced septa. The number of branches per compartment does not appear to be significantly different from the wild type, at least in the tip proximal intercalary compartments, which are readily visible. In addition, the apical compartment is shorter than in the wild-type grown under the same conditions, and often undergoes multiple branching in the absence of septa. Since hyphae do not have excessive branching per intercalary compartment, it appears that septation eventually 'catches

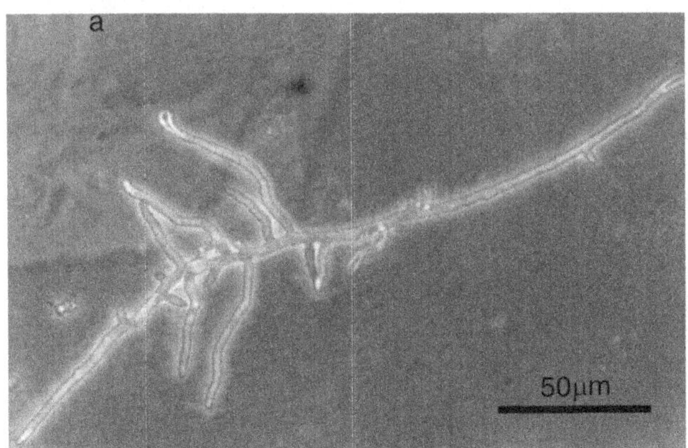
a
50μm

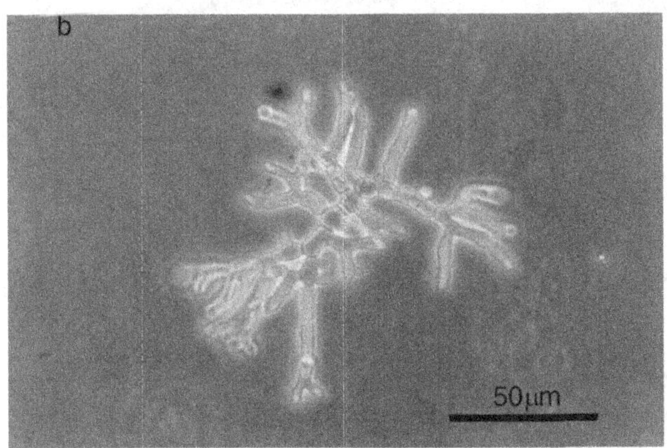
b
50μm

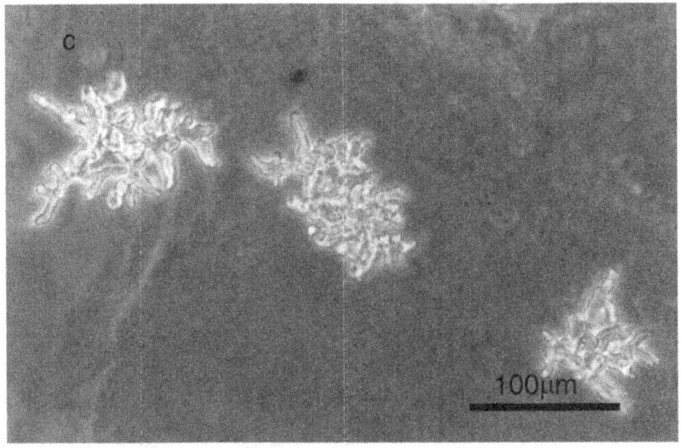
c
100μm

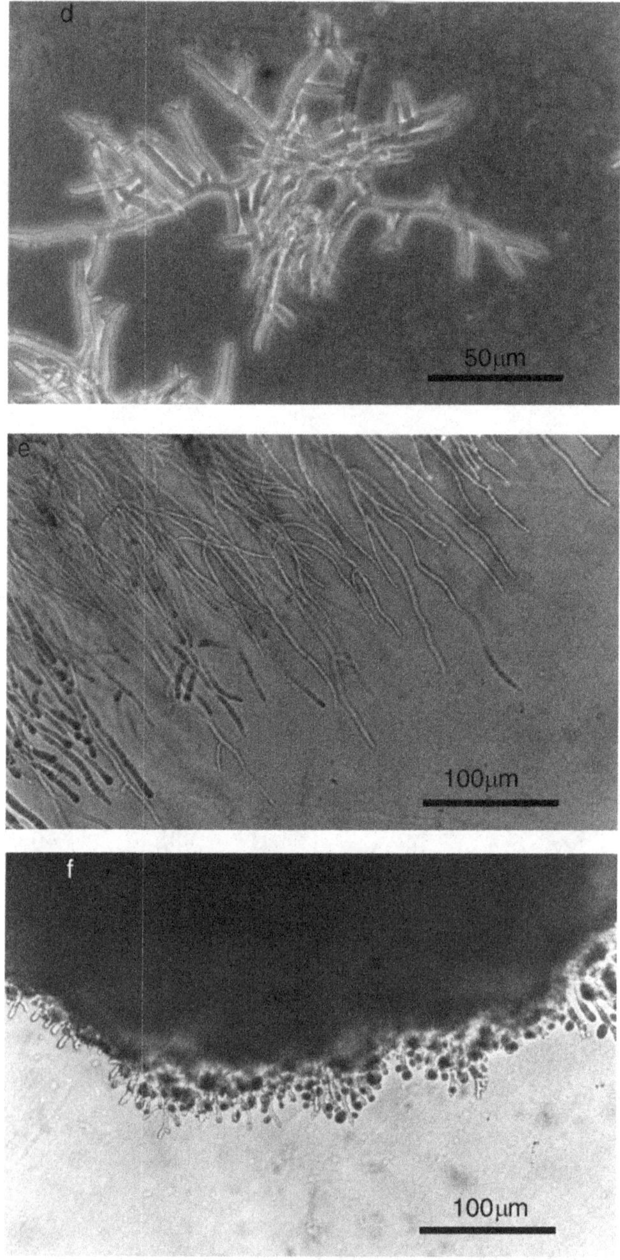

Fig. 10.5. Hyperbranching (*hbr*) mutants of *Aspergillus nidulans*. Wild-type and mutant phenotypes in 16–18 hour cultures at 42°C. Germlings: (a) wild-type, (b) *hbr-2*, (c) *hbr-3*, (d) *hbr-9*. Edge of two-day-old colonies: (e) wild-type, (f) *hbr-2*.

up' with branching. As mentioned previously, there appears to be a relationship between branching and septation in *A. nidulans* and other filamentous fungi. Indeed, Steele & Trinci (1977*b*) were able to demonstrate a close relationship between the hyphal growth unit and intercalary compartment length. Measurements of apical and intercalary compartment lengths in certain *hbr* mutants have confirmed this visual impression. For example, *hbr-2* and *hbr-9* mutations (Fig. 10.5) result in apical and first subapical compartments with less than half the length of the wild type (R.N. Turner, unpublished). Although G values are a convenient measure to categorize mutant phenotypes, Trinci *et al.* (1994) have pointed out that there is an assumption of uniform mycelial diameter. Visually, some of the *hbr* mutants appear to have rather thicker hyphae, though accurate measurements have not yet been made. It is possible that septation and branching are linked to mycelial volume, and only a small increase in diameter would be needed to double the volume of the intercalary compartment. Increased branching and septation frequency might thus result from an increase in hyphal diameter in some cases. *hbr-3* gives the most extreme phenotype (Fig. 10.5), and measurement of compartment size is difficult because of its irregularity. Most observations and measurements have been made on undifferentiated mycelium (16–20 hours on complete medium at 42°C), where the whole microcolony can be observed clearly, but some mutants exhibit their hyperbranching phenotype most clearly at the growing edge of older colonies (differentiated mycelium; Fig. 10.5). Nuclei have also been examined using DAPI staining. Nuclear distribution, strongly affected in *nud* mutants of *A. nidulans* and *ropy* mutants of *N. crassa*, does not appear to be markedly affected in any of the mutants examined to date.

Isolation of **hbr** *genes*

In order to learn more about the nature of the *hbr* mutations, gene isolation and molecular characterization is essential. The usual approach in *A. nidulans* is to assign the desired mutation to a particular linkage group using the parasexual cycle, then use cosmid DNA pools from a chromosome specific library (Brody *et al.*, 1991) to transform the mutant to a wild-type phenotype (assuming that the mutation is recessive). In the case of the severe mutants, it is possible to detect wild-type transformants in a mutant background when transformed protoplasts are regenerated at 42°C. Using this approach, we have now identified a cosmid which complements the *hbr-2* mutation (S. Memmott & G. Turner, unpublished), and subcloning is currently in progress to identify DNA sequence responsible for complementation. Some of the mutations are less extreme, and background growth at 42°C would be too great to permit this approach. In this case, cotransformation using a selectable marker (e.g. *argB*) will be necessary, following construction of *hbr argB* strains by sexual crossing.

A final perspective: the role of molecular genetics in understanding hyphal morphogenesis

Studies of cellular morphogenesis in fungi have historically emanated from two viewpoints. One view is based on the premise that the key to understanding morphogenesis is to identify all the necessary molecular components and determine how they function within the context of a 'genetic programme'. This bottom-up approach is founded on the genetic analysis of morphogenetic mutants and the molecular characterization of their cognate genes. In contrast, the other view is centred on the notion that although genes are important, morphogenesis is ultimately controlled by epigenetic factors which reflect a cell's physiological state and its external environment. This top-down approach entails analysing the effects of factors such as intracellular pH and/or Ca^{2+} gradients in establishing spatial patterns within the cell. Despite the fact the each of these approaches has helped to advance the understanding of morphogenesis in fungal cells, the bottom-up molecular genetic approach has tended to receive greater emphasis in recent years. In response to this bias, an argument has been raised that this approach is limited and cannot possibly determine how a cell generates three-dimensional patterns (Harold, 1995). The argument states that although numerous gene products involved in morphogenesis have been individually characterized, the manner in which they function as an ensemble to generate cellular shape is not at all obvious. This argument clearly has merit. It cannot be disputed that morphogenesis involves complex interactions between many gene products that must be properly regulated in space and time, and that these interactions are subject to physiological and environmental influences (see Machin *et al.*, 1996). Furthermore, proponents of the bottom-up approach recognize its limitations, as, for example, the sequence of the entire budding yeast genome has not yet made it clear how a bud forms. Given these arguments, is the role of molecular genetics simply to identify the components needed for morphogenesis?

Will the characterization of the organizational functions required for the generation of spatial patterns within fungal cells be refractory to genetic analyses? Insights into the molecular mechanisms underlying the spatial organization of yeast cells which were recently obtained through the analysis of cell patterning mutants emphatically demonstrates that the answer to these questions is no (Chant *et al.*, 1995; Sanders & Herskowitz, 1996; Mata & Nurse, 1997). The relatively subtle phenotypes caused by the patterning mutations suggests that the apparent limitations of the molecular genetic approach primarily reflect the tendency of geneticists to chose mutants that are easy to recognize. This problem has been largely rectified by the development of increasingly sensitive genetic (Verde *et al.*, 1995; Zahner *et al.*, 1996) and molecular screens (Sawin & Nurse, 1996) for mutants defective in spatial patterning. These advances, coupled with established ability of genetics to identify genes without preconceived notion and to assess their function unambiguously, suggests that the molecular genetic approach will continue to play a significant role in understanding cellular morphogenesis in fungi.

References

Adams, A. E. M. & Pringle, J. R. (1984). Relationship of actin and tubulin distribution to bud growth in wild-type and morphogenetic mutant *Saccharomyces cerevisiae*. *Journal of Cell Biology*, **98**, 934–45.

Allen, E. D., Aiuto, R. & Sussman, A. S. (1980). Effects of cytochalasins on *Neurospora crassa* I. growth and ultrastructure. *Protoplasma*, **102**, 63–75.

Amberg, D. C., Zahner, J. E., Mulholland, J. W., Pringle, J. R. & Botstein, D. (1997). Aip3p/Bud6p, a yeast actin-interacting protein that is involved in morphogenesis and the selection of bipolar budding sites. *Molecular Biology of the Cell*, **8**, 729–53.

Bainbridge, B. W. (1966). Research Notes. *Aspergillus Newsletter*, **7**, 19–21.

Bainbridge, B. W. (1970). Genetic analysis of an unequal chromosomal translocation in *Aspergillus nidulans*. *Genetics Research, Cambridge*, **15**, 317–26.

Bainbridge, B. W. & Trinci, A. P. J. (1969). Colony and specific growth rates of normal and mutant strains of *Aspergillus nidulans*. *Transactions of the British Mycological Society*, **53**, 473–5.

Bartnicki-Garcia, S., Hergert, F. & Gleitz, G. (1989). Computer simulation of fungal morphogenesis and the mathematical basis for hyphal (tip) growth. *Protoplasma*, **153**, 46–57.

Bender, A. & Pringle, J. R. (1989). Multicopy supression of the *cdc24* budding defect in yeast by *CDC42* and three newly identified genes including the *ras*-related gene *RSR1*. *Proceedings of the National Academy of Sciences, USA*, **86**, 9976–80.

Bennett, J. W. & Lasure, L. L. eds. (1991). *More Gene Manipulations in Fungi*. San Diego: Academic Press.

Bergen, L. & Morris, N. R. (1983). Kinetics of the nuclear division cycle of *Aspergillus nidulans. Journal of Bacteriology*, **156**, 155–60.

Brody, H., Griffith, J., Cuticchia, A. J., Arnold, J. & Timberlake, W. E. (1991). Chromosome-specific recombinant DNA libraries from the fungus *Aspergillus nidulans. Nucleic Acids Research*, **19**, 3105–9.

Bruno, K. S., Aramayo, R., Minke, P. F., Metzenberg, R. L. & Plamann, M. (1996*b*). Loss of growth polarity and mislocalization of septa in a *Neurospora* mutant altered in the regulatory subunit of cAMP-dependent protein kinase. *The EMBO Journal*, **15**, 5772–82.

Bruno, K. S., Tinsley, J. H., Minke, P. F. & Plamann, M. (1996*a*). Genetic interactions among cytoplasmic dynein, dynactin, and nuclear distribution mutants of *Neurospora crassa. Proceedings of the National Academy of Sciences, USA*,. **93**, 4775–80.

Buhr, T. L., Oved, S., Truesdell, G. M., Huang, C., Yarden, O. & Dickman, M. B. (1996). A kinase-encoding gene from *Colletotrichum trifolii* complements a colonial growth mutant of *Neurospora crassa. Molecular and General Genetics*, **251**, 565–72.

Caldwell, I. Y. & Trinci, A. P. J. (1973). The growth unit of the mould *Geotrichum candidum. Archives of Microbiology*, **88**, 1–10.

Chang, E. C., Barr, M., Wang, Y., Jung, V., Xu, H.-P. & Wigler, M. H. (1995). Cooperative interaction of *S. pombe* proteins required for mating and morphogenesis. *Cell*, **79**, 131–41.

Chant, J. (1994). Cell polarity in yeast. *Trends in Genetics*, **10**, 328–33.

Chant, J., Corrado, K., Pringle, J. R., & Herskowitz, I. (1991). Yeast *BUD5*, encoding a putative GDP-GTP exchange factor, is necessary for bud site selection and interacts with bud formation gene *BEM1*. *Cell*, **65**, 1213–24.

Chant, J. & Herskowitz, I. (1991). Genetic control of bud site selection in yeast by a set of gene products that constitute a morphogenetic pathway. *Cell*, **65**, 1203–12.

Chant, J., Mischke, M., Mitchell, E., Herskowitz, I. & Pringle, J. R. (1995). Role of Bud3p in producing the axial budding pattern of yeast. *Journal of Cell Biology*, **129**, 767–78.

Clutterbuck, A. J. (1970). Synchronous nuclear division and septation in *Aspergillus nidulans. Journal of General Microbiology*, **60**, 133–5.

Collinge, A. J., Fletcher, M. H. & Trinci, A. P. J. (1978). Physiology and cytology of septation and branching in a temperature-sensitive colonial mutant (*cot-1*) of *Neurospora crassa. Transactions of the British Mycological Society*, **71**, 107–20.

Cross, F. R. (1995). Starting the cell cycle: what's the point? *Current Opinion in Cell Biology*, **7**, 790–7.

Cvrckova, F., De Virgilio, C., Manser, E., Pringle, J. R. & Nasmyth, K. (1995). Ste20-like protein kinases are required for normal localization of cell growth and for cytokinesis in budding yeast. *Genes and Development*, **9**, 1817–30.

Dayton, J. S. & Means, A. R. (1996). Ca^{2+}/calmodulin-dependent kinase is essential for both growth and nuclear division in *Aspergillus nidulans. Molecular Biology of the Cell*, **7**, 1511–19.

Dayton, J. S., Sumi, M., Nanthakumar, N. N. & Means, A. R. (1997). Expression of a constitutively active Ca^{2+}/calmodulin-dependent kinase in *Aspergillus nidulans* spores prevents germination and entry into the cell cycle. *Journal of Biological Chemistry*, **272**, 3223–30.

Dicker, J. W. & Turian, G. (1990). Calcium deficiencies and apical hyperbranching in wild-type and the *frost* and *spray* morphological mutants of *Neurospora crassa*. *Journal of General Microbiology*, **136**, 1413–20.

Doshi, P., Bossie, C. A., Doonan, J. H., May, G. S. & Morris, N. R. (1991). Two α-tubulin genes of *Aspergillus nidulans* encode divergent proteins. *Molecular and General Genetics*, **225**, 129–41.

Evangelista, M., Blundell, K., Longtine, M., Chow, C., Adames, N., Pringle, J., Peter, M. & Boone, C. (1997). Bni1p, a yeast formin linking Cdc42p and the actin cytoskeleton during polarized morphogenesis. *Science*, **276**, 118–22.

Fiddy, C. & Trinci, A. P. J. (1976*a*). Mitosis, septation, branching and the duplication cycle in *Aspergillus nidulans*. *Journal of General Microbiology*, **97**, 169–84.

Fiddy, C. & Trinci, A. P. J. (1976*b*). Nuclei, septation, branching and growth of *Geotrichum candidum*. *Journal of General Microbiology*, **97**, 185–92.

Frazier, J. A. & Field, C. M. (1997). Are FH proteins local organizers? *Current Biology*, **7**, 414–17.

Fujita, A., Oka, C., Arikawa, Y., Katagai, T., Tonouchi, A., Kuhara, S. & Misumi, Y. (1994). A yeast gene necessary for bud-site selection encodes a protein similar to insulin-degrading enzymes. *Nature*, **372**, 567–70.

Garnjobst, L. & Tatum, E. L. (1967). A survey of new morphological mutants in *Neurospora crassa*. *Genetics*, **57**, 579–604.

Girbardt, M. (1969). Die ultrastruktur der apikalregion von pilzhyphen. *Protoplasma*, **67**, 413–41.

Gold, S., Duncan, G., Barrett, K. & Kronstad, J. (1994). cAMP regulates morphogenesis in the fungal pathogen *Ustilago maydis*. *Genes and Development*, **8**, 2805–16.

Gow, N. A. R. (1994). Tip growth and polarity. In *The Growing Fungus*, ed. N. A. R. Gow & G. M. Gadd, pp. 277–99. London: Chapman & Hall.

Halme, A., Michelitch, M., Mitchell, E. & Chant, J. (1996). Bud10p directs axial cell polarization in budding yeast and resembles a transmembrane receptor. *Current Biology*, **6**, 570–9.

Harold, F. M. (1995). From morphogenes to morphogenesis. *Microbiology*, **141**, 2765–78.

Harris, S. D., Hamer, L., Sharpless, K. E. & Hamer, J. E. (1997). The *Aspergillus nidulans sepA* gene encodes an FH1/2 protein involved in cytokinesis and the maintenance of cellular polarity. *The EMBO Journal*, **16**, 3474–83.

Harris, S. D., Morrell, J. M. & Hamer, J. E. (1994). Identification and characterization of *Aspergillus nidulans* mutants defective in cytokinesis. *Genetics*, **136**, 517–32.

Howard, R. J. (1981). Ultrastructural analysis of hyphal tip growth in fungi: Spitzenkorper, cytoskeleton, and endomembranes after freeze substitution. *Journal of Cell Science*, **48**, 89–103.

Hyde, G. J., & Heath, I. B. (1997). Ca^{+} gradients in hyphae and branches of *Saprolegnia ferax*. *Fungal Genetics & Biology*, **21**, 238–51.

Igual, J. C., Johnson, A. L. & Johnston, L. H. (1996). Coordinated regulation of gene expression by the cell cycle transcription factor *SWI4* and the protein kinase C MAP kinase pathway for yeast cell integrity. *The EMBO Journal*, **15**, 5001–13.

Jackson, S. L. & Heath, I. B. (1993). Roles of calcium ions in hyphal tip growth. *Microbiological Reviews*, **57**, 367–82.

Jacobs, C. W., Adams, A. E. M., Szaniszlo, P. J. & Pringle, J. R. (1988). Functions of microtubules in the *Saccharomyces cervisiae* cell cycle. *Journal of Cell Biology*, **107**, 1409–26.

Johnson, D. I. & Pringle, J. R. (1990). Molecular characterization of *CDC42*, a *Saccharomyces cerevisiae* gene involved in the development of cell polarity. *Journal of Cell Biology*, **111**, 143–52.

Kilmartin, J. V. & Adams, A. E. M. (1984). Structural rearrangements of tubulin and actin during the cell cycle of the yeast *Saccharomyces. Journal of Cell Biology*, **98**, 922–33.

Kobori, H., Yamada, N., Taki, A. & Osumi, M. (1989). Actin is associated with the formation of the cell wall in reverting protoplasts of the fission yeast *Schizosaccharomyces pombe. Journal of Cell Science*, **94**, 635–46.

Kron, S. J. & Gow, N. A. R. (1995). Budding yeast morphogenesis: signalling, cytoskeleton and cell cycle. *Current Opinion in Cell Biology* **7,** 845–855.

Leberer, E., Dignard, D., Harcus, D., Thomas, D. Y. & Whiteway, M. (1992). The protein kinase homologue Ste20p is required to link the yeast pheromone response G protein bg subunits to downstream signalling components. *The EMBO Journal* **11,** 4815–4824.

Lew, D. J. & Reed, S. I. (1993). Morphogenesis in the yeast cell cycle: regulation by Cdc28 and cyclins. *Journal of Cell Biology*, **120**, 1305–20.

Li, R. (1997). Bee1, a yeast protein with homology to Wiscott-Aldrich syndrome protein, is critical for the assembly of cortical actin cytoskeleton. *Journal of Cell Biology*, **136**, 649–658.

Longtine, M. S., DeMarini, D. J., Valencik, M. L., Al-Awar, O. S., Fares, H., De Virgilio, C. & Pringle, J. R. (1996). The septins: roles in cytokinesis and other processes. *Current Opinion in Cell Biology*, **8**, 106–19.

Machin, N., Lee, J. M., Chamany, K. & Barnes, G. (1996). Dosage suppressors of a benomyl-dependent tubulin mutant: evidence for a link between microtubule stability and cellular metabolism. *Genetics*, **144**, 1363–73.

Madden, K., Sheu, Y.-J., Baetz, K., Andrews, B. & Snyder, M. (1997). SBF cell cycle regulator as a target of the yeast PKC-MAP kinase pathway. *Science*, **275**, 1781–4.

Markham, P. (1994). Organelles of filamentous fungi. In *The Growing Fungus*, ed. N. A. R. Gow & G. M. Gadd, pp. 75–98. London: Chapman & Hall.

Marks, J., Hagan, I. M. & Hyams, J. S. (1986). Growth polarity and cytokinesis in fission yeast: the role of the cytoskeleton. *Journal of Cell Science*, **5**(Suppl.), 229–41.

Marks, J. & Hyams, J. S. (1985). Localization of F-actin through the cell division cycle of *Schizosaccharomyces pombe. European Journal of Cell Biology*, **39**, 27–32.

Martinelli, S. D. & Kinghorn, J. R. (eds.) (1994). *Aspergillus: 50 years On.* Progress in Industrial Microbiology, vol. 29. Amsterdam: Elsevier.

Mata, J. & Nurse, P. (1997). *tea1* and the microtubular cytoskeleton are important for generating global spatial order within the fission yeast cell. *Cell*, **89**, 939–49.

McGoldrick, C. A., Gruver, C. & May, G. S. (1995). *myoA* of *Aspergillus nidulans* encodes an essential myosin required for secretion and polarized growth. *Journal of Cell Biology*, **128**, 577–87.

Michelitch, M. & Chant, J. (1996). A mechanism of Bud1p GTPase action suggested by mutational analysis and immunolocalization. *Current Biology*, **6**, 446–54.

Miller, P. J. & Johnson, D. I. (1994). Cdc42p GTPase is involved in controlling polarized cell growth in *Schizosaccharomyces pombe. Molecular and Cellular Biology*, **14**, 1075–83.

Mitchell, M. B. & Mitchell, H. K. (1954). A partial map of linkage group D in *Neurospora crassa. Proceedings of the National Academy of Sciences, USA*, **40**, 436–46.

Mitchison, J. M., & Nurse, P. (1985). Growth in cell length in the fission yeast *Schizosaccharomyces pombe. Journal of Cell Science*, **75**, 357–76.

Morris, N. R. (1976). Mitotic mutants of *Aspergillus nidulans. Genetics Research, Cambridge*, **26**, 237–54.

Murray, J. C. & Srb, A. M. (1962). The morphology and genetics of wild-type and seven morphological mutant strains of *Neurospora crassa. Canadian Journal of Botany*, **40**, 337–49.

Nasmyth, K. (1996). At the heart of the budding yeast cell cycle. *Trends in Genetics*, **12**, 405–12.

Novick, P. & Botstein, D. (1985). Phenotypic analysis of temperature-sensitive yeast actin mutants. *Cell*, **40**, 405–16.

Nurse, P. (1994). Fission yeast morphogenesis – posing the problems. *Molecular Biology of the Cell*, **5**, 613–16.

Oakley, B. R. & Morris, N. R. (1980). Nuclear movement is β-tubulin-dependent in *Aspergillus nidulans. Cell*, **19**, 255–62.

Oakley, B. R. & Rinehart, J. E. (1985). Mitochondria and nuclei move by different mechanisms in *Aspergillus nidulans. Journal of Cell Biology*, **101**, 2392–7.

O'Connell, M. J., Norbury, C. & Nurse, P. (1994). Premature chromatin condensation upon accumulation of NIMA. *EMBO Journal*, **13**, 4926–37.

Oliver, S. G. (1997). Yeast as a navigational aid in genome analysis. *Microbiology*, **143**, 1483–7.

Osmani, S. A., Engle, D. B., Doonan, J. H. & Morris, N. R. (1988*a*). Spindle formation and chromosome condensation in cells blocked at interphase by mutation of a negative cell cycle control gene. *Cell*, **52**, 241–51.

Osmani, S. A., Pu, R. T. & Morris, N. R. (1988*b*). Mitotic induction and maintenance by overexpression of a G2-specific gene that encodes a potential protein kinase. *Cell*, **53**, 237–44.

Ottilie, S., Miller, P. J., Johnson, D. I., Creasy, C. L., Sells, M. A., Bagrodia, S., Forsburg, S. L. & Chernoff, J. (1995). Fission yeast *pak1*$^{+}$ encodes a protein kinase that interacts with Cdc42p and is involved in the control of cell polarity and mating. *EMBO Journal*, **14**, 5908–19.

Park, H.-O., Bi, E., Pringle, J. R. & Herskowitz, I. (1997). Two active states of the Ras-related Bud1/Rsr1 protein bind to different effectors to determine yeast cell polarity. *Proceedings of the National Academy of Sciences, USA*, **94**, 4463–8.

Park, H.-O., Chant, J. & Herskowitz, I. (1993). *BUD2* encodes a GTPase-activating protein for Bud1/Rsr1 necessary for proper bud site selection in yeast. *Nature*, **365**, 269–74.

Perkins, D. D., Radford, A., Newmeyer, D. & Bjorkman, M. (1982). Chromosomal loci of *Neurospora crassa. Microbiological Reviews*, **46**, 426–570.

Plamann, M., Minke, P. F., Tinsley, J. H. & Bruno, K. (1994). Cytoplasmic dynein and actin-related protein Arp1 are required for normal nuclear distribution in filamentous fungi. *Journal of Cell Biology*, **127**, 139–49.

Pontecorvo, G., Roper, J. A., Hemmons, L. M., MacDonald, K. D. & Bufton, A. W. J. (1953). The genetics of *Aspergillus nidulans*. *Advances in Genetics*, **5**, 141–239.

Prosser, J. I. (1983). Hyphal growth patterns. In *Fungal Differentiation: A Contemporary Synthesis*, ed. J. E. Smith, pp. 357–96. New York: Marcel Dekker, Inc.

Prosser, J. I. & Trinci, A. P. J. (1979). A model for hyphal growth and branching. *Journal of General Microbiology*, **111**, 153–64.

Pu, R. T. & Osmani, S. A. (1995). Mitotic destruction of the cell cycle regulated NIMA protein kinase of *Aspergillus nidulans* is required for mitotic exit. *EMBO Journal*, **14**, 995–1003.

Reynaga-Pena, C. G., Gierz, G. & Bartnicki-Garcia, S. (1997). Analysis of the role of the Spitzenkorper in fungal morphogenesis by computer simulation of apical branching in *Aspergillus niger*. *Proceedings of the National Academy of Sciences, USA*, **94**, 9096–101.

Roemer, T., Vallier, L. G. & Snyder, M. (1996). Selection of polarized growth sites in yeast. *Trends in Cell Biology*, **6**, 434–41.

Sanders, S. L. & Herskowitz, I. (1996). The Bud4 protein of yeast, required for axial budding, is localized to the mother/bud neck in a cell cycle-dependent manner. *Journal of Cell Biology*, **134**, 413–27.

Sawin, K. E. & Nurse, P. (1996). Identification of fission yeast nuclear markers using random polypeptide fusions with green fluorescent protein. *Proceedings of the National Academy of Sciences, USA*, **94**, 15146–51.

Scott, W. A. (1976). Biochemical genetics of morphogenesis in *Neurospora crassa*. *Annual Review of Microbiology*, **30**, 85–104.

Simanis, V. (1995). The control of septum formation and cytokinesis in fission yeast. *Seminars in Cell Biology*, **6**, 79–87.

Snell, V. & Nurse, P. (1993). Investigations into the control of cell form and polarity: the use of morphological mutants in fission yeast. *Development* (suppl.) 1993, 289–99.

Som, T. & Kolaparthi, V. S. R. (1994). Developmental decisions in *Aspergillus nidulans* are modulated by Ras activity. *Molecular and Cellular Biology*, **14**, 5333–48.

Stearns, T. & Botstein, D. (1988). Unlinked noncomplementation: isolation of new conditional lethal mutations in each of the tubulin genes of *Saccharomyces cerevisiae*. *Genetics*, **119**, 249–60.

Steele, G. C. & Trinci, A. P. J. (1977*a*). Effect of temperature and temperature shifts on growth and branching of a wild type and a temperature sensitive colonial mutant (*cot-1*) of *Neurospora crassa*. *Archives of Microbiology*, **113**, 43–8.

Steele, G. C. & Trinci, A. P. J. (1977*b*). Relationship between intercalary compartment length and hyphal growth unit length. *Transactions of the British Mycological Society*, **69**, 156–8.

Steinberg, G. & Schliwa, M. (1993). Organelle movements in the wild type and wall-less fz;sg;os-1 mutants of *Neurospora crassa* are mediated by cytoplasmic microtubules. *Journal of Cell Science*, **106**, 555–64.

Toda, T., Umesono, K., Hirata, A. & Yanagida. M. (1983). Cold-sensitive nuclear division arrest mutants of the fission yeast *Schizosaccharomyces pombe*. *Journal of Molecular Biology*, **168**, 251–70.

Trinci, A. P. J. (1973). The hyphal growth unit of wild-type and spreading colonial mutants of *Neurospora crassa*. *Archives of Microbiology*, **91**, 127–36.

Trinci, A. P. J. (1974). A study of the kinetics of hyphal extension and branch initiation of fungal mycelia. *Journal of General Microbiology*, **81**, 225–36.

Trinci, A. P. J. (1979). The duplication cycle and branching. In *Fungal Walls and Hyphal Growth*, 2nd Symposium of the British Mycological Society, eds. J. H. Burnett & A. P. J. Trinci, pp. 319–58. Cambridge: Cambridge University Press.

Trinci, A. P. J. & Morris, N. R. (1979). Morphology and growth of a temperature-sensitive mutant of *Aspergillus nidulans* which forms aseptate mycelium at non-permissive temperatures. *Journal of General Microbiology*, **114**, 53–9.

Trinci, A. P. J., Wiebe, M. G. & Robson, G. D. (1994). The mycelium as an integrated entity. In *The Mycota I: Growth, Differentiation and Sexuality*, eds. K. Esser and P. A. Lemke, pp. 175–93. Berlin: Springer-Verlag.

Verde, F., Mata, J. & Nurse, P. (1995). Fission yeast cell morphogenesis: identification of new genes and analysis of their role during the cell cycle. *Journal of Cell Biology*, **131**, 1529–38.

Vinh, D. B. N., Welch, M. D., Corsi, A. K., Wertman, K. F. & Drubin, D. G. (1993). Genetic evidence for functional interactions between actin non-complementing (ANC) gene products and actin cytoskeletal proteins. *Genetics*, **135**, 275–86.

Wessels, J. G. H. (1993). Tansley review no. 45: wall growth, protein excretion and morphogenesis in fungi. *New Phytologist*, **123**, 397–413.

Wiebe, M. G., Robson, G. D., Trinci, A. P. J. & Oliver, S. G. (1992). Characterization of morphological mutants generated spontaneously in glucose-limited, continuous flow cultures of *Fusarium graminearum*. *Mycological Research*, **96**, 555–62.

Wolkow, T. D., Harris, S. D. & Hamer, J. E. (1996). Cytokinesis in *Aspergillus nidulans* is controlled by cell size, nuclear positioning, and mitosis. *Journal of Cell Science*, **109**, 2179–88.

Xiang, X., Beckwith, S. M. & Morris, N. R. (1994). Cytoplasmic dynein is involved in nuclear migration in *Aspergillus nidulans*. *Proceedings of the National Academy of Sciences, USA*, **91**, 2100–4.

Xiang, X., Roghi, C. & Morris, N. R. (1995). Characterization and localization of the cytoplasmic dynein heavy chain in *Aspergillus nidulans*. *Proceedings of the National Academy of Sciences, USA*, **92**, 9890–4.

Yang, S., Ayscough, K. R. & Drubin, D. G. (1997). A role for the actin cytoskeleton of *Saccharomyces cerevisiae* in bipolar bud-site selection. *Journal of Cell Biology*, **136**, 111–23.

Yarden, O., Plamann, M., Ebbole, D. J. & Yanofsky, C. (1992). *cot-1*, a gene required for hyphal elongation in *Neurospora crassa*, encodes a protein kinase. *EMBO Journal*, **11**, 2159–66.

Ye, X., Xu, G., Pu, R. T., Fincher, R. R., McGuire, S. L., Osmani, A. H. & Osmani, S. A. (1995). The NIMA protein kinase is hyperphosphorylated and activated downstream of $p34^{cdc2}$/cyclin B: coordination of two mitosis promoting kinases. *EMBO Journal*, **14**, 986–94.

Zahner, J. E., Harkins, H. A. & Pringle, J. R. (1996). Genetic analysis of the bipolar pattern of bud site selection in the yeast *Saccharomyces cerevisiae. Molecular and Cellular Biology*, **16**, 1857–70.

Zheng, Y., Bender, A. & Cerione, R. A. (1995). Interactions among proteins involved in bud-site selection and bud-site assembly in *Saccharomyces cerevisiae. Journal of Biological Chemistry*, **270**, 626–30.

Zheng, Y., Cerione, R. & Bender, A. (1994). Control of the yeast bud-site assembly GTPase Cdc42: catalysis of guanine nucleotide exchange by Cdc24 and stimulation of GTPase activity by Bem3. *Journal of Biological Chemistry*, **269**, 2369–72.

11

Mating and sexual interactions in fungal mycelia

G. W. GOODAY

Introduction

Sexual reproduction is a major factor aiding adaptability and fitness in organisms throughout the natural world, and the fungi are no exception in exploiting its potential. Fungal mycelia in natural environments, unless they are self-fertile, are faced with the problem of finding a compatible partner. Their major senses are chemical, i.e. taste and smell, so we can imagine each mycelium, in for example the soil, exuding its own specific repertoire of chemicals, to announce its presence to potential mates. These chemicals have to be at least reasonably specific to fungal species, and completely specific to mating type within that species, so that attempts at mating stand a good chance of being successful. Thus potentially there are probably as many different chemicals as there are species. Such specific chemicals can be termed 'hormones' used in the context as defined by Raper (1952) for fungi 'substances produced by the affected plant or by others of the same species . . . performing indispensable regulatory roles in the sexual process'. An alternative term, increasingly used as a synonym in the fungal literature, is 'pheromone' for a chemical acting at a distance (cf., insect sex attractants). The very small number of such compounds that have been identified to date fall into two classes: isoprenoids (derived from mevalonic acid) among the 'lower fungi' (a very diverse phylogenetic group), and hydrophobic peptides, mostly isoprenylated, among Ascomycetes and Basidiomycetes. Reviews of various aspects of these fungal hormones/pheromones include those of Raper (1952), Machlis (1972), Gooday (1974), Van den Ende (1984) Gooday & Adams (1993), Gooday (1994) and Duntze, Betz & Nientiedt (1994). As well as these very specific effectors, there are numerous reports throughout the fungi of apparently less-specific sex factors, which include growth substances regulating development by their overall concentration, and morphogens regulating localized differentiation by providing positional information. Such sex factors have been reviewed by Dyer, Ingram & Johnstone (1992).

Steps leading up to successful mating can be summarized as follows, based on the very few examples where this has been investigated in detail. As mature mycelia approach each other, one or both of the potential partners is constitutively releasing a pheromone. The other partner responds to this because it has a specific receptor which can trigger the appropriate signal transduction pathway. An initial response is to initiate or to greatly increase production and release of the complementary pheromone. Commonly there is then growth arrest, if this has not already occurred through starvation, and then initiation of sexual differentiation, leading to the formation of the specific mating hyphae that will undergo plasmogamy. An essential feature of the process is also specific destruction of the pheromone by the recipient, to maintain its receptivity.

The key to the mating process is thus chemical communication between the mating mycelia, and it is with this aspect that the rest of the chapter is concerned.

Oomycetes

Species of the genus *Achlya* are water moulds, living saprobically on submerged plant and animal material. Their vegetative hyphae are diploid, and sexual reproduction entails the production of antheridial and oogonial initials on the same mycelium or two mycelia that are growing close to each other. The antheridial initials grow towards the oogonial initials, and when in contact with each other, both mature by delimitation of the hyphal apices by septal formation and meiosis, giving rise to egg cells and antheridial fertilization tubes. Most strains are homothallic, but some are heterothallic, self-sterile with three types of behaviour: solely male, solely female or male–female according to the nature of the strain with which they are paired (Raper, 1951; Barksdale, 1967; Thomas & Mullins, 1969).

In a series of elegant experiments, Raper (1951) showed that sexual reproduction in these fungi is regulated by complementary hormones. We now know that these are sterols, antheridiol and oogoniol (Fig. 11.1; Table 11.1; McMorris, 1978; Mullins, 1994; Carlile, 1996*b*). Antheridiol and oogoniol are biosynthesized by two separate pathways from the common plant sterol fucosterol. The complementary actions of these hormones account for the developmental sequences observed by Raper (1951) in each mating partner during sexual reproduction. Female

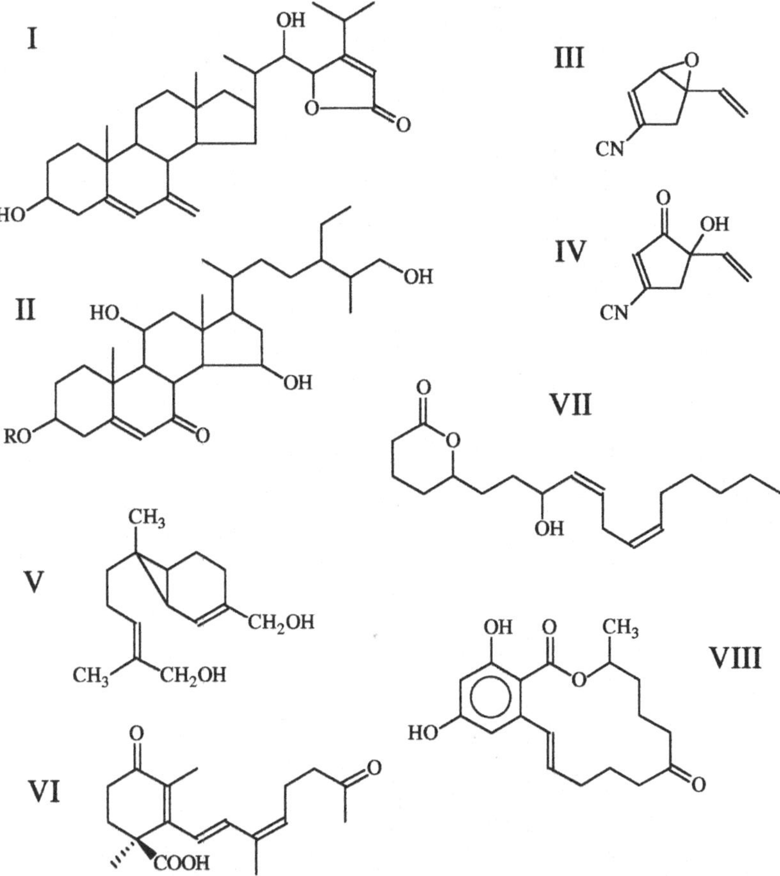

Fig. 11.1. Structures of some molecules that regulate sexual development in fungi. I, II: Antheridiol and oogoniol, sex hormones of *Achlya* species. For oogoniol-1, R is an isobutyrate residue. III, IV: Homothallins I and II, volatile metabolites from *Trichoderma koningii* that mimic hormone a1 of *Phytophthora* spp. V: Sirenin, sex attractant of *Allomyces* spp. VI: Trisporic acid B, sex hormone of the Mucorales. VII: Psi factor A from *Aspergillus nidulans*. VIII: Zearalenone from *Gibberella zeae*.

hyphae constitutively produce antheridiol, and under conditions of low nutrients male hyphae respond to increasing concentrations:

(a) They cease apical growth (Fig. 11.2; Gow & Gooday, 1987).
(b) They form antheridial branches (Fig. 11.2; Barksdale, 1967). Counting the number of these is the basis of the bioassay, which can detect 10 pg ml^{-1}.

Table 11.1. *Isoprenoid hormones of fungi*

Hormone	Molecular structure	Probable precursor	Site and specificity of synthesis	Optimal yield (M)	Sensitivity of bioassay (M)	Response
Antheridiol	Sterol ($C_{29}H_{42}O_5$)	Fucosterol	Female cells of *Achlya* species, constitutive	10^{-8}	10^{-11}	Antheridia and oogoniol production by males
Oogoniol	Sterol ($C_{33}H_{54}O_6$)	Fucosterol	Male cells of *Achlya* species, induced	–	10^{-7}	Oogonia by females
Sirenin	Sesquiterpene ($C_{15}H_{24}O_2$)	Farnesyl pyrophosphate	Female gametes of *Allomyces*	10^{-6}	10^{-10}	Chemotaxis of male gametes
Trisporic acid	Apocarotenoid ($C_{18}H_{26}O_4$)	Retinal	(+)/(–) Cells of Mucorales, collaborative	10^{-6} (*M. mucedo*), 10^{-3} (*B. trispora*)	10^{-8}	Zygophores by (+) and (–)

Adapted from Gooday & Adams (1993).

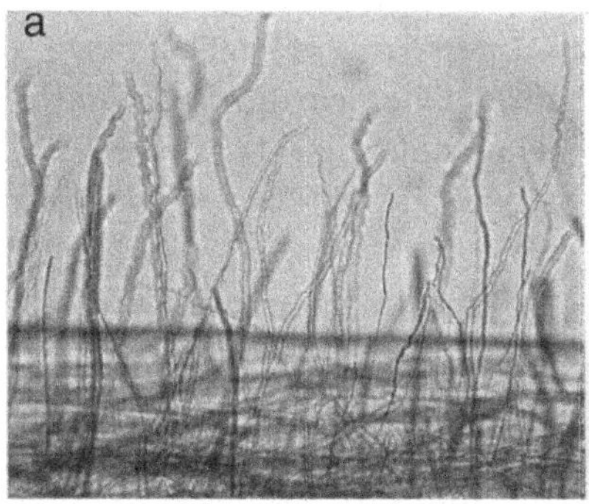

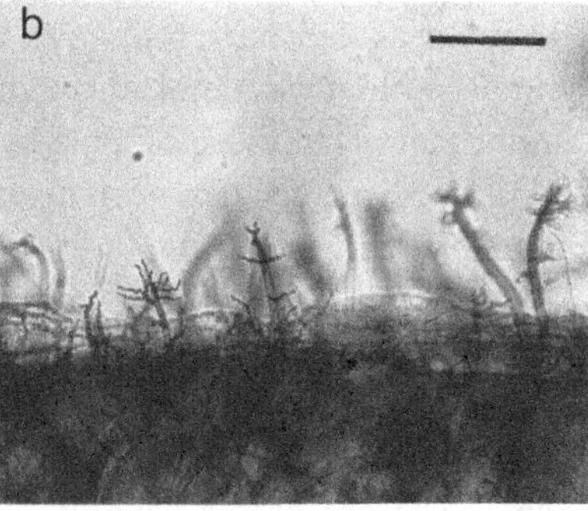

Fig. 11.2. Male hyphae of *Achlya ambisexualis* growing from a block of nutrient agar into 'artificial pond water': (a) control culture showing continued apical growth; (b) culture treated with 100 nM antheridiol, showing cessation of apical growth and profuse antheridial branching. Scale bar represents 200 μm. (From Gooday & Adams, 1993.)

(c) The antheridial branches grow chemotropically towards antheridiol, which is produced at increasing concentrations by the developing oogonial initials (Barksdale, 1967).
(d) The antheridia mature, become delimited by septa, and their nuclei undergo meiosis to form the male gametic nuclei (Barksdale, 1967).

Accompanying these morphological changes are a series of biochemical responses, which include the following:

(a) Synthesis and release of oogoniol (McMorris, 1978).
(b) Induced metabolism of antheridiol to inactive metabolites (Musgrave & Niewenhuis, 1975). This inactivation presumably plays a vital role in chemotropism, enabling the antheridial initials to remain responsive to the gradient of antheridiol.
(c) Increase in activity and secretion of cellulase (Thomas & Mullins, 1969). This may play a direct role in the antheridial branching, by softening the lateral cellulosic walls to allow pushing out of the new branches.
(d) Marked enhancement of synthesis of rRNA, mRNA and protein, and of histone acetylation (Timberlake & Orr, 1984).
(e) Up-regulation of heat-shock proteins correlated with branching and secretion of glycoproteins (Silver *et al.*, 1993).

In as far as they have been investigated, responses of female hyphae to the most potent oogoniol, the minor metabolite dehydro-oogoniol, are complementary to those of male hyphae to antheridiol (McMorris *et al.*, 1993). They include the formation of oogonial initials and enhanced synthesis of antheridiol.

There is good biological evidence for equivalent hormone systems in other oomycete species, but these remain to be characterized (Gooday & Adams, 1993). For example, formation of antheridia and oogonia occurred in male and female strains of *Pythium sylvaticum*, respectively, when they were grown separated by a permeable membrane (Gall & Elliott, 1985). There has been much interest in the possible occurrence of sterol hormones as regulators of sexual and asexual reproduction in the pythiaceous fungi, particularly in the plant pathogenic genera, *Phytophthora* and *Pythium* (Elliott, 1994). That sterols might be involved is an attractive idea, as these fungi are unable to synthesize sterols, but require them for reproduction, either as additions to the medium or from the host plant during pathogenesis (Elliott, 1983). A wide range of sterols stimulate reproduction (Elliott, 1983), suggesting that they may act as precursors to specific hormones, but to date these putative metabolites have not been characterized. They may be produced in tiny amounts and/or may be too unstable to have been identified. A confounding observation is that the same responses can be elicited in some cases by non-sterol lipids such as phospholipids (Ko, 1988), but there may have been trace sterol impurities or carry-over in these experiments. Kerwin & Duddles (1989) have investigated this in detail with *Pythium ultimum*, and conclude that trace levels of sterols can act synergistically with phospholipids containing unsaturated fatty acid moieties to induce oosporogenesis. There is also evidence for a hormonal system regulating mating in these fungi that is different from the putative sterol hormone system. This has been termed hormonal heterothallism, as both A1 and A2 mating types of heterothallic species form oospores when paired with the opposite mating type of the same or different species on the opposite side of a polycarbonate membrane. This shows that a heterothallic strain is induced to become self-fertile by a mating-type-specific hormone: A1 strains produce a1 hormone, which stimulates A2 but not A1 to produce oospores, while A2 strains produce a2 hormone with the complementary specificity (Ko, 1988; Guo & Ko, 1991). The effects of a1 hormone can be mimicked by volatile metabolites from *Trichoderma* species (Brasier, 1975) termed homothallins I and II (Fig. 11.1; Sakata & Rickards, 1980).

Chytridiomycetes

Sexual reproduction in the filamentous chytrids can take several forms, with fusion taking place between whole thalli, rhizoids, or motile anisogametes (Beakes, 1994). The best-studied system, that of *Allomyces* species, involves the fusion of motile male and female gametes. These are produced in male and female gametangia, often borne in pairs on the same hypha. Thus *Allomyces macrogynus* forms a pair of gametangia, the male being terminal and the female subterminal. Each is multinucleate and subtended by a complete septum separating it from the rest of the hypha. When starved of nutrients, gametogenesis occurs, with synchronous cleavage giving rise to small male gametes, bright orange with γ-carotene, and larger colourless female gametes, both of which swim free through exit pores that result from localized lysis of the gametangial walls (Pommerville, 1981). Their behaviours are different; the female cells swim sluggishly with frequent changes in direction so they tend to stay in the same place, and the males swim faster with fewer changes in direction so that they tend to cover larger distances (Pommerville, 1981). Karyogamy results in the formation of a biflagellate zygote, which eventually settles to germinate as a diploid vegetative thallus. Machlis (1972) showed that the female gametangium and swimming female gametes produce a potent attractant, which he termed sirenin (Carlile, 1996*a*). The response to sirenin is seen as a change in swimming pattern of the male gamete. In the vicinity of a female gamete it swims towards it in a helical path, occasionally re-orientating itself. When very near the female, the male swims in many very short runs and changes of direction until contact with the female occurs (Pommerville, 1981). Sirenin was characterized by growing large-scale cultures of a predominantly female hybrid (Machlis *et al*., 1966). It proved to be a bicyclic sesquiterpene with a cyclopropane ring (Fig. 11.1; Table 11.1). Bioassays can detect a threshold value of 10 pM. A range of analogues has been synthesized but only one has activity equivalent to that of the natural product (Pommerville *et al*., 1988). This demonstrates the specificity of the receptor. Sirenin is rapidly inactivated by male gametes (Machlis, 1973), which presumably aids their sensing its concentration gradient. Female cells are attracted by a complementary hormone produced by male cells, termed parisin (Pommerville & Olsen, 1987).

Zygomycetes

The system of sexual reproduction in the Mucorales was elucidated by Blakeslee (1904), who showed that there are two patterns of mating, homothallic and heterothallic. Most species are heterothallic, with sexual hyphae, zygophores, being formed only when two compatible strains grow together. The involvement of diffusible chemicals was shown by Burgeff (1924), who observed zygophore formation in both (+) and (−) mating types of *Mucor mucedo* when they were grown on the same agar medium separated by a collodion membrane (Fig. 11.3a). The hormone responsible for switching the mycelia from asexual to sexual differentiation in both mating types is trisporic acid (Fig. 11.1, Table 11.1). Its identification as a sex hormone followed from its characterization as a metabolite from mated cultures of *Blakeslea trispora*, which caused a massive stimulation of β-carotene biosynthesis when added to unmated cultures of this fungus (Gooday & Carlile, 1997). The trisporic acids are not species specific, either in production or effect. *M. mucedo*, which is used to bioassay trisporic acids, is especially sensitive, concentrations of 10 pM being detectable (Fig. 11.3b; Gooday, 1978; Van den Ende, 1984). Cultures of *Phycomyces blakesleeanus* need to be in a state of growth arrest before responding (Drinkard, Nelson & Sutter, 1982). The trisporic acids are apocarotenoids, C18 oxidative metabolites of β-carotene. They are biosynthesized by a remarkable collaborative metabolism of the two mating types (Fig. 11.4), whereby each possesses an incomplete enzymatic pathway, accumulating intermediates that can only be further metabolized by the opposite mating type. Only negligible amounts of trisporates are detectable in unmated cultures, but Sutter & Zawodny (1984) described the occurrence of C15 metabolites in cultures of (+) *B. trispora* that they interpreted as degradation products of trisporate precursors, preventing accumulation of trisporate in the unmated fungus. The mechanism whereby trisporate synthesis is enormously enhanced in mated colonies has been reviewed by Bu'Lock, Jones & Winskill (1976), Gooday (1983) and Van den Ende (1984) and is summarized in Fig. 11.4. The account that follows details only the probable major metabolites, but as befits pathways of secondary metabolism, the enzymes involved can transform a range of related substrates, and their action in a metabolic grid will give families of related products. Steps 1 to 5, from β-carotene via retinal to 4-dihydrotrisporin, are considered to be common to both (+) and (−) mycelia. Then the (+) strains oxidize the pro-*S*-methyl group on C-1 through to the carboxyl group and its methyl

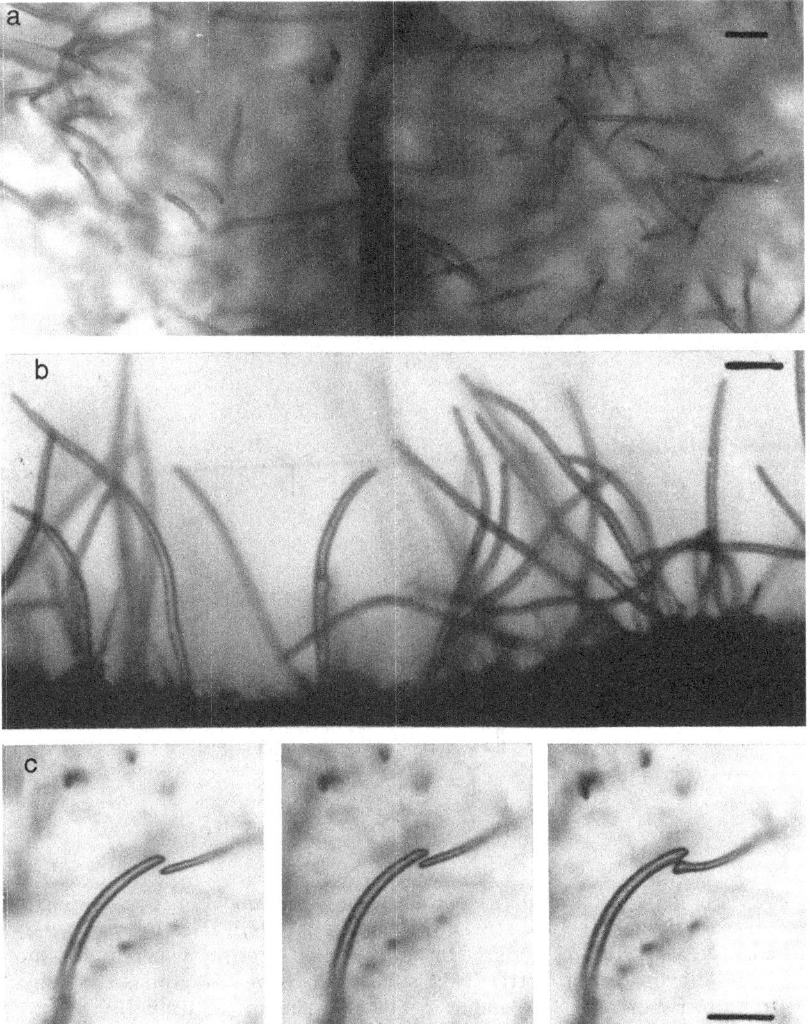

Fig. 11.3. Zygophore formation in *Mucor mucedo* in response to trisporic acid. (a) Reconstruction of experiment of Burgeff (1924), showing mutual induction of zygophore formation in cultures of (+) and (–) mating types (left and right respectively), separated by a cellophane membrane. (b) Zygophores of (–) strain formed on edge of bioassay well in response to trisporic acid. (c) Zygotropism between (+) and (–) zygophores (left and right respectively). Photographs at 4-min intervals. Scale bar represents 100 μm. From Gooday (1994).

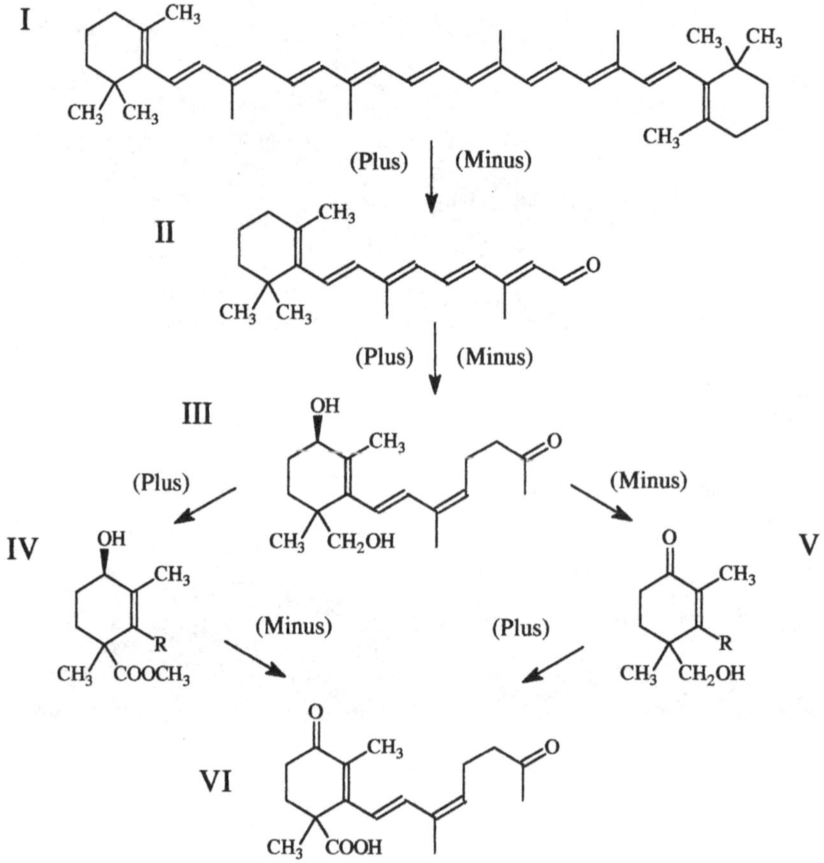

Fig. 11.4. Collaborative biosynthesis of trisporic acid by cross-feeding of intermediates between (+) and (−) mating types of *Blakeslea trispora*. *β*-Carotene (I) is metabolized by both (+) and (−) strains via retinal (II) to 4-dihydrotrisporol (III). This is metabolized by (+) strains to 4-dihydrotrisporic acid (IV) and its methyl ester, and by (−) strains to trisporol (V). These are metabolized to trisporic acid (VI) only after diffusing to the (−) and (+) strains, respectively. This scheme is an over-simplification since the process generates many cometabolites. R signifies the 9-carbon side chain as shown for the B series of metabolites (as in III and VI); the C series has the corresponding alcohol (Gooday & Adams, 1993).

ester, giving methyl-4-dihydrotrisporate. Meanwhile the (−) strains oxidize at C-4 to form the ketone, trisporin, and the pro-*S*-methyl group on C-1 to the alcohol to form trisporol. Thus methyl-4-dihydrotrisporate and trisporol form as two complementary prohormones, each inactive in its producing strain, but converted to active trisporic acid by diffusion into the complementary strain. Progress continues to be made in unravelling this fascinating story. The (–)-specific methyl-4-dihydrotrisporate dehydrogenase, with NADP as co-factor, shown to be localized in (–) zygophores of *M. mucedo* and in one of the two conjugating hyphae of the homothallic species *Zygorhynchus moelleri* by Werkman (1976), has been purified and its gene cloned by Czempinski *et al.* (1996). Sutter *et al.* (1996) have characterized (–) mating-type-specific mutants of *P. blakesleeanus* that are unable to carry out the initial oxidation of trisporin.

The trisporic acid then greatly stimulates carotene synthesis and the formation of its own prohormones, greatly amplifying its own synthesis by a 'cascade' mechanism (Bu'Lock *et al.*, 1976). At the same time there is a general stimulation of all isoprenoid biosynthesis, including sterols, especially ergosterol, but the mechanism of increased flux in these pathways remains unknown. The major developmental effect is to divert the mycelium from asexual sporulation to the formation of the characteristic sexual hyphae, zygophores. The dramatic increase in carotene production is readily observed when two compatible colonies of *M. mucedo* or *B. trispora* meet.

In *M. mucedo* the aerial zygophores grow towards one another in mated pairs, from distances up to 2 mm (equivalent to 100 m on our size scale). When they meet they fuse near their tips, leading to karyogamy and zygopore formation (Fig. 11.3c; Blakeslee, 1904; Gooday, 1973, 1978). This process of zygotropism must be in response to complementary volatile chemicals. These have not been definitively identified but are most likely to be the mating-type-specific prohormones, methyl-4-dihydrotrisporate and trisporol. Mesland, Huisman & Van den Ende (1974) demonstrated that zygophore formation could be induced when (+) and (–) mycelia of *M. mucedo* were separated by an air gap, showing that the trisporate precursors are volatile enough to have biological activity. The observation that zygophores cease to be attractive immediately they have fused with mating partners (Gooday, 1975) can be explained as the prohormones would immediately be metabolized through to trisporate.

Ascomycetes

There are many observations suggesting the activity of diffusible hormones during sexual interactions between and within colonies of filamentous ascomycetes, but as yet none of these compounds has been characterized. In contrast, pheromonal regulation of mating of hemi-ascomycetous yeasts, particularly *Saccharomyces cerevisiae* and *Schizosaccharomyces pombe*, has been defined in great detail (Gooday & Adams, 1993; Duntze *et al.*, 1994). For each species, there are complementary pairs of pheromones and pheromone receptors. In *S. cerevisiae*, α-cells secrete α-factor and have a membrane-bound **a**-factor receptor; **a**-cells secrete **a**-factor and have a membrane-bound α-factor receptor. In *S. pombe*, plus cells secrete P-factor and have a membrane-bound M-factor receptor, minus cells secrete M-factor and have a membrane-bound P-factor receptor. α-Factor and P-factor are unmodified peptides; **a**-factor and M-factor are peptides with COOH-terminal cysteines modified by *S*-farnesylation and methyl esterification. The receptors for these latter two lipopeptides are encoded by the genes *STE3* and $map3^{+}$, respectively. In each case, the yeast cells respond to the appropriate pheromone by producing a conjugation tube, which grows towards a cell of opposite mating type.

It is fairly certain that filamentous Ascomycetes have analogous pheromone systems, but direct evidence is still lacking. In *Neurospora crassa*, the mating-type loci, *A* and *a*, encode complementary regulatory genes (Staben, 1994). Presumably these control expression of genes encoding pheromones and pheromone receptors. There is good evidence for such hormones in *N. crassa*, in particular controlling chemotropic growth of the female filament (trichogyne) of each mating type (*A* and *a*) to the male fertilizing conidium (Bistis, 1983). Less specifically, chemical extracts have been reported to stimulate either fertility in poorly fertile crosses or development of protoperithecia in unmated strains: a factor described by Islam (1981) was lipid in nature, while Vigfusson & Cano (1974) described a factor that behaved as a protein. These factors may best be considered sexual growth substances. In *Arthroderma incurvatum*, A.S. Donald & G.W. Gooday (unpublished results) have obtained evidence for two reciprocal hormones, namely (+) factor and (–) factor, produced by (+) and (–) hyphae, respectively, which induce the formation of protocleistothecia in unmated cultures of (–) and (+) hyphae. These are small, diffusible, hydrophobic molecules, as yet uncharacterized. Other hormone systems in filamentous ascomycetes ripe for reinvestigation are those in the genera *Glomerella*, *Bombardia*, *Ascobolus* and *Nectria* (see Gooday, 1994 for references).

Champe & El-Zayat (1989) describe diffusible psi (precocious sexual induction) factors from the homothallic ascomycete *Aspergillus nidulans* which induce premature development of sexual structures. Psi activity was assayed by observing the response of mycelium growing from a uniform inoculum of conidia over an agar plate to filter discs impregnated with test samples. The activity resulted in formation of cleistothecia instead of conidiophores, and in the release of yellow pigment into the medium. A family of psi factors have been characterized as derivatives of linoleic acid, with minor components as the corresponding oleic acid derivatives (Fig. 11.1; Champe, Nagle & Yager, 1994). They appear to be species-specific. The lactone PsiA (Fig. 11.1) is interconvertible with its free acid, psiC. These two compounds appear to act in opposing fashion, with psiC favouring sexual sporulation and psiA favouring asexual sporulation (Champe *et al.*, 1994).

Chemical control of sexual morphogenesis in the heterothallic plant pathogen *Pyrenopeziza brassicae* is controlled by a class of sexual morphogens (SF, sexual factors) (Siddiq *et al.*, 1989). When lipid extracts from mated cultures were added to unmated cultures, they suppressed asexual sporulation and induced the formation of immature, sterile apothecia. When added to mated cultures, they greatly increased the speed of production and number of fertile apothecia. These activities were not species-specific, as effects on growth and sporulation were observed when the extracts were added to a wide range of fungi. Extracts from unmated cultures of *P. brassicae* had none of these effects. It is likely that this wide range of effects is due to several components acting as morphogens and growth substances. Enhancement of fertile perithecial development of *Nectria haematococca* by the addition of lipid extracts from mated cultures of this fungus has been described by Dyer, Ingram & Johnstone (1993). Linoleic acid had the same effect, and was detected as a metabolite of the mated cultures. Dyer *et al.* (1993) suggest that this and related fatty acids are produced during mating as sex factors enhancing the development of perithecia. There is evidence for linoleic acid being involved in sexual reproduction in several other fungi as an endogenous sexual growth substance (Dyer *et al.*, 1993). A well-characterized fungal metabolite whose role as a sexual regulator is unclear is zearalenone, an oestrogenic mycotoxin. This is a β-resorcylic acid lactone produced by *Gibberella zeae* and *Fusarium* species (Fig. 11.1: Wolf & Mirocha, 1973). At low concentrations it enhances perithecial production in *G. zeae*, but it inhibits at high concentrations. Its role as a specific sexual regulator, however, is put into question by the observation that there is no correlation between zearalenone production by different strains and their formation of perithecia (Windels *et al.*, 1989). Nelson (1971) reported that low concentrations of zearalenone stimulated sexual reproduction in a wide range of fungi, which suggests that it may act as a sexual growth substance.

Basidiomycetes

The best-characterized sexual signalling systems among the Basidiomycetes are those involving the lipopeptide sex pheromones produced by yeast cells of heterobasidiomycetous yeasts, smuts and jelly fungi (reviewed by Gooday & Adams, 1993; Caldwell, Naider & Becker, 1995; Vaillancourt & Raper, 1996). These pheromones are encoded by the mating type loci, along with genes for the complementary pheromone receptors. Mating in species of *Tremella*, *Rhodosporidium*, *Ustilago* and *Cryptococcus* occurs between compatible pairs of haploid yeast cells, which respond to cells of opposite mating type by developing conjugation hyphae which grow towards them.

Farnesylated peptide pheromones have been characterized from culture filtrates and/or their identity has been inferred from sequences of mating-type loci in these fungi. The characteristic amino acid sequence of the propheromones is a COOH-terminal 'CaaX box'; i.e. -cysteine-aliphatic amino acid-aliphatic amino acid-any other amino acid, and an NH_2-terminal methionine. By analogy with the *S. cerevisiae* mating pheromone **a**-factor, which has been studied in detail (Caldwell *et al.*, 1995; Chen *et al.*, 1997), each of these pheromone precursors is probably processed by initial farnesylation of the cysteine sulphur atom accompanied by proteolytic cleavage of the COOH-terminal 'aaX' tripeptide, and most likely by methylation of the new terminal carboxyl group. The resultant membrane-associated lipopeptide would then undergo one or more proteolytic cleavages at its NH_2-terminus, and the mature pheromone would be exported directly across the plasma membrane by an ATP-driven peptide pump of the ABC family (i.e. it contains an ATP-binding cassette). Variations in this pattern are shown by the pheromones rhodotorucine *A*, which is exported in an inactive form by *A* cells of *Rhodotorula toruloides*, and activated by a protease secreted by *a* cells; and tremerogen *A*-10, which appears to be secreted by the classical secretion pathway. The formation of *Ustilago maydis* mating hyphae is triggered by a complementary pair of such pheromones, *a1* (13 aa) and *a2* (9 aa) (Spellig *et al.*, 1994). The resultant mating hyphae then grow chemotactically towards the sources of the pheromone (Snetselaar, Bölker & Kahmann, 1996).

Mating among the Homobasidiomycetes is characterized by the extraordinary lengths to which some of them go in order to maximize their chances of outbreeding. The tetrapolar mating systems of *Schizophyllum commune* and *Coprinus cinereus* give rise to more than 20 000 and 12 000 mating types, respectively, giving outbreeding potentials better than 98% (Raper, 1966; Kües & Casselton, 1992; Kothe, 1996, 1997; Vaillancourt & Raper, 1996). In *S. commune*, mating type is collectively determined by two tightly linked multi-allelic loci, α and β, for *A*, and two, α and β, for *B* (Kothe, 1997; Wessels *et al.*, 1998). In the world-wide population, there are nine and 32 specificities, respectively, for $A\alpha$ and $A\beta$, and nine each for $B\alpha$ and $B\beta$. Compatible mating requires a difference at either of the two *A* loci and either of the two *B* loci. Thus α and β of each complex are functionally redundant. The complex *A* loci code for interacting homeodomain transcription factors, and the complex *B* loci code for complementary pheromones and their receptors. Thus in terms of sexual interactions between colonies, it is the *B* mating type which concerns us. Two complex *B* mating type loci, $B\alpha1$ and $B\beta1$, have been cloned (Wendland *et al.*, 1995; Specht, 1995; Vaillancourt & Raper, 1996; Vaillancourt *et al.*, 1997). Each contains a single pheromone receptor gene and three putative pheromone genes. The pheromone receptor proteins encoded by these genes, *Bar1* and *Bbr1*, and a further one, *Bar2*, have homology with the pheromone receptors Ste3p of *S. cerevisiae* (recognizing **a**-factor), Pra1 and Pra2 of *U. maydis* and Map3 of *S. pombe* (Vaillancourt & Raper, 1996; Vaillancourt *et al.*, 1997). These pheromone receptor proteins are part of the large family of receptor proteins with seven transmembrane domains, involved in signal transduction through G-proteins. The six putative pheromone precursors encoded by $B\alpha1$ and $B\beta1$ are recognized by having the 'CaaX' box (Table 11.2). There are no obvious differences between the α and β series, and each pheromone gene is unique in sequence.

DNA sequencing of the *B6* locus of *C. cinereus* shows six genes for putative pheromone precursors (O'Shea *et al.*, 1998). These genes code for peptides with 53–72 amino acids with the 'CaaX' box for farnesylation at their COOH-termini, but no conservation at their NH_2-termini. Four have a C-terminal sequence of Cys-Val-Ile-Ala, the other two of Cys-Val-Ile-Ser. There are two highly conserved amino acids, Glu-Arg or Asp-Arg, 14–16 residues from the C-terminal which are putative sites for

Table 11.2. *Pheromone precursors in* Schizophyllum commune

Pheromone	Amino acid pro-sequence
bap1(2)	Met-(42 aa)-Glys-Cys-Val-Arg-Gly
bbp1(2)	Met-(59 aa)-Tyr-Cys-Val-Val-Arg
bap1(1)	Met-(43 aa)-Arg-Cys-Val-Cys-His
bap1(3)	Met-(44 aa)-Trp-Cys-Val-Val-Arg
bbp1(1)	Met-(70 aa)-Trp-Cys-Val-Val-Arg
bbp1(3)	Met-(66 aa)-Trp-Cys-Val-Val-Arg
Mature forms?	(*n* aa)-X-(farnesyl-S)-Cys
	(*n* aa)-X-(farnesyl-S)-Cys-(methyl ester)

Adapted from Vaillancourt & Raper (1996).

proteolytic processing of the propheromones. With proteolytic cleavage of the C-terminal 'aaX', this predicts six mature peptide pheromones of 11–13 amino acids. Three other genes in the *B6* locus encode putative pheromone receptors with seven transmembrane domains, again with the homology to Ste3p of *S. cerevisiae*, and with Pra1 and Pra2 of *U. maydis*.

The discovery that the *B* complex encodes pheromones and pheromone receptors was a surprise, so what can their role be? Initial fusion between two hyphae of *S. commune* or of *C. cinereus* is independent of mating type. The *B* mating-type genes control two distinct processes in mating. Firstly, they regulate the remarkable phenomenon of nuclear migration. Following fusion of two monokaryons with different *B* genes, there is extensive bidirectional invasion of the compatible nuclei throughout the mycelia. This process involves enzymatic dissolution of the complex dolipore septa, and nuclear migration occurs at great speed, estimated at up to 3 mm/hour. This leads to very rapid establishment of dikaryotic mycelium. Northern blotting shows that pheromone genes are upregulated during this process (Vaillancourt *et al.*, 1997). This suggests that the pheromones have a direct role in nuclear migration. The pheromones may diffuse extracellularly (perhaps along the cell walls) ahead of the nuclei, preparing the hyphae for nuclear migration by triggering the activation of lytic enzymes involved in septal dissolution, and by regulating organization of cytoskeletal motor protein functions (Wendland *et al.*, 1995; Kothe, 1996, 1997; Vaillancourt & Raper, 1996; Vaillancourt *et al.*, 1997). The *B* mating type genes also control the hook-cell fusion during clamp connection formation (Raper, 1966). It has long been suspected that this process involves specific chemotropic attractants. Harder (1927) suggested that chemotropic hormones were involved in controlling the turning and re-fusion of the hook cell with the sub-apical wall; and Buller (1933) implies this '. . . the hook of a clamp-connexion does not fuse directly with the main hypha as hitherto has been supposed, but with a blunt process or peg sent out by the main hypha in response to a stimulus given by the apex of the hook'. This peg can clearly be seen in the photographs of Niederpruem, Jerslid & Lane (1971). The process of hook-cell fusion has two elements in common with initial dikaryon formation; nuclear movement and localized cell-wall dissolution, so it seems reasonable that both processes could share the same pheromone/pheromone-receptor systems. Wessels *et al.* (Chapter 13) suggest that the pheromones could be secreted locally into the wall adjacent to each nucleus, where they would regulate the spatial distribution of the nuclei, by diffusing laterally to interact with complementary receptors adjacent to the complementary nuclei.

It is unlikely that pheromones of *S. commune* and *C. cinereus* will become available for experimentation in the foreseeable future, thus investigations of their precise activities will have to be indirect. Experiments are underway localizing the pheromone receptors in the plasma membrane (Kothe, 1996) and it may be possible to localize the pheromone-export protein.

The relationship between chemotropic vegetative fusions between hyphae of the same species, the 'homing' chemotropic responses of monokaryotic hyphae to oidia, which are both independent of mating type, and the chemotropism of the hook cell has been discussed by Raper (1952) and Gooday (1975) but remains unclear.

Some of the sexual morphogens and growth factors that have effects on Ascomycetes, SF, zearalenone, mycosporines and linoleic acid, have been reported to stimulate sexual development in a range of Agaricales (Dyer *et al.*, 1992).

Conclusions

There are undoubtedly untold numbers of specific chemicals regulating all activities of fungal growth, differentiation and reproduction. Only a very few have been characterized. By definition they are difficult to characterize: they are produced in very small concentrations; they are produced only in certain physiological states; and they are unstable. A sensitive and discriminatory bioassay is thus essential for their study. In the few cases where they have been identified, use has been made of overproducing strains for their production and/or super-sensitive strains for their bioassay. In the case of biologically active peptides, the examples from *S. commune* and *C. cinereus* provide graphic examples of how the power of molecular biological techniques can lead to the discovery of novel pheromones.

These fungal systems clearly present evolutionary parallels to our own human hormone systems, and indeed among the human pathogenic fungi there are some specific interactions, which may represent co-evolution during development of pathogenesis, or mimicry between homologous systems (Gooday & Adams, 1993).

The metabolites described here as regulators of sexual development fall into two classes: very specific unique molecules, exemplified by the isoprenoids of the Oomycetes, chytrids and Zygomycetes (Table 11.1) and the lipopeptides of the Ascomycetes and Basidiomycetes; and the much less well characterized and much less specific sex factors and sexual growth factors. This latter group of compounds may turn out to be the fungal equivalents of the bacterial cell-density signalling molecules, the homoserine lactones. Many fungal activities, certainly including sexual reproduction, only take place when a colony has achieved an ill-defined maturity. For example, young colonies of *M. mucedo* do not produce trisporic acid, or respond to it, as do mature colonies.

There are large numbers of fungal species: Oomycetes ~700; Chytrids ~600; Zygomycetes ~900; Ascomycetes, ~32 000; Basidiomycetes, ~14 000, and so, extrapolating from the examples discussed here, we can predict that there are equally large numbers of hormones/pheromones waiting to be characterized, together with their receptors. Knowledge of some of these systems may have biotechnological value, for example by enabling us to divert pathogenic fungi from growth to abortive sexual development. We can, however, certainly predict that there will be many more surprises, which will provide stories that are at least as fascinating as the handful that have been revealed to date.

References

Barksdale, A. W. (1967). The sexual hormones of the fungus *Achlya*. *Annals of the New York Academy of Sciences*, **144**, 313–19.

Beakes, G. W. (1994). Sporulation of lower fungi. In *The Growing Fungus*, ed. N. A. R. Gow & G. A. Gadd, pp. 339–66. London: Chapman & Hall.

Bistis, G. N. (1983). Evidence for diffusible, mating-type-specific trichogyne attractants in *Neurospora crassa*. *Experimental Mycology*, **7**, 292–9.

Blakeslee, A. F. (1904). Sexual reproduction in the Mucorineae. *Proceedings of the National Academy of Arts and Science*, **40**, 205–319.

Brasier, C. M. (1975). Stimulation of sex organ formation in *Phytophthora* by antagonistic species of *Trichoderma*. 1. The effect in vitro. *New Phytologist*, **74**, 183–94.

Buller, A. H. R. (1933). *Researches on Fungi Vol. 5*. London: Longmans Green.

Bu'Lock, J. D., Jones, B. E. & Winskill, N. (1976). The apocarotenoid system of sex hormones and prohormones in Mucorales. *Pure and Applied Chemistry*, **47**, 191–202.

Burgeff, H. (1924). Untersuchungen über Sexualität und Parasitismus bei Mucorineen I. *Botanische Abhandlungen*, **4**, 1–135.

Caldwell, G. A., Naider, F. & Becker, J. M. (1995). Fungal lipopeptide mating pheromones: a model system for the study of protein prenylation. *Microbiological Reviews*, **59**, 406–22.

Carlile, M. J. (1996*a*). The discovery of fungal sex hormones: I. Sirenin. *Mycologist*, **10**, 3–6.

Carlile, M. J. (1996*b*). The discovery of fungal sex hormones: II. Antheridiol. *Mycologist*, **10**, 113–8.

Champe, S. P. & El-Zayat, A. A. E. (1989). Isolation of a sexual sporulation hormone from *Aspergillus nidulans*. *Journal of Bacteriology*, **171**, 3982–8.

Champe, S. P., Nagle, D. L. & Yager, L. N. (1994). Sexual sporulation. In *Aspergillus: 50 Years On*, ed. S. D. Martinelli & J. R. Kinghorn, *Progress in Industrial Microbiology, Vol. 29*, pp. 201–27. Amsterdam: Elsevier.

Chen, P., Sapperstein, S. K., Choi, J. D. & Michaelis, S. (1997). Biogenesis of the *Saccharomyces cerevisiae* mating pheromone a-factor. *Journal of Cell Biology*, **136**, 251–69.

Czempinski, K., Kruft, V., Wöstermeyer, J. & Burmester, A. (1996). 4-Dihydromethyltrisporate dehydrogenase from *Mucor mucedo*, an enzyme of the sexual hormone pathway: purification, and cloning of the corresponding gene. *Journal of General Microbiology*, **142**, 2647–54.

Drinkard, L. C., Nelson, G. E. & Sutter, R. P. (1982). Growth arrest: a prerequisite for sexual development in *Phycomyces blakesleeanus*. *Experimental Mycology*, **6**, 52–9.

Duntze, W., Betz, R. & Nientiedt, M. (1994). Pheromone in yeasts. In *The Mycota, Vol. I. Growth, Differentiation and Sexuality*, ed. J. G. H. Wessels & F. Meinhardt, pp. 381–99. Berlin: Springer-Verlag.

Dyer, P. S., Ingram, D. S. & Johnstone, K. (1992). The control of sexual morphogenesis in the Ascomycotina. *Biological Reviews*, **67**, 421–58.

Dyer, P. S., Ingram, D. S. & Johnstone, K. (1993). Evidence for the involvement of linoleic acid and other endogenous lipid factors in perithecial development of *Nectria haematococca* mating population VI. *Mycological Research*, **97**, 485–96.

Elliott, C. G. (1983). Physiology of sexual reproduction in *Phytophthora*. In *Phytophthora: its Biology, Taxonomy, Ecology and Pathology*, ed. D. C. Erwin, S. Bartnicki-Garcia & P. H. Tsao, pp. 71–80. St Paul: American Phytopathological Society.

Elliott, C. G. (1994). *Reproduction in Fungi. Genetical and Physiological Aspects*. London: Chapman & Hall.

Gall, A. M. & Elliott, C. G. (1985). Control of sexual reproduction in *Pythium sylvaticum*. *Transactions of the British Mycological Society*, **84**, 629–36.

Gooday, G. W. (1973). Differentiation in the Mucorales. *Symposium of the Society for General Microbiology*, **23**, 269–94.

Gooday, G. W. (1974). Fungal sex hormones. *Annual Review of Biochemistry*, **43**, 35–49.

Gooday, G. W. (1975). Chemotaxis and chemotropism in fungi and algae. In *Primitive Sensory and Communication Systems*, ed. M. J. Carlile, pp. 155–204. London: Academic Press.

Gooday, G. W. (1978). Functions of trisporic acid. *Philosophical Transactions of the Royal Society of London B*, **284**, 509–20.

Gooday, G. W. (1983). Hormones and sexuality in fungi. In *Secondary Metabolism and Differentiation in Fungi*, ed. J. W. Bennett & A. Ciegler, pp. 239–66, New York: Dekker.

Gooday, G. W. (1994). Hormones in mycelial fungi. In *The Mycota, Vol. I. Growth, Differentiation and Sexuality*, ed. J. G. H. Wessels & F. Meinhardt, pp. 401–11. Berlin: Springer-Verlag.

Gooday, G. W. & Adams, D. J. (1993). Sex hormones and fungi. *Advances in Microbial Physiology*, **34**, 69–145.

Gooday, G. W. & Carlile, M. J. (1997). The discovery of fungal sex hormones: III. Trisporic acid and its precursors. *Mycologist*, **11**, 126–30.

Gow, N. A. R. & Gooday, G. W. (1987). Effects of antheridiol on growth, branching and electrical currents in hyphae of *Achlya ambisexualis. Journal of General Microbiology*, **133**, 3531–5.

Guo, L. Y. & Ko, W. H. (1991). Hormonal regulation of sexual reproduction and mating type change in heterothallic *Pythium splendens. Mycological Research*, **95**, 452–6.

Harder, R. (1927). Zur Frage nach der Rolle von Kern und Protoplasma im Zellgeschehen und bei der Übertragung von Eigenschaften. *Zeitschrift für Botanik*, **19**, 337–407.

Islam, M. S. (1981). Sex pheromones in *Neurospora crassa*. In *Sexual Interactions in Eukaryotic Microbes*, ed. D. H. O'Day & P. A. Horgen, pp. 131–54. New York: Academic Press.

Kerwin, J. L. & Duddles, N. D. (1989). Reassessment of the role of phospholipids in sexual reproduction by sterol-auxotrophic fungi. *Journal of Bacteriology*, **171**, 3831–9.

Ko, W. H. (1988). Hormonal heterothallism and homothallism in *Phytophthora. Annual Review of Phytopathology*, **26**, 57–73.

Kothe, E. (1996). Tetrapolar fungal mating types: sexes by the thousands. *FEMS Microbiology Reviews*, **18**, 65–87.

Kothe, E. (1997). Solving a puzzle piece by piece: sexual development in the basidiomycetous fungus *Schizophyllum commune. Botanica Acta*, **110**, 208–13.

Kües, U. & Casselton, L. A. (1992). Fungal mating type genes – regulators of sexual development. *Mycological Research*, **96**, 993–1006.

Machlis, L. (1972). The coming of age of sex hormones in plants. *Mycologia*, **64**, 235–47.

Machlis, L. (1973). Factors affecting the stability and accuracy of the bioassay for the sperm attractant sirenin. *Plant Physiology*, **52**, 524–6.

Machlis, L., Nutting, W. H., William, M. W. & Rapoport, H. (1966). Production, isolation and characterisation of sirenin. *Biochemistry*, **5**, 2147–52.

McMorris, T. C. (1978). Antheridiol and the oogoniols, steroid hormones which control sexual reproduction in *Achlya*. *Philosophical Transactions of the Royal Society of London B*, **248**, 459–70.

McMorris, T. C., Toft, D. O., Moon, S. & Wang, W. (1993). Biological response of the female strain *Achlya ambisexualis* 734 to dehydro-oogoniol and analogues. *Phytochemistry*, **32**, 833–7.

Mesland, D. A. M., Huisman, J. G. & Van den Ende, H. (1974). Volatile sexual hormones in *Mucor mucedo*. *Journal of General Microbiology*, **80**, 111–7.

Mullins, J. T. (1994). Hormonal control of sexual dimorphism. In *The Mycota, Vol. I. Growth, Differentiation and Sexuality*, ed. J. G. H. Wessles & F. Meinhardt, pp. 413–21. Berlin: Springer-Verlag.

Musgrave, A. & Nieuwenhuis, D. (1975). Metabolism of radioactive antheridiol by *Achlya* species. *Archives of Microbiology*, **105**, 313–17.

Nelson, R. R. (1971). Hormonal involvement in sexual reproduction in the fungi with special reference to F-2, a fungal estrogen. In *Morphological and Biochemical Events in the Plant–Parasite Interaction*, ed. S. Akai & S. Ouchi, pp. 181–205. Tokyo: The Phytopathological Society of Japan.

Niederpruem, D. J., Jerslid, R. A. & Lane, P. L. (1971). Direct microscopic studies of clamp connection formation in growing hyphae of *Schizophyllum commune*. *Archiv für Mikrobiologie*, **78**, 268–80.

O'Shea, S. F., Chaure, P. T., Halsall, J. R., Olesnicky, N. S., Leibbrandt, A., Connerton, I. F. & Casselton, L. A. (1998). A large pheromone and receptor gene complex determines multiple *B* mating type specificities in *Coprinus cinereus*. *Genetics*, **148**, 1081–90.

Pommerville, J. (1981). The role of sexual pheromones in *Allomyces*. In *Sexual Interactions in Eukaryotic Microbes*, ed. D. H. O'Day & P. A. Horgen, pp. 53–92. New York: Academic Press.

Pommerville, J. & Olsen, L. W. (1987). Evidence for a male-produced pheromone in *Allomyces macrogynus*. *Experimental Mycology*, **11**, 245–8.

Pommerville, J. C., Strickland, J. B., Romo, D. & Harding, K. E. (1988). Effects of analogs of the fungal sex pheromone sirenin on male gamete motility in *Allomyces macrogynus*. *Plant Physiology*, **88**, 139–42.

Raper, J. R. (1951). Sexual hormones in *Achlya*. *American Scientist*, **39**, 110–20.

Raper, J. R. (1952). Chemical regulation of sexual processes in the thallophytes. *Botanical Reviews*, **18**, 447–545.

Raper, J. R. (1966). *Genetics of Sexuality in Higher Fungi*. New York: Ronald Press.

Sakata, K. & Rickards, R. W. (1980). Synthesis of homothallin II. In *Proceedings of 23rd Symposium of Natural Products*, pp. 165–72. Nagoya, Nagoya City University.

Siddiq, A. A., Ingram, D. S., Johnstone, K., Friend, J. & Ashby, A. M. (1989). The control of asexual and sexual development by morphogens in fungal pathogens. *Aspects of Applied Biology*, **23**, 417–26.

Silver, J. C., Brunt, S. A., Kyriakopolou, G., Borkar, M. & Nazarian-Armavil, V. (1993). Heat shock proteins in hyphal branching and secretion in steroid hormone induced fungal development. *Journal of Cellular Biochemistry Supplement*, **17**, 136.

Snetselaar, K. M., Bölker, M. & Kahmann, R. (1996). *Ustilago maydis* mating hyphae orient their growth towards pheromone sources. *Fungal Genetics and Biology*, **20**, 299–312.

Specht, C. (1995). Isolation of the *Bα* and *Bβ* mating-type loci of *Schizophyllum commune*. *Current Genetics*, **28**, 374–9.

Spellig, T., Bölker, M. Lottspeich, F., Frank, R. W. & Kahmann, R. (1994). Pheromones trigger filamentous growth in *Ustilago maydis*. *EMBO Journal*, **13**, 1620–7.

Staben, C. (1994). Sexual reproduction in higher fungi. In *The Growing Fungus*, ed. N. A. R. Gow & G. A. Gadd, pp. 383–402. London: Chapman & Hall.

Sutter, R. P., Grandin, A. B., Dye, B. D. & Moore, W. T. (1996). (–) Mating type-specific mutants of *Phycomyces* defective in sex pheromone biosynthesis. *Fungal Genetics and Biology*, **20**, 268–79.

Sutter, R. P. & Zawodny, P. D. (1984). Apotrisporin: a major metabolite of *Blakeslea trispora*. *Experimental Mycology*, **8**, 89–92.

Thomas, D. S. & Mullins, J. T. (1969). Cellulase induction and wall extension in the water mold *Achlya ambisexualis*. *Physiologia Plantarum*, **39**, 347–53.

Timberlake, W. E. & Orr, W. C. (1984). Steroid hormone regulation of sexual reproduction in *Achlya*. In *Biological Regulation and Development, Vol. 3b*, ed. R. F. Goldberger & K. Yamamoto, pp. 255–83. New York: Plenum Press.

Vaillancourt, L. J. & Raper, C. A. (1996). Pheromones and pheromone receptors as mating-type determinants in Basidiomycetes. In *Genetic Engineering, Principles and Methods, Vol. 18*, ed. J. K. Setlow, pp. 219–47. New York: Plenum Press.

Vaillancourt, L. J., Raudaskoski, M., Specht, C. A. & Raper, C. A. (1997). Multiple genes encoding pheromones and a pheromone receptor define the Bβ1 mating-type specificity in *Schizophyllum commune*. *Genetics*, **146**, 541–51.

Van den Ende, H. (1984). Sexual interactions in the lower filamentous fungi. In *Encyclopedia of Plant Physiology, vol. 17*, ed. H. F. Linskens & J. Heslop-Harrison, pp. 333–49. Berlin: Springer-Verlag.

Vigfusson, N. V. & Cano, R. J. (1974). Artificial induction of the sexual cycle in *Neurospora crassa*. *Nature*, **249**, 383–5.

Wendland, J., Vaillancourt, L. J., Hegner, B., Lengeler, K., Laddison, K. J., Specht, C. A., Raper, C. A. & Kothe, E. (1995). The mating-type locus Bβ1 of *Schizophyllum commune* contains a pheromone-receptor and putative pheromone genes. *EMBO Journal*, **14**, 5271–8.

Werkman, T. A. (1976). Localization and partial characterization of a sex-specific enzyme in homothallic and heterothallic Mucorales. *Archives of Microbiology*, **109**, 209–13.

Windels, C. E., Mirocha, C. J., Abbas, H. K. & Weiping, X. (1989). Perithecium production in *Fusarium graminearum* populations and lack of correlation with zearalenone production. *Mycologia*, **81**, 272–7.

Wolf, J. C. & Mirocha, C. J. (1973). Regulation of sexual reproduction in *Gibberella zeae* (*Fusarium roseum* 'Graminearum') by F-2 (zearalenone). *Canadian Journal of Microbiology*, **19**, 725–34.

12

Genetic stability in fungal mycelia

M. L. SMITH

Introduction

The genetic nature of the 'fungal individual' has intrigued biologists for over 50 years. To some extent this is because fungi have an indeterminate growth form and the physical shape of a 'fungus' can not usually be described *a priori* in the same way that a cat or a tree is characterized and identified. Most filamentous fungi are cryptic organisms whose visible portion consists of ephemeral fruit-bodies that occur in a temporally and spatially discontinuous manner and the boundaries of individuals are, therefore, not immediately evident to the casual observer (see Chapter 1). An early view of the fungal individual was recognized in the 'unit mycelium' concept (Buller, 1958; Raper, 1966). Within this view, a basidiomycete could be a mosaic of several different nuclear genotypes operating within a physiologically integrated individual. Studies that examine the distribution of genetic markers within fungal populations support an alternative view of fungal individualism (Todd & Rayner, 1980; Rayner, 1991), in which mycelia occupy discrete territories in space, have cellular non-self recognition systems and are genetically distinct. Reports that basidiomycete genotypes can occur throughout extensive geographic areas (Adams, 1974; Shaw & Roth, 1976; Anderson *et al.*, 1979; Dickman & Cook, 1989; Smith, Bruhn & Anderson, 1992) are surprising since some fungi must now be thought of as large, and consequently ancient, individuals. Intrinsic to this notion, such individuals must be genetically stable, at least to the extent that they are recognizable as genetic units, possibly to the extent that they are potentially immortal. In this chapter, genetic stability and factors that may influence genetic stability in fungi will be discussed.

Impacts of genetic stability in fungi

Why would we be at all interested in genetic stability of fungi? First, instability in some fungal strains can have tremendous negative economic consequences. On the one hand, genetic *stability* is necessary for fungi used as food and in pharmaceutical or industrial processes, where product quality and quantity are important. On the other hand, by generating variation within fungal species, genetic *instability* may bring forth new forms of plant and animal pathogens and, thereby, social and economic woes. Second, genetic stability of recombinant fungi must be evaluated prior to their release, for example as biological pest controls, for reasons of efficacy and safety. Third, towards a basic understanding of fungi, mutation events in fungal mycelia may be useful as markers to track growth patterns and dynamics of fungal mycelia in nature, just as colony sectors can be used in laboratory studies. Finally, an understanding of genetic stability in fungi provides another perspective into biological questions such as somatic stability of genomes, aging, genetic deterioration, mutation rates and the evolutionary processes of asexual organisms.

Genetic stability estimates

There is a dramatic range in the extent to which fungi are genetically stable. In laboratory-based studies, instability of some genotypes is evident by the appearance of colony sectors (Fig. 12.1) and these can arise frequently in some genotypes. For example, strains of *Neurospora crassa* and *Aspergillus nidulans* that carry chromosomal duplications can be extremely unstable. Rearrangements in these duplication-bearing strains result in the loss, or conversion of one of the duplicated segments, and can occur rather predictably for a given strain within one or a few weeks. Duplication instability is dealt with in more detail in this chapter under Maintenance of genetic identity within mycelia. Colony sectors are useful to fungal geneticists since they are easily identified, can be isolated into culture and the genetic basis of the underlying mutational process can be examined (for example, see Käfer, 1975). Based on laboratory studies, we might expect to detect sectors that correspond to specific mutational events in naturally occurring fungi. In doing so, we could extend the sector hunt to periods of decades or centuries and perhaps achieve a better appreciation for aspects of rate, type and the role of mutation events within mitotic cell lineages. It might also be expected that genetic analyses of naturally occurring fungal mycelia would provide a means of determining the upper limit to genetic stability in such lineages.

Understanding genetic processes in naturally occurring fungi requires both a clear conceptual and physical definition of the fungal individual. The conceptual definition used here is that of Smith, Bruhn & Anderson

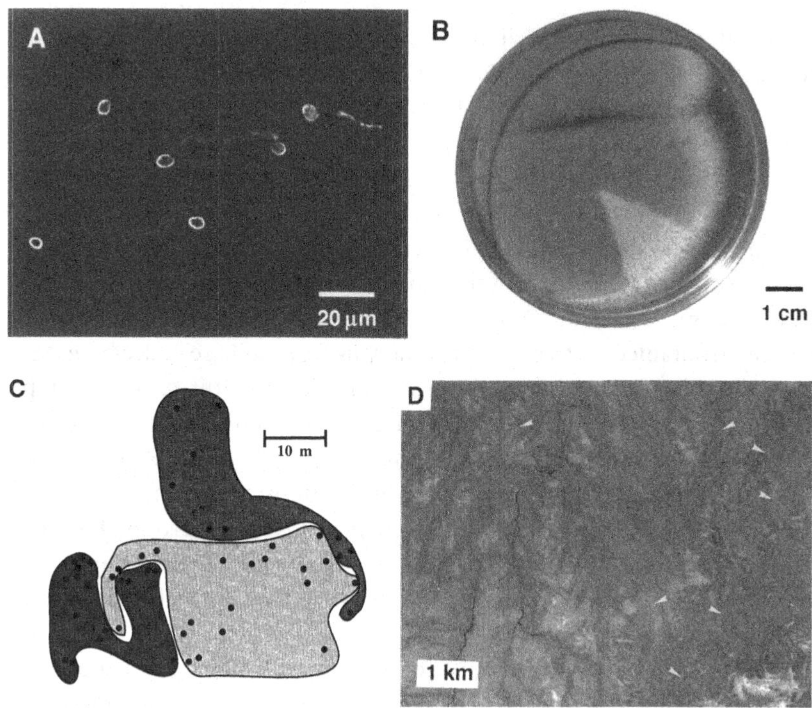

Fig. 12.1. The different aspects of the 'fungal mycelium'. (A) Basidiospore germlings of *Armillaria*. (B) Colour sector in *Aspergillus nidulans* diploid strain. Kindly provided by Dr. E. Käfer. (C) Distribution of *Armillaria ostoyae* genotypes in nature indicates irregular shaped genets result from antagonistic mycelial interactions. Dots represent mycelium sampling points. Lines are drawn around genets. Adapted from Smith *et al.*, 1994. (D) Forest rings (arrows) of thinned trees are evident in aerial photographs near Mammamattawa, Ontario, Canada. Forest rings may represent unimpeded expansion of infection foci by a tree pathogen such as *A. ostoyae.* A mutant search in large, concentric fungal colonies may provide a unique opportunity to study mutations in asexual cell lineages.

(1994), and corresponds to the use of 'genet' in mycological literature (Rayner, 1991), in which an individual constitutes a mitotic cell lineage that occupies a contiguous space. If there is the potential for physical connection and physiological cooperation across a genet, then a genet behaves as a mycelium. Means to physically define naturally occurring fungal genets will be illustrated here by studies with the basidiomycete *Armillaria*, a facultative tree-root pathogen. The genus *Armillaria* is an important component of forest ecosystems throughout the world with representative species on all continents except Antarctica. Members of the genus have several unusual and interesting attributes including bioluminescence, diploidy during the vegetative phase of the life cycle and the ability to form rather complex linear aggregates of hyphae, called rhizomorphs. Rhizomorphs are thought to be important as perennating structures, in 'foraging' through soil 'in search' of food sources, and for transport of materials within the genet (see Chapter 2). Several studies have examined the distribution of *Armillaria* genotypes in nature (reviewed by Anderson & Kohn, 1995). In these studies, fruit-bodies, mycelia from infected trees, or rhizomorphs from soil are collected within a geographic area and used to establish axenic cultures. Mating-type alleles, mitochondrial and nuclear DNA restriction fragment length polymorphisms (RFLPs), random amplification of polymorphic DNA (RAPD) and mycelial compatibility interactions, can then be used to infer the genotype of each isolate. A genetically identical group of isolates can be regarded to have arisen from the same genet if the isolates were collected from a contiguous area, that is, if they probably constitute different sampling points from within the same mycelium, and if the genetic markers used to identify the genets are sufficiently variable at the population level to exclude the possibility of identity by inbreeding (Smith *et al.*, 1992).

Genotype mapping has revealed several interesting biological attributes of *Armillaria*. Within a species, genets do not overlap, but apparently grow around each other to form complex shapes (Fig. 12.1). The shape of genets is probably determined by a complex interplay of nutrient distribution and intraspecific antagonistic interactions. Such interactions are evident as zone lines when genetically distinct mycelia are confronted in culture (Fig. 12.2). Genets of *Armillaria* that exceed one hundred metres on one axis are not uncommon. Minimum age estimates of these large genets are based on the following assumptions. First, genets arose from a unique sexual mating event. Second, this mating event occurred at a point in space from which the resultant diploid colony grew 'vegetatively' outward, that is, by mitotic cell division and hyphal extension. Third, growth rates determined in the field or in the laboratory are used to estimate the time required for a mycelium to grow, unimpeded, one-half the distance required to transect the genet. Based on these assumptions, some *Armillaria* genets were established several hundreds, or even thousands of years ago. For example, a single genet, 'Clone 1', of *A. bulbosa* (synonym: *A. gallica*), occupies about 15 hectares

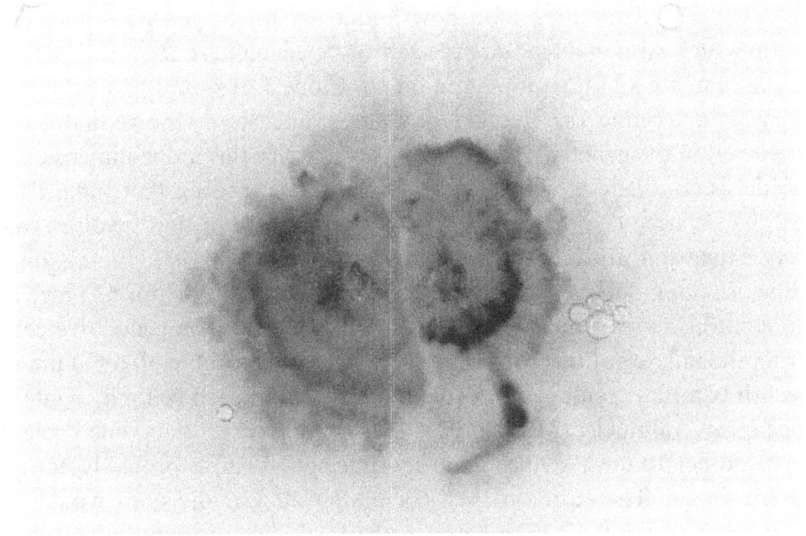

Fig. 12.2. Mycelial antagonism between two *Armillaria* genotypes is evident as a zone line in culture, compare to shapes in Figure 1C and 1D.

of hardwood forest in northern Michigan, and is estimated to be at least 1500 years in age (Smith *et al.*, 1992). (Note that 'Clone 1' is a genet. The name is apt since portions of this genet have been subsequently 'cloned' repeatedly by subculturing in the laboratory.) Remarkably, isolates taken throughout this genet were found to be identical for RFLPs associated with the mitochondrial DNA, and 20 RAPD and 27 DNA restriction fragments that were used as nuclear DNA markers. This was estimated to constitute a measure of identity for about 730 base pairs of nuclear DNA sequence, from several points across the genet. It should be emphasized that this is not based on DNA sequence data, but on the identity of genetic markers that probably represent several different regions of the genome. It should be also noted that this value is based on detecting a loss in either restriction enzyme recognition sites or RAPD primer annealing sites. It does not include possible gain-of-site mutations, that would have been evident as new RAPD or RFLP markers, also not observed. Further, the data did not reveal any large-scale mutational events such as insertions, deletions or loss of heterozygosity. The overall picture is one of a large fungal mycelium that has remained genetically stable during centuries of vegetative growth over an extensive area of the forest floor. Does this individual constitute an unusually stable genotype?

What types of mutation, and how frequently might genetic differences within such an individual be expected to be encountered?

To estimate a mitotic mutation rate in Clone 1, it would first be necessary to determine the number of mitotic nuclear divisions that have occurred in the genet: a difficult task. To simplify this a one-dimensional model is considered, while at the same time recognizing that fungi may exhibit an irregular three-dimensional growth pattern that involves colony expansion and contraction over time (Boddy, 1993). In this one-dimensional model, a rhizomorph grows in a straight line for 635 metres to connect the two most distant points in Clone 1. The generative cells from the interior of this rhizomorph are taken to be about 10 to 30 μm in length (Cairney, Jennings & Veltkamp, 1988) and each cell may contain one to several nuclei (Motta, 1969, 1982). On average, then, one nuclear division per 10 μm (Saville, Yoell & Anderson, 1996) is probably not an overestimate. Therefore, at least 63 million nuclear divisions would be required to span the 635 m between the most distant points sampled in Clone 1. A final estimate of mutation rates in this genet, then, would be less than 2×10^{-11} ($<$ 1 mutation/730 base pairs/6.3×10^{7} mitotic cell divisions).

How does this maximum mitotic mutation rate in an *Armillaria* genet compare to expected values? Mutation rates at the genomic level are remarkably similar in biological systems, at about 3.3×10^{-3} mutations/replication/genome (Drake, 1991). At this rate, many mutations should have occurred during the life of Clone 1, and those that were not deleterious to the cell in which they arose should be detectable. On a per base pair level, however, overall spontaneous mutation rates in *Saccharomyces cerevisiae* and *N. crassa* are estimated to be between 4×10^{-10} and 10^{-11} (Drake, 1991). Therefore, the DNA sequence stability of Clone 1 may not exceed expectations for fungal systems. It is remarkable to consider, however, that at a mutation rate of about 10^{-11}, detecting a single point mutation in a genet may require that several thousand base-pairs of DNA sequence is compared in distantly located cells, representing a mitotic lineage of over one thousand years. Could an understanding of potentially destabilizing genetic factors be used to increase the odds of detecting genetic variation within apparently stable genets?

Potentially destabilizing factors

Factors known to reduce genetic stability in fungi range from spontaneous mutations during DNA replication to induced changes to genome structure and content. Small-scale changes to DNA can occur spontaneously by misincorporation of nucleotides during replication, or as a result of natural, or anthropogenically derived DNA-damaging chemicals or ionizing radiation. The rate at which DNA replication errors occur can vary dramatically among different regions of the same genome. This variation may be due to a lack of constraint on the fidelity of a particular stretch of DNA or to specific DNA sequence characteristics of a region. Therefore, to increase the chances of locating spontaneous mutations within genets, analysis of non-coding regions or of repetitive tracts of DNA may be the best strategy. One possibility is simple tandemly repeated sequences of two, three or four nucleotides, called 'microsatellites', that are a common feature of eukaryotic genomes (Tautz & Renz, 1984). These sequences appear to be far less stable than other regions of DNA, perhaps due to DNA polymerase slippage within the repeated sequence during DNA replication. Mutation events in introduced poly-GT repeats in *S. cerevisiae*, for example, were found to range from about 10^{-7} to 10^{-4} per cell division, corresponding to an increase in repeat length from 15 to 105 base pairs, respectively (Wierdl, Dominska & Petes, 1997). The high rate of spontaneous alterations in microsatellite sequences could result in mutant sectors within naturally occurring fungal genets. These could be examined for by, first, screening an appropriate genomic library for specific microsatellite sequences and, second, sequencing into the unique flanking regions of the microsatellite. Third, oligonucleotide primers could be synthesized based on the flanking sequences for use in amplifying through the microsatellite region by polymerase chain reaction (PCR). Finally, these primers could be used to PCR amplify and compare a microsatellite region in multiple cultures taken from across a fungal genet.

In addition to changes at the nucleotide level that occur through errors during DNA replication, fungi are subject to mutations that alter chromosome structure or genome organization. Inter- and intrachromosomal rearrangements may occur spontaneously or through factors that cause chromosome breakage or recombination events. Large-scale mutations may be evident by alterations in restriction fragment patterns, electrophoretic karyotypes, or colony characteristics. Occasionally, isolates are recovered from nature that exhibit these types of minor differences (reviewed by Smith *et al.*, 1994), but it can be difficult to determine whether such isolates represent mitotic lineages that have accumulated minor mutations, or closely related genets that arose through sexual reproduction. Direct observations of morphological or physiological changes that occur during serial transfer of fungal strains are not uncommon, however, and reports of mitotic recombination in fungi under laboratory conditions are well-established (Ellingboe & Raper, 1962; Frankel, 1979). Similarly, 'strain degeneration' is a major concern to commercial ventures where consistent performance by fungi is required. For example, aberrant 'fluffy' sectors that arise periodically in spawn of *Agaricus bisporus*, the cultivated button mushroom, are attributed to a reduction in mushroom yield and quality. These sectors have been shown to exhibit a range of genetic differences which included a loss of heterozygosity attendant on deheterokaryotization, chromosome loss and mitotic recombination (Horgen *et al.*, 1996). Karyotypic changes are also reported for a diploid strain of *S. cerevisiae* (Longo & Vezinhet, 1993), and a strain of the rice blast fungus, *Magnaporthe grisea*, following prolonged serial transfers (Talbot *et al.*, 1993). It is notable in this latter case that cultures of the *M. grisea* strain were identical for DNA fingerprint patterns and pathological type. Evidently, karyotypic changes do not always result in noticeable alterations in strain characteristics. Rapid changes in *Candida albicans* and *S. cerevisiae* karyotypes have been observed to correlate to physiological stress during vegetative growth (reviewed by Zolan, 1995). This has led to the proposal that a hypermutable state can be induced in which the normally stringent control of karyotype is lost. This hypermutable state could arise either by a loss of suppression of ectopic recombination or induction of DNA repair enzymes and an increase in the frequency of ectopic recombination. Interestingly, mitotic hyperrecombination and chromosome instability is evident during mitotic cell division in *S. cerevisiae* strains with a deletion in the *SGS1* gene (Watt *et al.*, 1996). The product of *SGS1* is a member of a conserved DNA helicase family that includes those defective in Bloom's cancer-prone syndrome and Werner's premature aging syndrome. Mutations within *SGS1* are thought to interfere with normal helicase function whereby hydrogen bonds between strands of the double helix are disrupted during DNA replication. The dissection of recombination pathways in fungi may ultimately provide insights into how genetic stability may be modulated during the vegetative phase of the life cycle.

The frequency of karyotypic changes is unknown for fungal genets in nature. Colonial basidiomycetes would be an ideal point to initiate an examination of karyotypic stability. An analysis of electrophoretic karyotypes within *Armillaria* genets would be of particular interest for a number of reasons. First, *Armillaria* is one of the only known hymenomycetes to have a prolonged diploid stage to the life cycle (Hintikka, 1973; Anderson, 1983). One might expect an increased chance of karyotypic changes due to mitotic crossing-over in fungi with diploid nuclei. Second, the segregation of *Armillaria* diploids has been induced in laboratory experiments (Anderson, 1983). Third, the locations of several naturally occurring, well-characterized diploids of *Armillaria* are known. These genets will have encountered several rounds of freezing, local nutrient deprivation and periods of drought, in addition to chemicals associated with plant defence systems and soil microbes, and a host of potentially destabilizing genetic elements. Finally, as already mentioned, a spatial description of karyotypic differences within such genets would provide insight into chromosome mutation frequencies and fungal growth patterns in naturally occurring fungi.

Along with physiological stress, which is implicated in karyotypic rearrangement of fungal genomes, a wide array of potentially destabilizing genetic elements are known to occur in fungi. These elements vary in their ability to move about and insert into host genomes. Included here are transposable genetic elements such as those detected in *S. cerevisiae* (Boeke *et al.*, 1985), *N. crassa* (reviewed by Yeadon & Catcheside, 1995) and *M. grisea* (Shull & Hamer, 1996). Transposable elements can cause mutation in at least two ways. Their insertion into a coding region can result in gene disruption. In addition, multiple copies of transposon-derived DNA sequences, like other dispersed repetitive genetic elements, may be sites of ectopic recombination that yield chromosome translocations (reviewed by Smith & Glass, 1996). Mitochondrial plasmids are also widely distributed among fungal taxa (see review by Griffiths, 1995). For example, plasmids, or plasmid-related DNA sequences, were detected in mitochondria from 20 of 21 *Agaricus* species examined (Robison, Kerrigan & Horgen, 1997). Apparently, either these plasmid sequences occurred in a common ancestor to the genus, or significant horizontal transfer between species has occurred. Strain senescence in *Neurospora* species has been correlated with the insertion of plasmid DNA elements into the mitochondrial DNA, acting as insertional mutagens in the disruption of genes that have respiratory function. Genome changes associated with virus-like elements (reviewed by Nuss & Koltin, 1990), and group II introns (Mueller *et al.*, 1993; Sellem, Lecellier & Belcour, 1993), may also be common in fungi.

Mycelial interactions between individuals are also expected to alter the genetic constitution of individuals. For example, horizontal transfer of plasmids has been documented in unstable heterokaryon fusions within a *Neurospora* species (Collins & Saville, 1990), between *Neurospora* species (Griffiths *et al.*, 1990), and even between different genera of fungi (Kempken, 1995). Horizontal transfer of genetic elements can be regarded as an infectious process and may explain discontinuous distributions of mobile genetic elements across taxa.

There is also an acknowledged potential for intraspecific nuclear parasitism when mycelia of the same, or closely related, species meet (Rayner, 1991). Genetic destabilization may result from an inadequate non-self recognition response by one of the interacting partners. In basidiomycetes, mating interactions between heterokaryotic (or diploid, in the case of *Armillaria*) and haploid, homokaryotic mycelia may also have a destabilizing influence on established individuals. This type of mating interaction was first described by Buller (1930) and is known as the 'Buller phenomenon', or as 'he-ho' (heterokaryon–homokaryon) or 'di-mon' (dikaryon–monokaryon) mating, and is at least partly governed by mating type (reviewed by Carvalho, Smith & Anderson, 1995). There appear to be many variations of the process, but the general outcome is that of a nuclear component of the heterokaryon moving into the homokaryotic partner. Obviously, this would result in a major disruption to the genetic constitution of the homokaryotic partner, but little change to the heterokaryotic partner. However, mitotic recombination is required in some cases for the invading nucleus to become mating-type compatible with the recipient homokaryon. Regardless of whether mitotic recombination is induced, or an existing recombinant nucleus is recruited during this process, the selection of recombinant genotypes may have a destabilizing influence on the heterokaryotic partner in some cases. In some instances there may be complete replacement of nuclei in recipient cells by invading nuclei from the heterokaryotic partner. In *A. gallica*, mitochondria do not migrate along with the nuclei into the recipient cells (Carvalho *et al.*, 1995). One would expect this would result in heteroplasmic individuals since distinct mitochondrial genomes could become associated with a single, uniform nuclear genome following replacement. However, heteroplasmic genets of *Armillaria* do not appear to be common in nature (Smith *et al.*, 1990, 1994; Saville *et al.*, 1996), although there is strong inference that mitochondrial mixes and recombination occurs in

Armillaria (Smith & Anderson, 1994; Saville *et al.*, 1996) and in other fungi.

Maintenance of genetic identity within mycelia

The necessarily incomplete, yet impressive, array of potentially destabilizing factors listed above overshadows the small number of mechanisms that may play a role in maintaining genetic stability within clonal lineages. Probably a major stabilizing influence in naturally occurring mycelia is selection pressure. It is unlikely that variation introduced into an organism at the genetic level will confer an immediate benefit. Consequently, a mutant cell, or sector of mycelium, will most likely not be able to compete with surrounding regions of the same individual, and will be lost. In the rare event that a 'super sector' develops, either by mutation or by invasion of an aggressive nucleus, mycelial overgrowth or migration of the 'improved' nuclear genotype through a pre-existing hyphal network may occur during a relatively brief time. Whether selection is a mechanism 'employed' by a fungus is questionable. Below are several examples of mechanisms by which fungi may maintain genetic stability.

Recognition and correction of small-scale mutations that arise during mitotic growth occur in all life forms, including fungi. Because of the fundamental importance of mutation, and the concern we share over cancer as a genetic disease of somatic cells, the process of DNA repair has been actively studied in 'model eukaryotic systems'. Among these systems, the biochemical basis of DNA repair is well characterized in *S. cerevisiae* and *Schizosaccharomyces pombe* (review by Friedberg, Walker & Siede, 1995). Cellular response to DNA damage in these yeasts ranges from tolerance to recognition and precise reversal of damage. Replication fidelity is dependent on a multi-step process. First, in cell free systems correct base pairing of nucleotides along the double helix is energetically favoured. Second, polymerases and accessory proteins increase the accuracy of semiconservative replication of DNA through nucleotide selection and editing. Finally, after DNA is replicated error rates may be further reduced by other cellular factors that detect and replace mismatched nucleotides. In *S. cerevisiae* there are more than 30 *RAD* genes involved in nucleotide excision repair and the cellular response to DNA damage (Friedberg *et al.*, 1995). This illustrates the complexity of post-replicative repair systems. The *RAD* genes constitute three groups, each corresponding to a multimeric complex or to an independent repair pathway.

The degree to which organisms tolerate or correct changes in DNA sequence probably depends upon their physiological condition and life history strategy. Unfavourable growth conditions may, in some instances, promote a relaxation in the control of DNA replication fidelity and result in an increase in mutation frequency. Some of the resulting mutations may be beneficial and provide for an adaptive advantage. Asexually reproducing fungi may likewise be more tolerant of mitotic mutations.

In addition to cellular mechanisms that correct small-scale mutations, several processes have been described in fungi that may promote large-scale genome stability. The following section briefly describes six such processes known to occur in fungi: RIP (repeat induced point mutation), MIP (methylation induced premeiotically), Quelling, spontaneous loss of duplications, non-random segregation of mitotic sister chromatids, and non-self recognition by vegetative incompatibility systems. Some of these and additional means of maintaining genome stability are likely to occur in diverse fungi but have been examined in a few well-studied species.

RIP, MIP and Quelling are all processes in which introduced duplications are detected and genes within the duplication are 'silenced' (review by Irelan & Selker, 1996). In *Neurospora crassa*, RIP causes G:C → A:T transition mutations within, and in the vicinity of, duplications introduced into the genome through DNA transformation or by crosses. Methylation of cytosine residues at these duplications is also associated with RIP and may be maintained during vegetative growth. RIP occurs between fertilization and karyogamy, and therefore may not be directly involved in maintenance of genetic stability during vegetative growth. However, relatively little repetitive DNA, and evidence of putatively RIPped remnants of transposable genetic elements in the genomes of *N. crassa* strains indicate that RIP may function in maintaining genome stability of the species. By significantly altering introduced repeated sequences, RIP may reduce the homology among these sequences and thus suppress chromosome rearrangements due to ectopic mitotic cross-overs. Similarly, introduced duplications are recognized in some way, and silenced in *Ascobolus immersus* and *Coprinus cinereus* by MIP. Unlike RIP, MIP does not result in transition mutations, but epigenetically inactivates duplications by methylation of cytosine residues. The pattern of cytosine methylation can be maintained subsequently during vegetative growth in these fungi. Quelling, or transgene-induced silencing, of some introduced duplicated segments of DNA occurs during vegetative growth in *N. crassa*. Quelling does not apparently involve cytosine methylation, although other forms of DNA modification have not been ruled out. Models of gene silencing by Quelling include interaction between RNA and DNA or between homologous RNA molecules (Cogoni *et al.*, 1996). Gene silencing of introduced repeated sequences by RIP, MIP, Quelling or by other processes, may effectively reduce the proliferation of replicative, parasitic DNA elements during vegetative growth of fungi. For example, gene silencing may inactivate transposable genetic elements that contain and require transcription of a transposase to move within and between nuclei.

Silencing duplicated tracts of DNA may prevent further genetic destabilization by infectious DNA elements but does not result in a return to euploidy; the infected cells and subsequent daughter cells will contain the inserted DNA element or a remnant of that element. Restoration of genome balance by spontaneous loss of duplicated DNA (review by Perkins, 1997) has been investigated in both *N. crassa* (review by Perkins & Barry, 1977) and *A. nidulans* (Nga & Roper, 1969; Case & Roper, 1981). Haploid strains of *N. crassa* that carry a duplicated segment can be produced by a cross between translocation-bearing and wild-type strains. One-quarter of the meiotic products from such a cross contain a duplication of the translocated segment and give rise to partial diploid progeny. Unstable partial diploids can be recognized as visible sectors under non-selective conditions, or by increased growth rates under selective conditions. To select for the spontaneous breakdown of a duplication-bearing strain, partial diploids are synthesized that are self-incompatible due to heteroallelism at a heterokaryon incompatibility locus located on the translocated segment. The partial diploid carries two different alleles of a heterokaryon incompatibility gene (*het* gene), and is self-incompatible, as evident by slow growth and/or abnormal colony characteristics. Spontaneous loss of one of the duplicated segments can occur within a few weeks and is evident by an increase in growth rate, a return to wild-type colony characteristics or, sometimes, by full fertility (partial diploids are usually barren). This ‘escape’ from self-incompatibility appears to be associated with deletion of one the duplicated copies, usually with a bias toward deletion of the copy in the translocated position (Turner, 1977; Newmeyer & Galeazzi, 1977; Smith *et al.*, 1996). Similarly, the spontaneous breakdown of partial diploids that are heteroallelic for the recessive trait conferring cycloheximide resistance can be observed on cycloheximide-containing medium (Turner, 1976). Loss of duplicated segments has also been described in haploid and diploid strains of *A. nidulans* (review by Case & Roper, 1981). Diploid strains of *A. nidulans* that carry an additional segment of DNA in a translocated position can be very unstable, losing one of the excess segments, usually that in the translocated position. As in *Neurospora*, the loss of the translocated genetic material results in an euploid or nearly euploid condition. These observations suggest that there are general mechanisms in place by which duplicated tracts of DNA are recognized and deleted in fungi. Such mechanisms may be induced by physiological stress. A preference to delete the ectopic, or improperly placed, duplicated segment of DNA suggests that there is either a strong selection for the euploid condition, or a mechanism to preferentially delete DNA that is out of place in the genome. This might involve a temporary gene silencing, with the result being a release from growth inhibition, followed by permanent removal of the silenced DNA.

Non-random sister chromatid segregation during mitotic division has been demonstrated in *A. nidulans* (Rosenberger & Kessel, 1968). Briefly, DNA in germinating conidia was pulse labelled and followed through subsequent mitotic nuclear divisions by autoradiography. Labelled DNA was not evenly distributed among nuclei after the second nuclear division. At the four- and eight-nuclei stage, 75% of labelled DNA was in two nuclei, and the greatest proportion of radioisotope was in nuclei located near the tips of growing hyphae. This suggests that the original template strands cosegregate toward the outer margin of an expanding colony and that the growing front contains what may be considered 'germ nuclei'. Extending this further, DNA located near actively growing areas of a fungal colony may have undergone relatively few replications and thereby have accumulated relatively few mutations.

Finally, nearly all filamentous fungi have well-developed vegetative incompatibility systems (also referred to as heterokaryon, cytoplasmic and somatic incompatibility). Exceptions apparently include some fungal associates of grasses, such as symbiotic species of *Epichloë*, which have either poorly developed, or no vegetative compatibility groups (Chung & Shardl, 1997). One function of incompatibility systems may be to help maintain genetic stability by reducing the incidence of transmission of genetic elements between fungal colonies (Caten, 1972; Hartl, Dempster & Brown, 1975). Vegetative incompatibility in fungi has been reviewed

(Glass & Kuldau, 1992; Leslie, 1993; Bégueret, Turcq & Clavé, 1994; Esser & Blaich, 1994; Worrall, 1997). Here, vegetative incompatibility is understood as any cellular-based process that prevents stable fusion of genetically distinct hyphae during the assimilative, or vegetative, phase of the life cycle. Incompatibility could conceivably act prior to, or following, hyphal fusion. Among the best-characterized is vegetative incompatibility of *Podospora anserina* and *N. crassa*. In both species, cell fusion can occur within and between mycelia during the vegetative phase of the life cycle. In *Neurospora*, a single allelic difference at any one of several *het* loci in fusing cells gives rise to an incompatibility response, characterized by cell lysis or inhibited growth of the resultant heterokaryon. In *Podospora*, vegetative incompatibility is governed by both allelic interactions, as in *Neurospora*, and non-allelic interactions in which specific combinations of alleles at distinct loci that are brought together in the same cell result in an incompatible interaction. In many species of fungi there are multiple vegetative incompatibility loci, and multiple alleles may occur at some of these loci making compatibility unlikely among independently derived mycelia. Vegetative incompatibility may thus reduce the transfer of invasive genetic elements between mycelia by reducing successful cell fusions. Genetic elements restricted in this way could include exploitive, or parasitic, nuclei and mitochondria, virus-like elements, transposable genetic elements or plasmids. Actual examples of this process are few but convincing. The vegetative incompatibility system in *Cryphonectria parasitica*, the causal agent of chestnut blight, appears to be an effective barrier to transmission of hypovirulence-associated double-stranded RNA virus (Anagnostakis, 1982). In *Neurospora* and *Aspergillus*, near isogenic strains have been used to show that allelic differences at single and multiple *het* loci can be effective in reducing the transmission of cytoplasmic elements (Caten, 1972; Debets, Yang & Griffiths, 1994). Territoriality exhibited by fungal genets further indicates that a well-developed non-self recognition system exists among fungi in general, and that at least one function of vegetative incompatibility is to help maintain genetic stability in well-adapted individuals.

Concluding remarks

The study of genetic stability in fungi has sufficient practical and theoretical importance to merit further investigation. Large basidiomycete genets probably represent mitotic cell lineages that are centuries old. The detection of mutations in such genets would enhance our understanding of mutation rates and processes in eukaryotes in general, and growth patterns in naturally occurring fungi, specifically. One would expect to encounter genetic variation within such genets, given (a) these genets are comprised of vast numbers of cells, (b) spontaneous and induced mutations are relatively common in cultured fungi, (c) spontaneous mutation frequencies are estimated to be about 3.3×10^{-3} mutations/replication/genome, and (d) there are numerous potentially destabilizing genetic elements known to occur in fungi. In addition, the introduction of genetic variation into a mitotic lineage may be beneficial under some circumstances and fungi may have, therefore, developed cellular systems that modulate genetic instability. Preliminary analysis did not reveal genetic differences in a limited number of samples taken from an *Armillaria* genet that covers approximately 15 hectares. DNA sequencing in non-coding regions, karyotype analysis and changes in microsatellite or rDNA repeat length appear to be good areas to begin a search for mutations within similarly large genets. The risk associated with a search for mutations in such colonies is that no mutations are identified. However, if such is the case, a careful experimental design should provide a conservative estimate of mutation frequencies that are far lower than expected and an indication that some fungal individuals have well-developed means of promoting genetic stability, and thereby, may indeed be practically immortal.

References

Adams, D. H. (1974). Identification of clones of *Armillaria mellea* in young-growth Ponderosa pine. *Northwest Science*, **48**, 21–8.

Anagnostakis, S. L. (1982). Biological control of chestnut blight. *Science*, **215**, 466–71.

Anderson, J. B. (1983). Induced somatic segregation in *Armillaria mellea* diploids. *Experimental Mycolology*, **7**, 141–7.

Anderson, J. B. & Kohn, L. M. (1995). Clonality in soilborne, plant-pathogenic fungi. *Annual Reviews of Phytopathology*, **33**, 369–91.

Anderson, J. B., Ullrich, R. C., Roth L. F. & Filip, G. M. (1979). Genetic identification of clones of *Armillaria mellea* in coniferous forests in Washington. *Phytopathology* **69**, 1109–11.

Bégueret, J., Turcq, B. & Clavé, C. (1994). Vegetative incompatibility in filamentous fungi: *het* genes begin to talk. *Trends in Genetics*, **10**, 441–6.

Boddy, L. (1993). Saprotrophic cord-forming fungi: warfare strategies and other ecological aspects. *Mycological Research*, **97**, 641–55.

Boeke, J. D., Garfinkel, D. J., Styles, C. A. & Fink, G. R. (1985). Ty elements transpose through an RNA intermediate. *Cell*, **40**, 491–500.

Buller, A. H. R. (1930). The biological significance of conjugate nuclei in *Coprinus lagopus* and other hymenomycetes. *Nature*, **126**, 686–89.

Buller, A. H. R. (1958). *Researches on Fungi, Vol. IV*. New York: Hafner.

Cairney, J. W. G., Jennings, D. H. & Veltkamp, C. J. (1988). Structural differentiation in maturing rhizomorphs of *Armillaria mellea* (Tricholomatales). *Nova Hedwigia*, **46**, 1–25.

Carvalho, D. B., Smith, M. L. & Anderson, J. B. (1995). Genetic exchange between diploid and haploid mycelia of *Armillaria gallica*. *Mycological Research*, **99**, 641–7.

Case, B. L. & Roper, J. A. (1981). Mitotic processes which restore genome balance in *Aspergillus nidulans*. *Journal of General Microbiology*, **124**, 9–16.

Caten, C. E. (1972). Vegetative incompatibility and cytoplasmic infection in fungi. *Journal of General Microbiology*, **72**, 221–9.

Chung, K.-R. & Shardl, C. L. (1997). Vegetative compatibility between and within *Epichloë* species. *Mycologia*, **89**, 558–65.

Cogoni, C., Irelan, J. T., Schumacher, M., Schmidhauser, T. J., Selker, E. U. & Macino, G. (1996). Transgene silencing of the *al-1* gene in vegetative cells of *Neurospora* is mediated by a cytoplasmic effector and does not depend on DNA–DNA interactions or DNA methylation. *EMBO Journal*, **15**, 3153–63.

Collins, R. A. & Saville, B. J. (1990). Independent transfer of mitochondrial chromosomes and plasmids during unstable vegetative fusion in *Neurospora*. *Nature*, **345**, 177–9.

Debets, F., Yang, X. & Griffiths, A. J. F. (1994). Vegetative incompatibility in *Neurospora*: its effect on horizontal transfer of mitochondrial plasmids and senescence in natural populations. *Current Genetics*, **26**, 113–19.

Dickman, A. & Cook, S. (1989). Fire and fungus in a mountain hemlock forest. *Canadian Journal of Botany*, **67**, 2005–16.

Drake, J. W. (1991). A constant rate of spontaneous mutation in DNA-based microbes. *Proceedings of the National Academy of Sciences, USA.*, **88**, 7160–64.

Ellingboe, A. H. & Raper, J. R. (1962). Somatic recombination in *Schizophyllum commune*. *Genetics*, **47**, 85–98.

Esser, K. & Blaich, R. (1994). Heterogenic incompatibility in fungi. In *The Mycota I: Growth, Differentiation and Sexuality*, ed. J. Wessels & F. Meinhardt, pp. 212–32. Berlin: Springer.

Frankel, C. (1979). Meiotic-like recombination in vegetative dikaryons of *Schizophyllum commune*. *Genetics*, **92**, 1121–6.

Friedberg, E. C., Walker, G. C. & Siede, W. (1995). *DNA Repair and Mutagenesis*. Washington, DC: American Society for Microbiology.

Glass, N. L. & Kuldau, G. A. (1992). Mating type and vegetative incompatibility in filamentous ascomycetes. *Annual Review of Phytopathology*, **30**, 201–24.

Griffiths, A. J. F. (1995). Natural plasmids of filamentous fungi. *Microbiological Reviews*, **59**, 673–85.

Griffiths, A. J. F., Kraus, S. R., Barton, R., Court, D. A., Meyers, C. J. & Bertrand, H. (1990). Heterokaryotic transmission of senescence plasmid DNA in *Neurospora*. *Current Genetics*, **17**, 139–45.

Hartl, D. L., Dempster, E. R. & Brown, S. W. (1975). Adaptive significance of vegetative incompatibility in *Neurospora crassa*. *Genetics*, **81**, 553–69.

Hintikka, V. (1973). A note on the polarity of *Armillariella mellea*. *Karstenia*, **13**, 32–9.

Horgen, P. A., Carvalho, D., Sonnenberg, A., Li, A. & Van Griensven, L. J. L. D. (1996). Chromosomal abnormalities associated with strain degeneration in the cultivated mushroom, *Agaricus bisporus*. *Fungal Genetics and Biology*, **20**, 229–41.

Irelan, J. T. & Selker, E. U. (1996). Gene silencing in filamentous fungi: RIP, MIP and quelling. *Journal of Genetics*, **75**, 313–24.

Käfer, E. (1975). Reciprocal translocations and translocation disomics of Aspergillus and their use for genetic mapping. *Genetics*, **79**, 7–30.

Kempken, F. (1995). Horizontal transfer of a mitochondrial plasmid. *Molecular and General Genetics*, **248**, 89–94.

Leslie, J. F. (1993). Fungal vegetative compatibility. *Annual Review of Phytopathology*, **31**, 127–50.

Longo, E. & Vezinhet, F. (1993). Chromosomal rearrangements during vegetative growth of a wild strain of *Saccharomyces cerevisiae*. *Applied and Environmental Microbiology*, **59**, 322–6.

Motta, J. J. (1969). Cytology and morphogenesis in the rhizomorph of *Armillaria mellea*. *American Journal of Botany*, **56**, 610–19.

Motta, J. J. (1982). Rhizomorph cytology and morphogenesis in *Armillaria tabescens*. *Mycologia*, **74**, 671–4.

Mueller, M. W., Allmaier, M., Eskes, R. & Schweyen, R. J. (1993). Transposition of group II intron *al1* in yeast and invasion of mitochondrial genes at new locations. *Nature*, **366**, 174–6.

Newmeyer, D. & Galeazzi, D. R. (1977). The instability of *Neurospora* duplication *Dp(IL→IR)H4250*, and its genetic control. *Genetics*, **85**, 461–87.

Nga, B. H. & Roper, J. A. (1969). A system generating spontaneous intrachromosomal changes at mitosis in *Aspergillus nidulans*. *Genetics Research, Cambridge*, **14**, 63–70.

Nuss, D. L. & Koltin, Y. (1990). Significance of dsRNA genetic elements in plant pathogenic fungi. *Annual Review of Phytopathology*, **28**, 37–58.

Perkins, D. D. (1997). Chromosome rearrangements in *Neurospora* and other filamentous fungi. *Advances in Genetics*, **36**, 239–398.

Perkins, D. D. & Barry, E. G. (1977). The cytogenetics of *Neurospora*. *Advances in Genetics*, **19**, 133–285.

Raper, J. R. (1966). *Genetics of Sexuality in Higher Fungi*. New York: The Ronald Press.

Rayner, A. D. M. (1991). The challenge of the individualistic mycelium. *Mycologia*, **83**, 48–71.

Robison, M. M., Kerrigan, R. W. & Horgen, P. A. (1997). Distribution of plasmids and a plasmid-like mitochondrial sequence in the genus *Agaricus*. *Mycologia*, **89**, 43–7.

Rosenberger, R. F. & Kessel, M. (1968). Nonrandom sister chromatid segregation and nuclear migration in hyphae of *Aspergillus nidulans*. *Journal of Bacteriology*, **96**, 1208–13.

Saville, B. J., Yoell, H. & Anderson, J. B. (1996). Genetic exchange and recombination in populations of the root-infecting fungus *Armillaria gallica*. *Molecular Ecology*, **5**, 485–97.

Sellem, C. H., Lecellier, G. & Belcour, L. (1993). Transposition of a group II intron. *Nature*, **366**, 176–8.

Shaw III, C. G. & Roth, L. F. (1976). Persistence and distribution of a clone of *Armillaria mellea* in a ponderosa pine forest. *Phytopathology*, **66**, 1210–13.

Shull, V. & Hamer, J. E. (1996). Rearrangements at a DNA fingerprint locus in the rice blast fungus. *Current Genetics*, **30**, 263–71.

Smith, M. L. & Anderson, J. B. (1994). Mitochondrial DNAs of the fungus *Armillaria ostoyae*: restriction map and length variation. *Current Genetics*, **25**, 545–53.

Smith, M. L., Bruhn, J. N. & Anderson, J. B. (1992). The fungus *Armillaria bulbosa* is among the largest and oldest living organisms. *Nature*, **356**, 428–31.

Smith, M. L., Bruhn, J. N. & Anderson, J. B. (1994). Relatedness and spatial distribution of *Armillaria* genets infecting red pine seedlings. *Phytopathology*, **84**, 822–9.

Smith, M. L., Duchesne, L. C., Bruhn, J. N. & Anderson, J.B. (1990). Mitochondrial genetics in a natural population of the plant pathogen *Armillaria*. *Genetics*, **126**, 575–82.

Smith, M. L. & Glass, N. L. (1996). Mapping translocation breakpoints by orthogonal field agarose-gel electrophoresis. *Current Genetics*, **29**, 301–5.

Smith, M. L., Yang, C. J., Metzenberg, R. L. & Glass, N. L. (1996). Escape from *het-6* self-incompatibility in *Neurospora crassa* partial diploids involves preferential deletion within the ectopic segment. *Genetics*, **144**, 523–31.

Talbot, N. J., Salch, Y. P., Ma, M. & Hamer, J. E. (1993). Karyotypic variation within clonal lineages of the rice blast fungus, *Magnaporthe grisea*. *Applied and Environmental Microbiology*, **59**, 585–93.

Tautz, D. & Renz, M. (1984). Simple sequences are ubiquitous components of eukaryotic genomes. *Nucleic Acids Research*, **12**, 4127–38.

Todd, N. K. & Rayner, A. D. M. (1980). Fungal individualism. *Science Progress*, **66**, 331–54.

Turner, B. C. (1976). Dominance of the wild-type (sensitive) allele of *cyh-1*. *Neurospora Newsletter*, **23**, 24.

Turner, B. C. (1977). Euploid derivatives of duplications from a translocation in *Neurospora*. *Genetics*, **85**, 439–60.

Watt, P. M., Hickson, I. D., Borts, R. H. & Louis E. J. (1996). *SGS1*, a homologue of the Bloom's and Werner's syndrome genes, is required for maintenance of genome stability in *Saccharomyces cerevisiae*. *Genetics*, **144**, 935–45.

Wierdl, M., Dominska, M. & Petes, T. D. (1997). Microsatellite instability in yeast: dependence on the length of the microsatellite. *Genetics*, **146**, 769–79.

Worrall, J. J. (1997). Somatic incompatibility in basidiomycetes. *Mycologia*, **89**, 24–36.

Yeadon, P. J. & Catcheside, D. E. A. (1995). *Guest:* a 98 bp inverted repeat transposable element in *Neurospora crassa*. *Molecular and General Genetics*, **247**, 105–9.

Zolan, M. E. (1995). Chromosome-length polymorphism in fungi. *Microbiological Reviews*, **59**, 686–98.

13

Nuclear distribution and gene expression in the secondary mycelium of *Schizophyllum commune*

J. G. H. WESSELS, T. A. SCHUURS, H. J. P. DALSTRA AND J. M. J. SCHEER

Introduction

Germinating meiospores of *Schizophyllum commune* grow into primary mycelia which are monokaryons, i.e. mycelia in which each hyphal compartment contains one nucleus only. When two primary mycelia are grown together so that hyphal contacts can occur, hyphae may fuse, irrespective of mating type, and produce heterokaryotic cells. The fate of these heterokaryotic cells depends on the mating-type genes carried by the opposing mycelia. If they carry different *MAT*A and different *MAT*B genes, a stable heterokaryotic dikaryon (the secondary mycelium) is formed (Fig. 13.1). First, nuclei from each mate, rapidly travel into the other mate, inducing the dissolution of the complex dolipore septa which otherwise block nuclear translocation (Raper, 1966; Wessels, 1978). This process of septal dissolution and nuclear migration is apparently controlled by the presence of different *MAT*B genes in the heterokaryon, because it occurs when the *MAT*B genes only are different (*MAT*A= *MAT*B≠). However, if the *MAT*A genes are also different (*MAT*A≠ *MAT*B≠) the foreign nucleus associates with a resident nucleus in a tip cell and from now on septal dissolution is switched off and the two nuclei (carrying different *MAT*A and *MAT*B genes) remain closely associated with each other and divide synchronously. During this division the spindle of one of the dividing nuclei is oriented obliquely so that one of its daughter nuclei is positioned in a small lateral branch, the hook cell. This branch is subsequently cut off from the apical compartment by a septum. The other nucleus divides with a longitudinally oriented spindle followed by formation of a division septum at the position of the hook cell. This process of hook-cell formation and septation is obviously controlled by the *MAT*A genes, since it also occurs in heterokaryotic cells containing nuclei with different *MAT*A genes but the same *MAT*B genes (*MAT*A≠

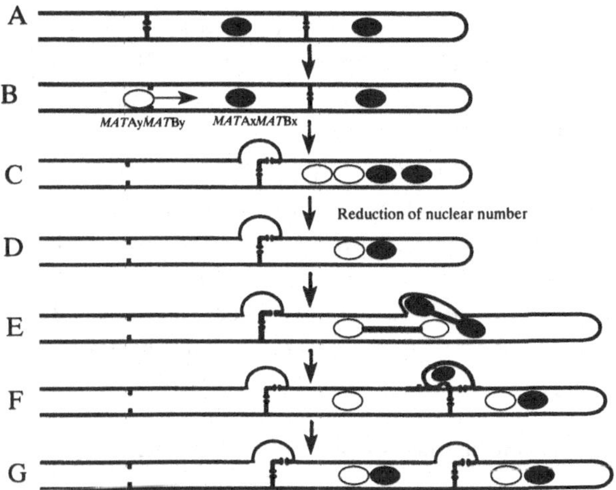

Fig. 13.1. Formation of a dikaryon of *Schizophyllum commune* (*MAT*A≠ *MAT*B≠) by mating two compatible monokaryon (*MAT*Ax *MAT*Bx + *MAT*Ay *MAT*By). After fusion of hyphae, nuclei from each mate migrate into the opposing mycelium. The compatible invading nucleus (white in B) interacts with resident nuclei and this causes septal dissolution and nuclear migration (function of *MAT*B≠) resulting in multinucleate tip cells in which septa are no longer eroded and clamp connections are formed (C). The number of nuclei in these tip cells is now gradually reduced by mechanisms not shown but described in detail by Niederpruem (1980) till two compatible nuclei remain (D). These henceforth divide synchronously accompanied by formation of a hook cell (functions of MATA≠) and subsequent fusion of the hook cell (function of *MAT*B≠) resulting in a clamp connection at each division (E–F).

*MAT*B=). The final fusion of the hook cell with the subapical compartment is clearly orchestrated by the presence of different *MAT*B genes in the *MAT*A≠ *MAT*B≠ heterokaryon since hook-cell fusion does not occur in the *MAT*A≠ *MAT*B= heterokaryon.

The role described for the mating-type genes is strengthened by the effects of rare, so-called constitutive, mutations in these genes that switch on the respective morphogenetic pathways (Parag, 1962; Raper, Boyd & Raper, 1965; Raper, 1966, 1983). The phenotypes of these mutants (homokaryons, containing only one type of nucleus) mimics heterokaryons in which sexual morphogenesis is dependent on heterogenic mating-type genes, that is, the *MAT*A^{con} *MAT*Bx homokaryon mimics the *MAT*A≠ *MAT*B= heterokaryon, the *MAT*Ax *MAT*B^{con} homokaryon mimics the *MATA*= *MATB*≠ heterokaryon, and the *MATA*^{con} *MATB*^{con} homokaryon mimics the *MATA*≠ *MATB*≠ heterokaryon.

An additional phenotypic feature of the *MATA*≠ *MATB*≠ heterokaryon and the *MATA*^{con} *MATB*^{con} homokaryon is the propensity of these mycelia to form fruiting bodies, a property not shared by any of the other mycelia. In these fruiting bodies compatible nuclei fuse and immediately undergo meiosis leading to independent reassortment in the four basidiospores of the different *MATA* and *MATB* genes of the *MATA*≠ *MATB*≠ heterokaryon and to the formation of exclusively *MATA*^{con} *MATB*^{con} spores in the *MATA*^{con} *MATB*^{con} homokaryon.

Structure and function of the mating-type genes

Rapid progress has recently been made in elucidating the molecular structure of the *MAT* genes of both *Schizophyllum commune* and *Coprinus cinereus*. Extensive reviews of this work have appeared (Kües & Casselton, 1992; Casselton & Kües, 1994; Kothe, 1996) and therefore this research will not be detailed here.

Although for simplicity the genetic entities governing mating are referred to as *MATA* and *MATB* genes, these entities are actually part of complex genetic loci called *A* and *B* incompatibility factors, respectively. The *A* factor (subdivided in two equivalent subloci *Aα and Aβ*) is a locus containing a series of different alleles each containing two genes, called *Y* and *Z* in *S. commune* (Stankis *et al.*, 1992) and *HD1* and *HD2* in *C. cinereus* (named after the homeodomain sequences they contain), which are divergently transcribed. The effective interaction was shown to be between a *Y* (*HD1*) gene in one nucleus and a *Z* (*HD2*) gene in the other nucleus belonging to the same allelic series. *Y* (*HD1*) and *Z* (*HD2*) genes within the same nucleus do not interact because they belong to different allelic series. Interactions between *Y*(*HD1*) and *Z*(*HD2*) gene products are thought to produce a gene-activating regulator because of the presence of homeodomains in the proteins. Specific interactions between HD1 and HD2 proteins of *C. cinereus* have been demonstrated *in vitro* (Banham *et al.*, 1995). In *S. commune*, the two-hybrid system was used to demonstrate interaction between the products of the *Y* and *Z* genes (Magae, Novotny & Ullrich, 1995). The *MATA*^{con} allele was shown to be due to a deletion, effectively fusing an *HD1* and an *HD2* gene within a *MATA* locus in such a way that the recognition sequence is deleted and a presumed DNA-binding domain (HD2) from one gene becomes associated with the activating domain of the other gene (Kües *et al.*, 1994) Also in *S. commune* the presence of only one homeodomain is sufficient for *MAT*A-activated development (Luo, Ullrich & Novotny, 1994).

Although the genetic structure of the *B*-incompatibility factor appears similar to that of the *A*-incompatibility factor (they also contain equivalent α and β alleles), interactions between different *B*-incompatibility factors seem to involve quite different processes since genes within the *B*-incompatibility factor (in both $B\alpha$ and $B\beta$) code for a pheromone receptor and multiple pheromones (Wendland *et al.*, 1995; Kothe, 1996; Vaillancourt *et al.*, 1997). This was quite unexpected because in these homobasidiomycetes the mating-type genes do not control fusion between hyphae; this occurs irrespective of mating-type. Nevertheless, the hypothesis was forwarded that the specific pheromone receptors are located in the plasmalemma and the pheromone secreted into the medium by one mate induced the other mate with the dissimilar *MAT*B genes to prepare for *B*-regulated processes occurring after hyphal fusion, namely septal dissolution and nuclear migration (Kothe, 1996). However, as will be shown below, there is also some evidence that *MAT*B genes act as nuclear identity genes and that it is the differential interaction between nuclei with different *MAT*B genes within the heterokaryon that determines the pattern of gene expression and morphogenesis.

Nuclear identity and nuclear interactions

There is ample evidence that nuclei with different mating-type genes maintain their identity in a heterokaryon. In a *MAT*B≠ interaction the invading nucleus rapidly migrates through the recipient mycelium (Snider & Raper, 1958) with concomitant dissolution of septa. Since nuclear migration is controlled by the presence of different *MAT*B genes, interactions between the products of invading and resident nuclei results in a rapid induction of the migratory machinery. In a *MAT*A≠ *MAT*B≠ heterokaryon, nuclear migration is just as rapidly switched off again once the migrating nucleus reaches a tip cell and associates with a resident nucleus, probably an effect of interactions between different *MAT*A genes.

A significant discovery related to recognition of nuclear type was made by Niederpruem and co-workers (Niederpruem, 1980; Nguen & Niederpruem, 1984). They found that after nuclear migration had occurred many tip cells were multinucleate, containing 3–12 nuclei. Such tip cells formed normal clamp connections, proving the presence in these cells of nuclei of opposite mating type (Fig. 13.1). Then a reduction of the number of nuclei took place by various mechanisms, including intercalary septation and outgrowth of pseudoclamps in which two nuclei were entrapped, ultimately leading to exclusively binucleate tip cells and complete dikaryotization (Fig. 13.1). Apparently, nuclei of opposite mating type recognized each other and were sorted out to combine in a binucleate state. Possibly, this recognition phenomenon is also a function of the *MAT*B genes. A similar phenomenon was noted in the ascomycete *Podospora anserina* (Zickler *et al.*, 1995; Arnaise, Debuchy & Picard, 1997). After fertilization of the female organ, the parental nuclei of opposite mating type undergo a number of mitotic divisions leading to a multinucleate cell from which dikaryotic ascogenous hyphae arise, each containing two nuclei of opposite mating type. Mutations in either of the two mating-type loci, *mat*+ or *mat*-, led to loss of recognition and the appearance of uniparental progeny. It was concluded that these mating-type loci contain genetic elements that determine nuclear identity. This was confirmed by transformation experiments involving internuclear complementation tests (Arnaise *et al.*, 1997).

In *S. commune*, experiments on dedikaryotization have also indicated that nuclei in the dikaryon are marked. After regeneration of a cell wall, protoplasts derived from the secondary mycelium revert to either dikaryotic or monokaryotic hyphae (Wessels, Hoeksema & Stemerding, 1976). Apparently, in the homokaryotic revertants, the two nuclear types became separated in individual protoplasts. However, the homokaryotic revertants behaved differently from those derived from the parental primary mycelia (monokaryons). For a number of cell divisions they formed pseudoclamps (hook cells that fail to fuse with penultimate cells) indicating that hook-cell fusion, controlled by the presence of different *MAT*B genes in the secondary mycelium, was immediately switched off after separation of the nuclei but that the morphogenetic event controlled by the presence of different *MAT*A genes (pseudoclamp formation) persisted for some time in the absence of a nucleus carrying a different *MAT*A gene. Only after about the thirtieth septum was formed did the hyphal morphology become typically monokaryotic (Wessels *et al.*, 1976). Similar observations were made much earlier by Harder (1927) who, by microsurgery of a *S. commune* dikaryon, obtained homokaryotic cells that exhibited transient pseudoclamp formation. Significantly, among the reverting protoplasts of the secondary mycelium the two nuclear types were often recovered with different frequencies despite the fact that in the highly structured *MAT*A≠ *MAT*B≠ heterokaryon they occur in a strict 1:1 ratio. Raper (1985) showed that the survival of the nuclei depended on the *MAT*B genes. The *Bα* and *Bβ* loci could be put in an hierarchical order with respect to survival of nuclei. It is as if in the dikaryon one of the two nuclei becomes entrained by the other nucleus in its mitotic cycle and loses the ability to divide on its own in the absence of the nucleus with *MAT*B genes higher up in the hierarchical order. Asymmetry of recovery of nuclear types was also noted using sodium cholate to dedikaryotize the *MAT*A≠ *MAT*B≠ heterokaryon (Leonard, Gaber & Dick, 1978) and thus was not an artifact of the method of dedikaryotization. Curiously, the hierarchical order of *B* loci was reversed in homokaryotic revertants of protoplasts obtained from a *MAT*A= *MAT*B≠ heterokaryon (Raper, 1985).

Genes under control of mating-type genes

Although no genes have been isolated that function in the monokaryon–dikaryon transition (Hoge *et al.*, 1982*a*), a number of genes have been found that are differentially expressed in the primary and secondary mycelium (Hoge, Springer & Wessels, 1982*b*; Mulder & Wessels, 1986). For instance, the genes *SC1*, *SC4*, *SC6*, *SC7*, and *SC14* have been found to be expressed in the secondary mycelium only, whereas the *SC3* gene is prevalently expressed in the primary (monokaryotic) mycelium but to a variable extent also in the secondary mycelium. The *SC1*, *SC3*, *SC4* and *SC6* genes have been shown to encode members of a family of small (about 10 kDa) cysteine-rich proteins, called hydrophobins (Schuren & Wessels, 1990; Wessels *et al.*, 1991*a*, 1995), which were found to be ubiquitous in fungi and fulfil a variety of functions (Wessels, 1994, 1996, 1997). The biophysical properties of the SC3 hydrophobin, which may be exemplary of all so-called class I hydrophobins (Wessels, 1994), have been best investigated. SC3 hydrophobin monomers, which are quite soluble in water, assemble when confronted with a hydrophilic–hydrophobic interface such as between water and air (Wösten, de Vries & Wessels, 1993) or water and oil or water and a hydrophobic solid (Wösten *et al.*, 1994*a*,*b*; Wösten, Schuren & Wessels, 1994*c*; Wösten *et al.*, 1995). During assemblage, the hydrophobin undergoes a conformational change producing an SDS-insoluble amphipathic membrane with a hydrophobic surface (i.e. water contact angles, θ, 110°) that exhibits a characteristic mosaic pattern of parallel rodlets, whereas the opposite surface becomes hydrophilic (θ 45°). SC3 hydrophobin monomers secreted at tips of hyphae thus, by self-assembly, cover these hyphae with a hydrophobic rodlet layer when they grow into the air (Wösten, *et al.*, 1993, 1994*a*). Because SC3 assembles on hydrophobic solids such as Teflon, making the surface completely wettable, SC3-secreting hyphae can tightly adhere to Teflon. This is borne out by the fact that adhesion no longer occurs when the *SC3* gene is disrupted (Wösten. *et al.*, 1994*c*).

Another remarkable property of the SC3 hydrophobin is that upon assemblage at a water–air interface it lowers the surface tension of the aqueous solution (100 $\mu g\ ml^{-1}$) from 72 to as low as 32 $mJ\ m^{-2}$, thus being one of the most powerful surface-active substances known (van der Vegt *et al.*, 1996). Disruption of the *SC3* gene in the primary (monokaryotic) mycelium greatly reduces the capacity to form aerial hyphae and those that are formed (in well-aerated cultures) have a hydrophilic surface (van Wetter *et al.*, 1996). During cultivation, SC3 accumulation in the medium lowers the surface tension of the medium to values as low as 26 $mJ\ m^{-2}$ and this probably enables hyphae to breach the water–air interface. This is corroborated by the finding that added SC3 facilitates formation of aerial hyphae in the SC3-knock-out mutant though the aerial hyphae that are formed have a hydrophilic surface (H.A.B. Wösten, M.-A. van Wetter & J.G.H. Wessels, unpublished). Thus secretion of SC3 into the medium may help hyphae to breach the water–air interface by virtue of the capacity of SC3 to lower the surface tension of the medium. This property can therefore be complemented in the SC3-knock-out mutant by adding SC3 to the medium. However, formation of a hydrophobic surface on a hypha growing into the air can only be accomplished by this very hypha secreting SC3 at its apex, and therefore cannot be complemented in the mutant by adding SC3 to the medium.

In common-A (*MATA*= *MAT*B≠) heterokaryons (Ásgeirsdóttir, van Wetter & Wessels, 1995) and *MAT*Ax *MAT*B^{con} homokaryons (Ruiters, Sietsma & Wessels, 1988) the *SC3* gene is suppressed and therefore few aerial hyphae are produced (flat mycelium). Apparently, the presence of different *MAT*B genes or constitutive activity of the *MAT*B gene negatively regulates the *SC3* gene. Also the *thn* mutation, when present in a monokaryon, down-regulates the *SC3* gene and prevents formation of aerial hyphae. Moreover, when present in double dose in the *MAT*A≠ *MAT*B≠ heterokaryon, this mutation not only prevents accumulation of SC3 mRNA but also accumulation of all identified dikaryon-expressed mRNAs, including *SC4* mRNA, and prevents formation of fruiting bodies (Wessels *et al.*, 1991*b*). However, in a *MAT*A≠ *MAT*B≠ heterokaryon with a disrupted *SC3* gene in both nuclear types, the disruption only affects formation of aerial hyphae; apparently normal sporulating fruiting bodies are produced (van Wetter *et al.*, 1996).

The *SC4* gene is only abundantly expressed in the *MATA*≠ *MATB*≠ heterokaryon and the *MATA*^{con} *MATB*^{con} homokaryon, particularly in the fruiting bodies (Mulder & Wessels, 1986; Ruiters *et al.*, 1988). The SC4 hydrophobin is found as an SDS-insoluble membrane lining air spaces within the plectenchyma of the fruiting bodies, providing these with a hydrophobic rodlet surface (Wessels *et al.*, 1995; L.A. Lugones & J.G.H. Wessels, unpublished). The product of the *SC7* gene, a hydrophilic protein with homology to PR1 proteins of plants (Schuren *et al.*, 1993), is found in the fruiting bodies within the extracellular matrix that binds hyphae together in the plectenchyma but was also found secreted into the medium (Ásgeirsdóttir *et al.*, 1995).

It is important to note that all the genes discussed above and implicated in emergent growth are expressed and similarly regulated by mating-type genes when the mycelia are grown in shaking cultures in which no emergent structures are formed (Wessels *et al.*, 1987). Expression of these genes is thus not a consequence of the formation of specific emergent structures but may be a prerequisite for their formation.

Regulation of *SC3* and dikaryon-expressed genes in the MATA≠ MATB≠ heterokaryon (the secondary mycelium)

As already mentioned, *SC3* is highly expressed in the primary mycelium (monokaryon) and to a variable extent in the secondary mycelium (*MATA*≠ *MATB*≠ heterokaryon). In fact, the gene (then named *1D10*) was originally isolated as a monokaryon-specific sequence (Dons *et al.*, 1984). Under some conditions, emergent growth in the secondary mycelium only encompasses the appearance of fruiting bodies and no aerial hyphae. In that case, *SC3* is hardly expressed in the secondary mycelium (Mulder & Wessels, 1986) and of the hydrophobins only SC4 is detected in the culture medium (Wessels *et al.*, 1991*a*). However, mostly both aerial hyphae and fruiting bodies are formed on the secondary mycelium and then both *SC3* and *SC4* (and other dikaryon-expressed genes) are active. In fact, an inverse relationship between accumulation of *SC3* mRNA and dikaryon-specific mRNAs was generally observed; when SC3 mRNA was high, the dikaryon-specific mRNAs were low, and vice versa (Wessels *et al.*, 1987). For instance, as can be seen in Fig. 13.2, in 4-day-old surface cultures, an increase in atmospheric CO_2 con-

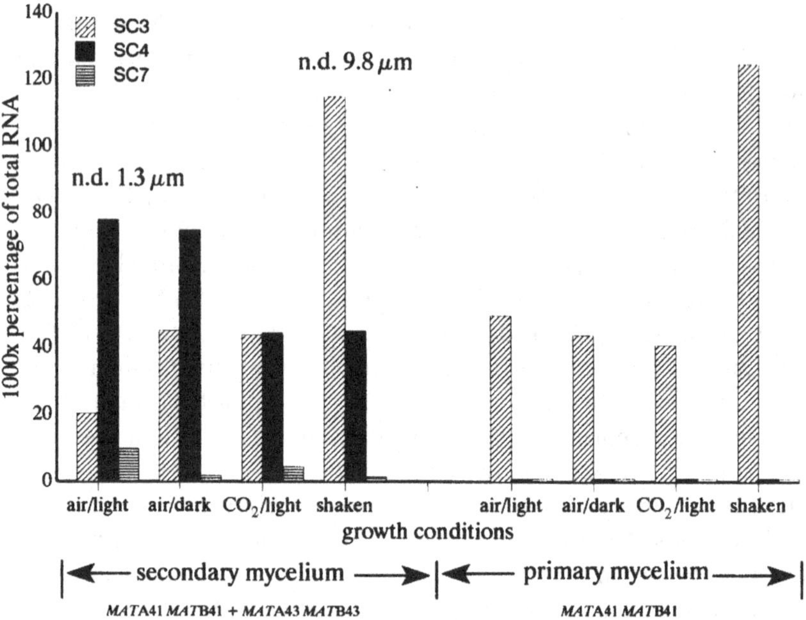

Fig. 13.2. Expression of the *SC3*, *SC4*, and *SC7* genes in 4-day-old secondary and primary mycelia of *Schizophyllum commune*, grown in surface culture, as influenced by light, darkness and increased ambient CO_2 concentration and by growth in shaken culture. Nuclear distance is indicated as n.d. (Expression data from Wessels *et al.*, 1987).

centration from ambient to 5% led to an increase in *SC3* mRNA concentration but a decrease in the concentration of dikaryon-specific mRNAs in the secondary mycelium, resulting in a shift in the *SC3* mRNA/*SC4* mRNA ratio from 0.26 to 1 (these ratios being 13.75 and 19.43, respectively, in the primary monokaryotic mycelium). Cultivating the homogenized mycelium in shaken cultures instead of surface culture raised the *SC3* mRNA/*SC4* mRNA ratio from 0.26 to 2.56 (13.75 and 27.8, respectively, in the primary mycelium). It is evident that the effects of enhanced CO_2 concentrations are much more pronounced in the secondary than in the primary mycelium.

There is good evidence that 'monokaryon-specific' mRNAs (exemplified by *SC3* mRNA) and 'dikaryon-specific' mRNAs (exemplified by abundant mRNAs such as *SC4* and *SC7* mRNAs) are not produced by the same hyphae of the secondary mycelium (Ásgeirsdóttir *et al.*, 1995). When secreted proteins produced by a colony of a fruiting secondary mycelium were trapped on a PVDF membrane and localized with specific antibodies, it was found that the dikaryon-specific SC7 protein, but not the monokaryon-specific SC3 hydrophobin, was detected in a peripheral ring of light-grown mycelium where fruiting bodies were being formed whereas the rest of the colony, formed in the dark and making aerial hyphae, released SC3 but not SC7 (Fig. 13.4). Immuno-localization then detected SC3 on these aerial hyphae and on aerial hyphae covering fruiting bodies but no SC3 was found in hyphae that make up the inner tissue (plectenchyma) of the fruiting bodies. These produced SC7 (Ásgeirsdóttir *et al.*, 1995) and SC4 (Wessels *et al.*, 1995). Apparently, in the secondary mycelium there is a spatial differentiation leading to hyphae that produce either monokaryon-specific or dikaryon-specific mRNAs.

Exclusion of *SC3* expression in dikaryotic hyphae of the *MAT*A≠ *MAT*B≠ heterokaryon would be expected if the presence of different *MAT*B genes suppresses activity of the *SC3* gene as evident in the *MAT*A= *MAT*B≠ heterokaryon (Ásgeirsdóttir *et al.*, 1995). Conversely, it would appear that in hyphae of the secondary mycelium that express *SC3* but not genes normally expressed in the dikaryon, the interaction between different *MAT*B genes is somehow disturbed. However, there is a caveat here. If the interaction of different *MAT*B genes represses the SC3 gene in the *MAT*A≠ *MAT*B≠ heterokaryon, one would also expect silencing of the *SC3* gene in the *MAT*A^{con} *MAT*B^{con} homokaryon. However, although no activity of the *SC3* gene is found in the *MAT*Ax *MAT*B^{con} homokaryon (Ruiters *et al.*, 1988), the gene is expressed in the *MAT*A^{con} *MAT*B^{con} homokaryon together with the dikaryon-expressed genes (Springer & Wessels, 1989). The reason for this behaviour is not clear but it should be emphasized that in other respects also the *MAT*A^{con} *MAT*B^{con} homokaryon is not a perfect mimic of the *MAT*A≠ *MAT*B≠ heterokaryon. Although binucleate clamped cells are found, many cells bear pseudoclamps (non-fused hook cells) and fruiting and sporulation are less regular than in the *MAT*A≠ *MAT*B≠ heterokaryon.

Correlation of nuclear distance and gene expression

One way for the secondary mycelium to express 'monokaryotic' mRNAs would be to segregate its nuclei into separate hyphae (dedikaryotization). This was observed during re-growth of homogenized secondary mycelium in which pseudoclamps containing one nucleus grew out to form monokaryotic hyphae (Nguyen & Niederpruem, 1984). Gabriel (1967, cited by

Kühner, 1977) observed in *Panellus serotinus* and *Phlebia radiata* that poor aeration was responsible for a transition of the secondary mycelium to pseudo-primary mycelium without clamp connections but still containing the two types of nuclei in multicellular tip cells. A similar observation was made by Ross, Loftus & Foster (1991) who observed that an arginine-requiring mutant heterokaryon of *Coprinus congregatus* reversibly switched from the dikaryotic to the multinucleate homokaryotic phenotype when deprived of exogenous arginine. Such a process was also suggested in *S. commune*, where the hyphae covering fruiting bodies showed simple septa and no clamp connections and manufactured the SC3 hydrophobin but none of the dikaryon-expressed proteins (Ásgeirsdóttir *et al.*, 1995). Unfortunately, nuclei could not be visualized in these hyphae. Similarly, the vegetative aerial hyphae formed by the *MAT*A≠ *MAT*B≠ heterokaryon of *S. commune* were covered by the SC3 hydrophobin, like aerial hyphae of the primary mycelium containing only one nuclear type. By isolating from the secondary mycelium an early stage in aerial hyphae formation, in which nuclei could be stained by 4,6-diamidino-2phenylindole (DAPI), it was found (Ásgeirsdóttir *et al.*, 1995) that in these hyphae the two nuclei were separated by a distance of 8.3 ± 4.8 µm compared to the binucleate substrate hyphae in which the nuclei mostly occurred in close proximity (0.9 ± 0.4 µm). Since the expression of the *SC3* gene is suppressed by the presence of different *MATB* genes in the *MAT*A= *MAT*B≠ heterokaryon, this may also be the case in the presence of different *MAT*B genes in the *MAT*A≠ *MAT*B≠ heterokaryon but only if the nuclei are closely together to allow for interactions between the products of *MAT*B genes. However, in combination with interactions between the products of different *MAT*A genes, this interaction between the products of different MATB genes would activate the dikaryon- expressed genes. Spatial separation of the nuclei and their cytoplasmic domains could then lead to disruption of interaction between the products of the *MAT*B genes and thus to a monokaryotic type of gene expression.

To extend this correlation between gene expression and nuclear distance the aforementioned observation (Fig. 13.2) that a monokaryotic type of gene expression is enhanced in the secondary mycelium by growing it in submerged shaken culture instead of in still culture (Wessels *et al.*, 1987) was revisited. It was found that the accumulation of the *SC3* mRNA in shaken cultures was paralleled by an increase in the distance of nuclei in apical cells, from around 1.5 µm in still culture and young shaken culture, which predominantly produced *SC4* mRNA, to about 10 μm in shaken cultures, which predominantly produced *SC3* mRNA. It appears that when the nuclear distance exceeds 3 to 4 μm, gene expression shifts towards the monokaryotic type (Fig. 13.3).

The spatial differentiation between secretion of SC3 and SC7 in a fruiting colony of the secondary mycelium (Ásgeirsdóttir *et al.*, 1995) could also be correlated with a difference in the distance between the compatible nuclei (Fig. 13.4). Substrate mycelium grown in the dark showed a considerable distance between the two nuclei in hyphal compartments but in hyphae grown in the light the nuclei were very close. Apparently light did not restore the binucleate state in hyphae grown in the dark and these continued to secrete SC3 but not SC7 and formed aerial hyphae only. The light-grown hyphae on which the fruiting bodies developed did not secrete SC3 but produced SC7 and presumedly expressed all dikaryon-specific mRNAs. Colonies returned to darkness continued to form hyphae with widely separated nuclei; only the nuclei in the most peripheral leading apical compartments were in close proximity.

If the distance between the nuclei with different mating-type genes is responsible for these spatial differences in gene expression, one would not expect this to occur in a fruiting *MATA*^{con} *MATB*^{con} homokaryon in

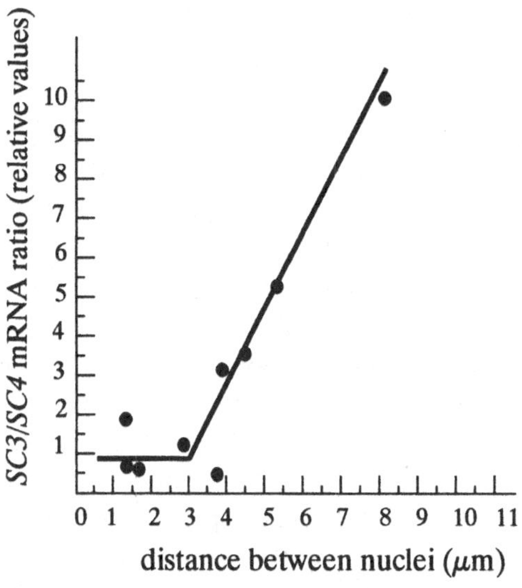

Fig. 13.3. Correlation between the distance between nuclei in apical cells of the secondary mycelium of *Schizophyllum commune* and the ratio of *SC3* mRNA and *SC4* mRNA.

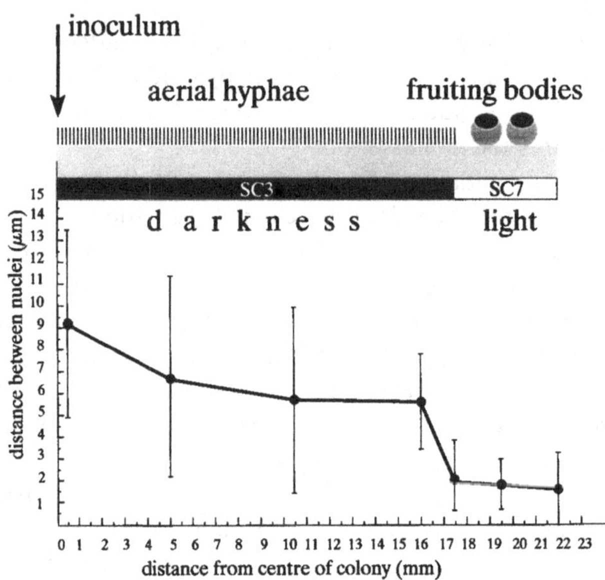

Fig. 13.4. Correlation between secretion of SC3 and SC7 into the medium and formation of aerial hyphae and fruiting bodies on the one hand and distance between nuclei in substrate hyphae of the secondary mycelium of *Schizophyllum commune* on the other hand. The *MAT*A≠ *MAT*B≠ heterokaryon was first grown as a colony in the dark for 4 d (radius 16 mm) and then transferred to light, resulting in formation of fruiting bodies in the light-grown periphery of the colony.

which the nuclei in a pair are the same. Indeed when the experiment shown in Fig. 13.4 was done with such a homokaryon it was found that SC3 and SC7 secretion was not spatially separated and both occurred throughout the colony. Moreover, fruiting bodies did not arise in a peripheral ring of light-grown mycelium but were also formed on dark-grown mycelium in the centre of the colony.

To correlate nuclear distance and pattern of gene expression at the level of an individual hypha, advancing hyphae at the mycelial front of a colony of *MAT*A≠ *MAT*B≠ mycelium were forced to grow over hydrophilic glass (θ 15°) or over glass made hydrophobic by treatment with dimethyldichlorosilane (θ 100°). As shown by specific antibodies, hyphae of the secondary mycelium growing over the hydrophobic surface mostly produced the SC3 hydrophobin and little or no SC4 hydrophobin on their surfaces and tightly adhered to the hydrophobic surface as was noted earlier for hyphae of the primary mycelium growing over Teflon

(Wösten, 1994*c*). The preparations were then treated with DAPI to reveal the position of the nuclei in the same hyphae in which the hydrophobin was localized. As summarized in Fig. 13.5, in all hyphae that strongly reacted with the SC3 antibody, the binucleate state typical for the dikaryon was disrupted; the average distance of the nuclei being 9.4 μm. However, hyphae growing over the hydrophilic glass mostly produced SC4 hydrophobin and little or no SC3 hydrophobin. In these hyphae the binucleate state was maintained; the average distance between nuclei being 1.2 μm.

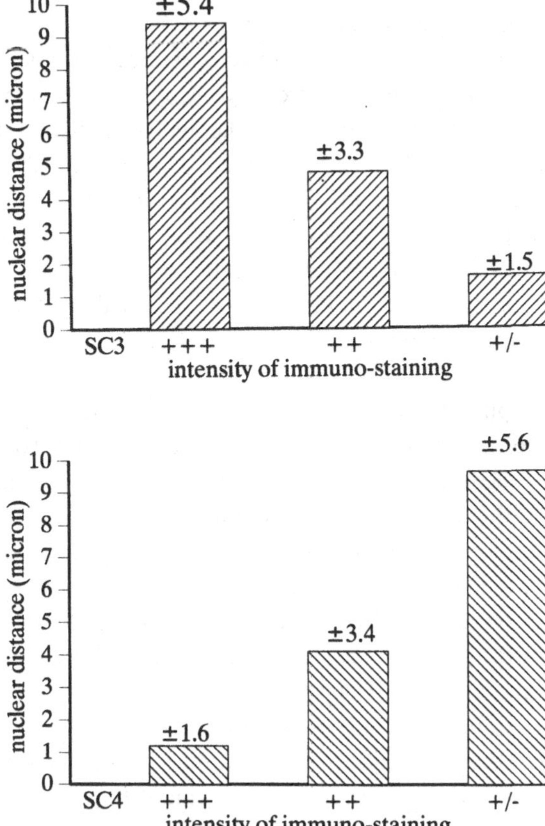

Fig. 13.5. Correlation between the presence of SC3 and SC4 hydrophobin on apical cells of the secondary mycelium of *Schizophyllum commune* growing on hydrophobic and hydrophilic surfaces as revealed by immuno-cytochemistry, and the distance between nuclei in these cells.

Although these data only concern correlations, they suggest that it is the alteration of the distance between the compatible nuclei in the hyphae of the secondary mycelium that determines their pattern of gene expression. In this case the nuclear distance is thought to be controlled by environmental factors. The possibility that the disruption of the binucleate state was brought about by the adhesion of the hyphae to the hydrophobic substrate mediated by SC3 assembly was ruled out by showing that this disruption also occurred when a *MATA*≠ *MATB*≠ heterokaryon with knocked-out *SC3* genes in both nuclei was grown on a hydrophobic surface.

Disruption of the binucleate state also generally occurs in secondary mycelia carrying a homozygous *fbf* mutation (Springer & Wessels, 1989). In these mycelia the hook cells fail to fuse with the penultimate cells so that the cells derived from binucleate tip cells receive one nucleus only (Ásgeirsdóttir *et al.*, 1995). These heterokaryons, in spite of the presence of different *MATB* genes, produce none of the dikaryon-specific mRNAs but abundantly express the *SC3* gene (Springer & Wessels, 1989). It thus appears that dikaryon-specific mRNAs are only expressed when the nuclei are in close proximity (the binucleate state); monokaryon-specific mRNAs are expressed when the nuclei drift apart, thereby probably interfering with the nuclear signalling system encoded by the *MATB* genes. A hypothetical scheme involving these controls is given in Fig. 13.6.

It should be emphasized that in the scheme given in Fig. 13.6, the solid arrows do not necessarily indicate direct interactions between controlling and controlled genes. On the contrary, it is probable that interactions between different *MATB* genes entails activation of a complex reaction cascade (Kothe, 1996). Also the *THN* gene is probably involved in a signalling cascade. With respect to formation of hydrophobic aerial hyphae, it was found that the *thn* mutation can be complemented by a diffusible substance secreted into the culture medium by the wild-type monokaryon (F.H.J. Schuren, personal communication). Apparently the *thn* mutant is unable to produce this signal substance. Since at least one other non-allelic *thn* mutation has been mapped in *S. commune* (Raper, 1988) it would not be surprising if the *THN* genes are part of a signalling cascade analogous to that proposed for the *bld* genes which are involved in the formation of a surface-active polypeptide, sapB, and aerial hyphae in *Streptomyces coelicolor* (Nodwell, McGovern & Losick, 1996).

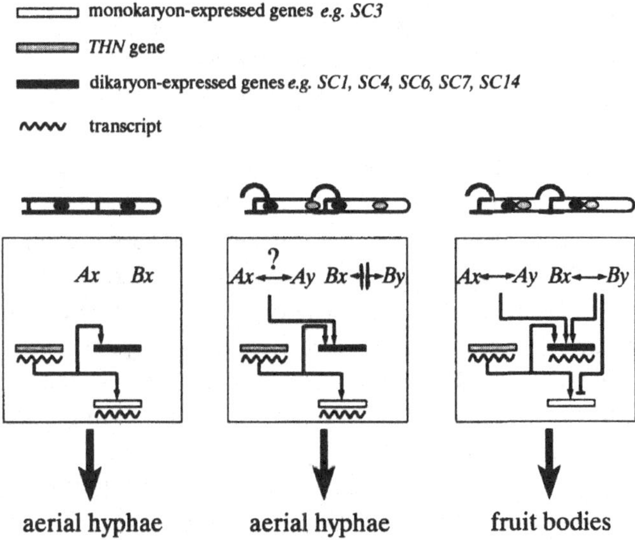

Fig. 13.6. Hypothetical scheme for the regulation of target genes by the mating-type genes in primary and secondary mycelium of *Schizophyllum commune* and the effects of disruption of the binucleate state on the expression of these target genes. The *THN* gene is necessary for activity of both the monokaryon- and dikaryon-expressed genes. Interaction between the *MAT*B genes prevents accumulation of monokaryon-specific mRNAs but is necessary, together with interaction between the *MAT*A genes for activity of the dikaryon-expressed genes. Failure of the *MAT*B genes to interact, presumably by increased distance between compatible nuclei in the secondary mycelium, leads to a monokaryon-type of gene expression. The arrows do not indicate direct interactions between genes. Probably they represent cascades of interactions involving intermediary genes.

It is unknown how environmental conditions such as light, the hydrophobicity of the substrate, or shaking of the culture medium can alter the distance between nuclei in the secondary mycelium. Possibly, these conditions have an influence on cytoskeletal elements that hold compatible nuclei together in the binucleate state. Kamada, Hirai & Fujii (1993) showed that, in the *MAT*A≠ *MAT*B≠ heterokaryon of *Coprinus cinereus*, homozygous mutations in α- and β-tubulin genes disturb the pairing of compatible nuclei causing an increase in the distance between nuclei. A similar effect was seen when applying the anti-tubular drug benomyl to wild-type mycelia. Runeberg & Raudaskoski (1983) have shown, with indirect immunofluorescence microscopy, that intersecting microtubule tracks surround the nuclear pair in *S. commune*.

Does nuclear distance affect communication between nuclei through the pheromone-receptor system?

The results described show that the expression of monokaryotic and dikaryotic mRNAs in the secondary mycelium of *S. commune* correlates with the distance between the two compatible nuclei in a hyphal compartment. Only when the nuclei are close together (the binucleate state) do dikaryon-specific mRNAs accumulate, while monokaryotic sequences are suppressed. There are at least three possibilities to explain this correlation. First, nuclear distance and gene expression may be causally unrelated to each other and both may change in response to the same environmental factor. This does not seem likely because the correlation holds with changes in a multitude of environmental factors. Second, the changes in environmental factors may primarily influence the activities of mating-type genes or the functioning of their products. In turn this would result in changing the distance between nuclei carrying different mating-type genes. For instance, the presence of different *MAT*A genes or the mutation *MAT*A^{con} leads to pairing and conjugate division of nuclei. Interference with the activities of the *MAT*A genes or their products could thus lead to the nuclei drifting apart. However, as shown, it is the interaction between the *MAT*B genes that appears primarily responsible for switching off the activity of the *SC3* gene in the secondary mycelium with closely paired nuclei. Third, environmental factors may primarily determine the position of the nuclei in the secondary mycelium and this may cause failure of the compatible nuclei to interact properly. For instance, it seems possible that signalling between the two nuclei occurs via products of the *MAT*B genes, which also mark the identity of each nucleus.

In the absence of any information at that time on the molecular structure of the mating-type loci, Raper (1983) has extensively speculated on the possibility that the *MAT*B genes may code for specific proteins in the nuclear envelope which play a role in nuclear migration and/or mitosis or in recognition of other nuclei carrying mating-type genes of compatible type. At first sight the operation of such a system of inter-nuclear communication would seem improbable since the two nuclei are supposed to be in a common cytoplasm and thus would be expected to obtain the same complement of proteins from the cytoplasm. However, evidence is emerging from work on mammalian cells that certain mRNAs carry localization signals that target them to particular regions of the cytoplasm, for instance the perinuclear cytoplasm and cytoskeletal-bound ribosomes (Veyrune *et al.*, 1996 and references cited therein). In this way a protein synthesized in a heterokaryon could be specifically incorporated in the nuclear envelop of the nucleus encoding the mRNA for this particular protein or in any cytoplasmic domain, including plasmalemma, dominated by that nucleus.

Since the *MAT*B genes have now been shown to encode pheromones and pheromone receptors (Wendland *et al.*, 1995; Vailancourt *et al.*, 1997), signalling between compatible nuclei may be mediated by this system, assuming that the products of the *MAT*B genes are synthesized and located within the cytoplasmic domain belonging to the nucleus that contains these genes. The concentration of specific pheromone(s) encoded in a nucleus may fall off at some distance of this nucleus and thus fail to interact with the receptor encoded by a compatible nucleus positioned at some distance. This model could be accommodated either by the receptors located in the nuclear envelop and pheromones diffusing within the cytoplasm (intracytoplasmic communication) or by receptors located in the plasmalemma and pheromones secreted within the wall domain (extracytoplasmic communication). Intracytoplasmic communication would perhaps best explain the marking of the nuclei but calls for an unprecedented mechanism of pheromone signalling. Extracytoplasmic communication would be more analogous to the yeast system (Duntze, Betz & Nientiedt, 1994) with the exception that the pheromones may be primarily present in the wall domain and not in the culture medium, so that a gradient can become established. Topical secretion of the pheromone into the wall is conceivable because these putative lipopeptides, without obvious signal peptides for secretion, are probably secreted via ABC transporters, like the *Saccharomyces cerevisiae* STE6 **a**-factor peptide export system (Kuchler, Sterne & Thorner, 1989; McGrath & Varshavsky, 1989). Also, apically directed secretory vesicles would deliver the pheromones at the outside of the wall from where they would be expected to diffuse easily into the medium (Wessels, 1994) whereas ABC transporters would deliver the pheromones at the inside of the mature wall. It may also be noted that mycelial fungi like *S. commune*, unlike yeasts, do not normally grow in liquid but rather on solid surfaces or into the air. Under these circumstances only the wall domain (apoplast) would be available for extracytoplasmic signalling. Within the tubular cell a pheromone secreted into the wall by a particular nucleus-dominated cytoplasmic domain could establish a concentration gradient in the wall (particularly if degradation of the pheromones occurred) with no pheromone present at the site where the responsive receptor encoded by the compatible nucleus resides in the plasmalemma. Only if the two nuclei are close together would their cytoplasmic domains overlap and permit binding of the pheromone to the responsive receptor. In either model, positioning of the compatible nuclei at some distance would lead to loss of interaction between the different *MAT*B genes and thus to expression of a gene like *SC3* and cessation of expression of genes normally expressed in the dikaryon such as *SC4* and *SC7*. Since normal clamp connections are nevertheless formed in these hyphae, it must be assumed that separation of the nuclei only occurs in mitotic interphase and that the binucleate state is restored during mitosis. Exclusive location of the pheromones in the hyphal wall would also explain why the effect of different *MAT*B genes is not manifested at the level of hyphal fusions. Important information concerning a choice between models will come from unequivocal localization of the pheromones and pheromone receptors.

It is difficult to say whether an increased distance between nuclei in the secondary mycelium would also interfere with the interaction of the products of different *MAT*A genes because experimental dedikaryotization by protoplast formation has shown the persistence of pseudoclamp formation in the derived homokaryons (Wessels *et al.*, 1976) and loss of such an interaction would thus not be expected to lead to immediate cessation of the *MAT*A≠ phenotype. On the other hand, the formation of protoplasts in these dedikaryotization experiments would have resulted in loss of the pheromones if they were located in the wall and thus to immediate loss of interaction between the different *MAT*B genes, which is what was observed; pseudoclamps but no true clamps were detected in the derived homokaryons. Also, during reversion of heterokaryotic protoplasts into dikaryons, initially pseudoclamps were observed instead of true clamps.

Whatever the mechanisms involved, the observation that the distance between compatible nuclei in heterokaryons may determine the pattern of gene expression, if of general occurrence in fungi, may largely explain why these organisms maintain heterokaryons during vegetative growth and development rather than forming diploids like animals and plants. It apparently provides for additional regulatory possibilities not possible in haploid–diploid organisms.

Summary

In this Chapter spatial differences in gene expression in the secondary mycelium of *Schizophyllum commune* are described in relation to the distribution of nuclei within hyphae. Hyphae of this secondary mycelium containing compatible nuclei (different *MATA* and *MATB* mating-type genes) were found to have a dikaryotic type of gene expression (formation of the proteins SC4 and SC7) when the two compatible nuclei were in close proximity. Certain environmental conditions led to an increase in the distance between these nuclei and this was accompanied by a decrease in the dikaryotic type of gene expression and enhancement of the expression of a gene (*SC3*) also highly expressed in the primary mycelium containing one nuclear type only. This leads to spatial differentiation in colonies of the secondary mycelium resulting in areas that form aerial hyphae or fruiting bodies, respectively. It is speculated that the nuclear interaction that regulates this differential gene expression is mediated by products of the *MATB* genes and that an increased distance between compatible nuclei, brought about by environmental factors, interferes with this interaction, causing a switch from a dikaryotic type to a monokaryotic type of gene expression in the secondary mycelium. This represents a novel mechanism of gene regulation and cell differentiation, probably confined to the fungi.

Note added in proof: Details of the work on nuclear distance have now been published (Schuurs *et al.*, 1998).

Acknowledgements

The authors are grateful to Drs E.M. Kothe, M. Raudaskoski, S.A. Ásgeirsdóttir and H.A.B. Wösten for critically reading a draft of the manuscript and suggesting improvements.

References

Ásgeirsdóttir, S. A., van Wetter, M. A. & Wessels, J. G. H. (1995). Differential expression of genes under control of the mating-type genes in the secondary mycelium of *Schizophyllum commune*. *Microbiology*, **141**, 1281–8.

Arnaise, S., Debuchy, R. & Picard, M. (1997). What is a *bona fide* mating-type gene? Internuclear complementation of *mat* mutants in *Podospora anserina*. *Molecular and General Genetics*, **256**, 169–78.

Banham, A. H., Asante-Owusu, R. N., Göttgens, B., Thompson, S. A. J., Kingsnorth, C. S., Mellor, E. J. C. & Casselton, L. A. (1995). An *N*-terminal dimerization domain permits homeodomain proteins to choose compatible partners and initiate sexual development in the mushroom *Coprinus cinereus*. *The Plant Cell*, **7**, 733–83.

Casselton, L. A. & Kües, U. (1994). Mating-type genes in homobasidiomycetes. In *The Mycota Vol. I Growth, Differentiation and Development*, ed. J. G. H. Wessels & F. Meinhardt, pp. 307–21. Berlin: Springer.

Dons, J. J. M., Springer J., de Vries S. C. & Wessels, J. G. H. (1984). Molecular cloning of a gene abundantly expressed during fruit-body initiation in *Schizophyllum commune. Journal of Bacteriology*, **157**, 802–5

Duntze. W., Betz, R. & Nientiedt, M. (1994). Pheromones in yeasts. In *The Mycota Vol. I Growth, Differentiation and Development*, ed. J. G. H. Wessels & F. Meinhardt, pp. 381–99. Berlin: Springer.

Harder, R. (1927). Zur Frage der Rolle von Kern und Zytoplasma im Zellgeschehen und bei der Übertragung von Eigenschaften. *Zeitschrift für Botanik*, **19**, 337–407.

Hoge, J. H. C., Springer, J., Zantinge B. & Wessels, J. G. H. (1982*a*). Absence of differences in polysomal RNAs from vegetative monokaryotic and dikaryotic cells of the fungus *Schizophyllum commune. Experimental Mycology*, **6**, 225–32.

Hoge, J. H. C., Springer, J. & Wessels, J. G. H.(1982*b*). Changes in complex RNA during fruit-body initiation in the fungus *Schizophyllum commune. Experimental Mycology*, **6**, 233–43.

Kamada, T., Hirai, K. & Fujii, M. (1993). The role of the cytoskeleton in the pairing and positioning of the two nuclei in the apical cell of the dikaryon of the basidiomycete *Coprinus cinereus. Experimental Mycology*, **17**, 338–44.

Kothe, E. M. (1996). Tetrapolar fungal mating types: sexes by the thousands. *FEMS Microbiology Reviews*, **18**, 65–87.

Kuchler, K,. Sterne, R. E. & Thorner, J. (1989). *Saccharomyces cerevisiae STE6* gene product: a novel pathway for protein export in eukaryotic cells. *EMBO Journal*, **8**, 3973–84.

Kües, U. & Casselton L. A. (1992) Fungal mating type genes – regulators of sexual development. *Mycological Research*, **96**, 993–1006.

Kües, U., Göttgens, B., Stratmann, R., Richardson, W., O'Shea, S. & Casselton, L. A. (1994). A chimeric homeodomain protein causes self-compatibility and constitutive sexual development in the mushroom *Coprinus cinereus. EMBO Journal*, **13**, 4054–9.

Kühner, R. (1977). Variation of nuclear behaviour in the homobasidiomycetes. *Transactions of the British Mycological Society*, **68**, 1–16.

Leonard, T. J., Gaber, R. F. & Dick, S. (1978). Internuclear genetic transfer in dikaryons of *Schizophyllum commune*. II Direct recovery and analysis of recombinant nuclei. *Genetics*, **89**, 685–93.

Luo, Y. H., Ullrich, R. C. & Novotny, C. P. (1994). Only one of the paired *Schizophyllum commune Aα* mating-type, putative homeobox genes encodes a homeodomain essential for *Aα*-regulated development. *Molecular and General Genetics*, **244**, 318–24.

Magae, Y., Novotny, C. P. & Ullrich, R. C. (1995). Interaction of the *Aα Y* and *Z* mating type homeodomain proteins of *Schizophyllum commune* detected by the two-hybrid system. *Biochemical and Biophysical Research Communications*, **211**, 1071–6.

McGrath, J. P. & Varshavsky, A. (1989). The yeast *STE6* gene encodes a homologue of the mammalian multidrug resistance P-glycoprotein. *Nature*, **340**, 400–4.

Mulder, G. H. & Wessels, J. G. H. (1986). Molecular cloning of RNAs differentially expressed in monokaryons and dikaryons of *Schizophyllum commune* in relation to fruiting. *Experimental Mycology*, **10**, 214–27.

Nguyen, T. T. & Niederpruem, D. J. (1984). Hyphal interactions in *Schizophyllum commune*: the di-mon mating. In *The Ecology and Physiology of the Fungal Mycelium*, ed. D. H. Jennings & A. D. M. Rayner, pp. 73–102.Cambridge: Cambridge University Press.

Niederpruem, D. J. (1980). Direct studies of dikaryotization in *Schizophyllum commune*. II. Behavior and fate of multikaryotic hyphae. *Archives of Microbiology*, **128**, 172–8.

Nodwell, J. R., McGovern, K. & Losick, R. (1996). An oligopeptide permease responsible for the import of an extracellular signal governing aerial mycelium formation in *Streptomyces coelicolor*. *Molecular Microbiology*, **22**, 881–93.

Parag Y. (1962). Mutations in the *B* incompatibility factor of *Schizophyllum commune*. *Proceedings of the National Academy of Sciences, USA*, **48**, 743–50.

Raper, C. A. (1983). Controls for development and differentiation of the dikaryon in basidiomycetes. In *Secondary Metabolism and Differentiation in Fungi*, ed. J. W. Bennett & A. Ciegler, pp. 195–236. New York: Marcel Dekker.

Raper, C. A. (1985). *B*-mating type genes influence survival of nuclei separated from heterokaryons of *Schizophyllum commune*. *Experimental Mycology*, **9**, 149–60.

Raper, C. A. (1988). *Schizophyllum commune*, a model for genetic studies of the basidiomycetes. In *Genetics of Plant Pathogenic Fungi*, ed. G. S. Sidhu, pp. 511–22. London: Academic Press.

Raper, J. R. (1966). *Genetics of Sexuality in Higher Fungi*. New York: The Ronald Press.

Raper, J. R., Boyd, D. H. & Raper, C. A. (1965). Primary and secondary mutations at the incompatibility loci in *Schizophyllum*. *Proceedings of the National Academy of Sciences, USA*, **53**, 1324–32.

Ross, I. K., Loftus, M. G. & Foster, L. M. (1991). Homokaryon–dikaryon phenotypic switching in an arginine requiring mutant of *Coprinus congregatus*. *Mycological Research*, **95**, 776–81.

Ruiters, M. H. J., Sietsma, J. H. & Wessels, J. G. H. (1988). Expression of dikaryon-specific mRNAs of *Schizophyllum commune* in relation to incompatibility genes, light, and fruiting. *Experimental Mycology*, **12**, 60–9.

Runeberg, P. & Raudaskoski, M. (1983). Cytoskeletal elements in the hyphae of the homobasidiomycete *Schizophyllum commune* visualized with indirect immunofluorescence and NBD-phallacidin. *European Journal of Cell Biology*, **41**, 25–32.

Schuren, F. H. J., Ásgeirsdóttir, S. A., Kothe, E. M., Scheer, J. M. J. & Wessels, J. G. H. (1993). The *Sc7/Sc14* gene family of *Schizophyllum commune* codes for extracellular proteins specifically expressed during fruit-body formation. *Journal of General Microbiology*, **139**, 2083–90.

Schuren, F. H. J. & Wessels, J. G. H. (1990). Two genes specifically expressed in fruiting dikaryons of *Schizophyllum commune*: homologies with a gene not regulated by mating-type genes. *Gene*, **90**, 199–205.

Schuurs, T.A., Dalstra, H. J. P., Scheer, J. M. J. & Wessels, J. G. H. (1998). Positioning of nuclei in the secondary mycelium of *Schizophyllum commune* in relation to differential gene expression. *Fungal Genetics and Biology*, **23**, 150–61.

Snider, P. J. & Raper J. R. (1958). Nuclear migration in the basidiomycete *Schizophyllum commune*. *American Journal of Botany*, **45**, 538–46.

Springer, J. & Wessels, J. G. H. (1989). A frequently occurring mutation that blocks the expression of fruiting genes in *Schizophyllum commune*. *Molecular and General Genetics*, **219**, 486–8.

Stankis, M. M., Specht, C. A., Yang, H., Giasson, L., & Ullrich, R. C. (1992). The *Aα* mating locus of *Schizophyllum commune* encodes two dissimilar multiallelic homeodomain proteins. *Proceedings of the National Academy of Sciences, USA*, **89**, 7169–73.

Vaillancourt, L. J., Raudaskoski, M., Specht, C. A. & Raper, C. A. (1997). Multiple genes encoding pheromones and a pheromone receptor define the Bβ1 mating-type specificity of *Schizophyllum commune*. *Genetics*, **146**, 541–51.

van der Vegt, W., van der Mei, H. C., Wösten, H. A. B., Wessels, J. G. H. & Busscher, H. J. (1996). A comparison of the surface activity of the fungal hydrophobin SC3p with those of other proteins. *Biophysical Chemistry*, **51**, 253–60.

van Wetter, M.-A., Schuren, F. H. J., Schuurs, T. A. & Wessels, J. G. H. (1996). Targeted mutation of the *SC3* hydrophobin gene of *Schizophyllum commune* affects formation of aerial hyphae. *FEMS Microbiology Letters*, **140**, 265–70.

Veyrune, J.-L., Campbell, G. P., Wiseman, J., Blanchard, J.-M. & Hesketh, J. E. (1996). A localization signal in the 3′ untranslated region of c-myc mRNA targets c-myc mRNA and β-globin reporter sequences to the perinuclear cytoplasm and cytoskeletal-bound ribosomes. *Journal of Cell Science*, **109**, 1185–94.

Wendland, J., Vaillancourt, L. J., Hegner, J., Lengeler, K., Laddison, K. J., Specht, C. A., Raper, C. A. & Kothe, E. (1995). The mating type locus *Bα1* of *Schizophyllum commune* contains a pheromone receptor and putative pheromone genes. *EMBO Journal*, **14**, 5271–8.

Wessels, J. G. H. (1978). Incompatibility factors and the control of biochemical processes. In *Genetics and Morphogenesis in the Basidiomycetes*, ed. M. N. Schwalb & P. G. Miles, pp. 81–104. New York, San Francisco, London: Academic Press.

Wessels, J. G. H. (1994). Developmental regulation of fungal cell wall formation. *Annual Review of Phytopathology*, **32**, 413–37.

Wessels, J. G. H. (1996). Fungal hydrophobins: proteins that function at an interface. *Trends in Plant Science*, **1**, 5–9.

Wessels, J. G. H. (1997). Hydrophobins: proteins that change the nature of the fungal surface. *Advances in Microbial Physiology*, **38**, 1–45.

Wessels, J. G. H., Ásgeirsdóttir, S. A., Birkenkamp K. U., de Vries, O. M. H., Lugones, L. G., Schuren, F. H. J., Schuurs, T. A., van Wetter, M.-A. & Wösten, H. A. B. (1995). Genetic regulation of emergent growth in *Schizophyllum commune*. *Canadian Journal of Botany*, **73** (Supplement 1), S273–81.

Wessels, J. G. H., de Vries, O. M. H., Ásgeirsdóttir, S. A. & Schuren, F. H. J. (1991*a*). Hydrophobin genes involved in formation of aerial hyphae and fruit bodies in *Schizophyllum*. *The Plant Cell*, **3**, 793–9.

Wessels, J. G. H, de Vries, O. M. H., Ásgeirsdóttir, S. A. & Springer, J. (1991*b*). The *thn* mutation of *Schizophyllum commune*, which suppresses formation of aerial hyphae, affects expression of the *Sc3* hydrophobin gene. *Journal of General Microbiology*, **137**, 2439–45.

Wessels, J. G. H., Hoeksema H. L. & Stemerding, D. (1976). Reversion of protoplasts from dikaryotic mycelium of *Schizophyllum commune*. *Protoplasma*, **89**, 317–21.

Wessels, J. G. H., Mulder, G. H. & Springer, J. (1987). Expression of dikaryon-specific and non-specific mRNAs of *Schizophyllum commune* in relation to environmental conditions and fruiting. *Journal of General Microbiology*, **133**, 2557–61.

Wösten, H. A. B., Ásgeirsdóttir, S. A., Krook J. H., Drenth J. H. H. & Wessels, J. G. H. (1994*a*). The fungal hydrophobin Sc3p self-assembles at the surface of aerial hyphae as a protein membrane constituting the hydrophobic rodlet layer. *European Journal of Cell Biology*, **63**, 122–9.

Wösten, H. A. B., de Vries, O. M. H. & Wessels, J. G. H. (1993). Interfacial self-assembly of a fungal hydrophobin into a hydrophobic rodlet layer. *The Plant Cell*, **5**, 1567–74.

Wösten, H. A. B., de Vries, O. M. H., van der Mei, H. C., Busscher, H. J. & Wessels, J. G. H. (1994*b*). Atomic composition of the hydrophobic and hydrophilic sides of self-assembled Sc3p hydrophobin. *Journal of Bacteriology*, **176**, 7085–6.

Wösten, H. A. B., Ruardy, T. G., van der Mei, H. C., Busscher, H. J. & Wessels, J. G. H. (1995). Interfacial self-assembly of a *Schizophyllum commune* hydrophobin into an insoluble amphipathic membrane depends on surface hydrophobicity. *Colloids and Surfaces B: Biointerfaces*, **5**, 189–95.

Wösten, H. A. B., Schuren F. H. J. & Wessels, J. G. H. (1994*c*). Interfacial self-assembly of a hydrophobin into an amphipathic membrane mediates fungal attachment to hydrophobic surfaces. *EMBO Journal*, **13**, 5848–54.

Zickler, D., Arnaise, S., Coppin, E., Debuchy, R. & Picard, M. (1995) Altered mating-type identity in the fungus *Podospora anserina* leads to selfish nuclei, uniparental progeny, and haploid meiosis. *Genetics*, **140**, 493–503.

Index

Printed in the United States
By Bookmasters